General Radiotelephone Operator's License Handbook

5th Edition

No. 3156
$23.95

GENERAL RADIOTELEPHONE OPERATOR'S LICENSE HANDBOOK

5th Edition

HARVEY F. SWEARER and JOSEPH J. CARR

TAB BOOKS Inc.
Blue Ridge Summit, PA

FIFTH EDITION
FIRST PRINTING

Copyright © 1989 by TAB BOOKS Inc.
Earlier editions copyright © 1971, 1974, 1975, and 1982 by TAB BOOKS Inc.
Printed in the United States of America

Reproduction or publication of the content in any manner, without express permission of the publisher, is prohibited. The publisher takes no responsibility for the use of any of the materials or methods described in this book, or for the products thereof.

Library of Congress Cataloging in Publication Data

Swearer, Harvey F.
   General radiotelephone license handbook / by Harvey F. Swearer and
 Joseph J. Carr. — 5th ed.
    p.   cm.
   Includes index.
   ISBN 0-8306-3156-9 (pbk.)   ISBN 0-8306-1456-7
   1. Radio—Examinations, questions, etc. 2. Radio operators-
-Licenses—United States. I. Carr, Joseph J. II. Title.
TK6554.5.S93   1989
621.3841'076—dc19   88-36659
                                                        CIP

TAB BOOKS Inc. offers software for sale. For information and a catalog, please contact TAB Software Department, Blue Ridge Summit, PA 17294-0850.

Questions regarding the content of this book should be addressed to:

   Reader Inquiry Branch
   TAB BOOKS Inc.
   Blue Ridge Summit, PA 17294-0214

CONTENTS

	Introduction	vii
1	**BASIC LAW**	**1**

Basic Law ♦ What's the Exam Like? ♦ Element 1: Questions and Answers ♦ Element 1: Sample Questions (Basic Law) ♦ Element 2: Questions and Answers ♦ Element 2: Sample Questions (Series O and Series M)

2	**BASIC RADIOTELEPHONE: PART I**	**25**

Electron Theory ♦ Properties of Components ♦ Vectors ♦ Circuit Components and Properties ♦ Questions and Answers

3	**SOLID-STATE ELECTRONICS**	**71**

Basic Semiconductor Theory ♦ The pn Junction ♦ Basic Transistor Theory ♦ Amplifier Basics ♦ Transistor Amplifiers ♦ Field-Effect Transistors ♦ Designing with FETs ♦ Bipolar Transistor Power Amplifiers ♦ Transistor RF Power Amplifiers

4	**BASIC RADIOTELEPHONE: PART II**	**135**

Indicating Instruments ♦ Oscillators ♦ Audio Amplifiers ♦ Microphones ♦ Radio Frequency Amplifiers ♦ Transmitters and AM Modulation ♦ Single-Sideband Suppressed Carrier (SSBSC) ♦ Questions and Answers

5 BASIC RADIOTELEPHONE: PART III — 179
Frequency Modulation Systems ♦ Antennas ♦ Transmission Lines ♦ Frequency Measurements ♦ Formulas

6 BASIC RADIOTELEPHONE: PART IV — 207
Motors and Generators ♦ Microwave Equipment ♦ Troubleshooting ♦ Two-Way Radio ♦ Questions and Answers

7 SAMPLE QUESTIONS — 225
Notes about Two-Way Radio ♦ Sample Questions (Basic Radiotelephone)

8 RADAR ENDORSEMENT — 243
Rules and Regulations ♦ Questions and Answers

9 TECHNICIAN CERTIFICATION PROGRAMS — 269
About Study Guides ♦ Good Study Habits ♦ Test Taking

APPENDIX: ANSWERS TO SAMPLE QUESTIONS — 275

Index — 277

INTRODUCTION

Many years ago, someone said, "Radio is the field of the future." A little later, the word "electronics" was substituted for "radio" to better cover the expanded growth of the radio art. There can no longer be doubt in anyone's mind regarding the veracity of the statement. There are dozens of jobs from serviceman to broadcast station engineer at locations all over the world. In addition to broadcasting, there are hosts of other jobs available in communications, cable TV, and related areas.

The FCC licenses discussed in this book are not Amateur Radio (or "ham") licenses. Amateur licenses permit the operation of radio communications equipment only in designated amateur bands. Commercial licenses, on the other hand, allow the operator to operate, install, adjust, and repair commercial, marine, public service and broadcast radio transmitters.

An FCC license is a valuable asset. You might get by uninitiated sales or administrative personnel with a few prepared statements, but if you are face-to-face with an experienced electronic technician, you might need to convince him or her that you know the job. This is where the FCC license can help even though the license, or "ticket," might not even be required. You have a good understanding of electronics to get a passing grade from that tough FCC examining officer. Some applicants try to memorize 500 or 600 answers. A few "sneak by," but they are not usually ready to handle the job and might be carrying the toolbox for the regular technician until they learn what they should have mastered before taking the exam. If you want to be a commercial radio operator or even a transmitter engineer, roll up your sleeves and learn the theory and formulas to figure out the answers to the test.

There have been some recent changes in the FCC rules and regulations concerning the commercial radiotelephone operator's license. Under the old system, there were three classes of licenses (1st, 2nd, and 3rd). Under the new system, the name of the license has been changed to *General Radiotelephone Operator License* (GROL). Broadcast stations are no longer required to hire FCC-licensed technicians, but some still look for a GROL as evidence of lessons learned.

Taking up the slack left by deregulating the old system of FCC licenses are private and industry groups who certify electronic technicians. Although once limited to the International Society of Certified Electronics Technicians (ISCET) and the Electronics Technicians Association International (ETAI), today there are several groups active with programs that vary from broad to very narrow in focus.

Regardless of whether licenses or certification is required, however, this book can assist any aspirant to employment in radio, communications, or any of the subfields to learn what it takes.

1

Basic Law

Radio communications and broadcasting must operate under certain laws, rules, and regulations of the Federal Communications Commission (FCC). The Commission has the force of law, but there are some commonly used practices that are historically rather than legally mandated. This chapter looks at some of these factors.

At one time, all transmitters (except the few very weak ones) required a licensed "operator" to service (i.e. adjust). There were once four categories of radiotelephone permit. The *Restricted Radiotelephone Permit* required no examination but did require registration with the FCC. Holders of this permit were operators of marine radios, aircraft radios, and radio station announcers (and other personnel) who turned the station transmitter on and off at the beginning and end of the broadcast day. The *Third Class Radiotelephone Permit* allowed a little more authority and was used by operators of certain communications stations. The *Second Class Radiotelephone Permit* allowed technical adjustments of radio transmitters in the communications services. Finally, the *First Class Radiotelephone Permit* allowed all the privileges of the lower grades plus the authority to operate broadcast station transmitters (and make technical adjustments on them).

The legacy of these licenses is still with us in some industries, but deregulation has taken its toll. Today, the only offering is the *General Radiotelephone Permit* (GRP), which is roughly analogous to the old Second Class Permit, including the examination. The Marine Radiotelephone Operators Permit (MROP) is analogous to a maritime version of the old Third Class Radiotelephone Permit and is used for merchant marine operators on the high seas and the Great Lakes. The Restricted Permit was once deleted from the line-up but is once again with us, and most of the same people as before are required to have it. For the license examinations (you take them at FCC offices), a passing score is 75 percent, and the questions are multiple choice. The permit is required for making technical adjustments of certain communications equipment. No license is required in the broadcast industry. In some industries where licenses were once required (including broadcasting), certain substitutes have sprung up in the form of technician and/or engineer certification programs.

The remainder of this chapter discusses typical questions from the FCC examination. These questions are generated from a knowledge of past examinations, knowledge of the requirements of jobs covered by the license, and information on the examinations published by the FCC. Elements I and II are broken into two sections. Category "O" is for General Radiotelephone Permit applicants, and Category "M" is for the Marine Radiotelephone Operator applicants. In some cases, questions pertain to both, so they are so indicated with a bracketed [M] and [O] in the text.

BASIC LAW

During your course of study, some words or phrases might seem foreign or even somewhat removed from the intended meaning, such as "ticket," which refers to the FCC license. "Ticket" means *label*, *tag*, *certificate*, *license*, or *permit*. Actually, nothing could be truer than the reference to the general radiotelephone license (or Marine Radio Operator License) as a "ticket." The doors to opportunity in the field of radio communications are many, but those permitting passage without a ticket are few. Even where licenses ("tickets") are not legally required, employers nonetheless look with greater favor on license holders because they know the applicant has a certain basic level of knowledge.

There are several exam elements dealing with commercial operator licenses, and each is a complete examination. The required elements in each case must be taken for the specific license or permit desired. If you fail to pass an element, you are finished for that day, but you may take it again.

You may not leave the room during an element—only between elements. Because Element 3 consists of 100 questions, you are probably going to be working on it for a couple of hours. There is no time limit on any of the examinations (except FCC office hours), so allow yourself plenty of time, and don't have your stomach growling at you through lunch hour.

The elements are (1) Basic Law, (2) Basic Operating Practice (O and M series), (3) Basic Radiotelephone, (4) Radiotelegraph Operating Practice, (5) Advanced Radiotelegraph, (6) Aircraft Radiotelegraph (Endorsement), and (7) Ship Radar (Endorsement). Elements 1 and 2 are offered in Category "O" for General Radiotelephone Applicants and in Category "M" for Marine Radiotelephone Operator Applicants. All elements are covered in this volume except those pertaining to radiotelegraph (5,6, and 7), which are of no practical interest to most applicants because of the dwindling number of jobs in this area.

WHAT'S THE EXAM LIKE?

None of the questions require an essay-type answer; rather, each offers a choice of answers (multiple choice). Only one is correct, of course, even though one or more of the others might *seem* to fit. You might also be asked to draw a few simple diagrams or correct incomplete or incorrect diagrams. Always sign every sheet of paper, even those for figuring. No books, notes, or paper of any kind may be taken into the examination room. You may, however, take a calculator or "slip-stick" (sliderule) to the exam. Calculator users might be required to demonstrate that internal memories are erased.

The basic law in Element 1 covers simple FCC rules and regulations and should be understood for proper retention, rather than memorized. If you rely solely on memorization, you'll probably forget the answers by tomorrow, so reason them out instead. The language might seem stiff, but it is professional and quite important. Remember that the rules are for the common good and protection of all our people.

After reviewing the answers to each question as suggested in the study guide, reason each out and determine the logic involved. The common sense behind each is apparent, and this analysis helps you to keep the information handy in your mind for future use, not only in your exams but later on in your work as well. When you feel confident that you know the material in Elements 1 and 2, try the sample test questions and see how you are doing. If you are a little weak in an area, go over it again until you are confident.

ELEMENT 1: QUESTIONS AND ANSWERS

The following questions and answers should acquaint you more with the licensing process and other information contained in Element 1.

Q1: Where and how are FCC licenses and permits obtained?
A: Bring the application on the prescribed form along with any specified documents in person or by mail to the FCC regional office where you plan to take the test. A license or permit is issued upon satisfactory completion of the exam. You might be advised of the outcome before leaving the office in order to make any desired preparations, but the actual license or permit will be mailed to the applicant's home address by the regional office. Check with the FCC [(202)632-7240] for exam schedules, as exams are only given at certain times of the year.

Q2: When a licensee qualifies for a higher grade FCC license or permit, what happens to the lesser grade license?
A: The lesser grade license is canceled upon issuance of the higher grade and, therefore, must be submitted to the examining officer upon passing the higher grade test. It will be returned by mail (cancelled) along with new (higher grade) license.

Q3: Who may apply for an FCC license?
A: Although commercial licenses are issued only to citizens of the United States, an alien holding an Aircraft Pilot Certificate issued by the Federal Aviation Administration may have the requirement waived by the FCC if it finds that the public interest can be served thereby.

Q4: If a license or permit is lost, what action must be taken by the operator?
A: The licensee must notify the FCC immediately and properly file an application to the office of original issue for a duplicate, informing them of the circumstances involved in the loss of the original license or how it was destroyed. Include a statement that a reasonable search has been made for the original license and that if found later will be returned to the FCC office for cancellation. Documentary evidence or a sworn statement of service performed under the original license must also be submitted. While awaiting receipt of

the duplicate license, an operator can continue his/her duties; however, a signed copy of the application for the duplicate license must be exhibited.

Q5: What is the usual license term of radio operators?
A: Commercial radio operator license terms are 5 years from the date of issuance.

Q6: What government agency inspects radio stations in the U.S.?
A: The Federal Communications Commission (FCC).

Q7: When may a license be renewed?
A: Renewal application can be made at any time during the final year of the license term or during a one-year period following expiration. However, the expired license may not be used during the grace period.

Q8: Who keeps the station logs?
A: Each log is kept by a competent person or persons having actual knowledge of the facts required. Program logs and maintenance logs must be signed before going on duty and again when going off duty.

Q9: Who corrects errors in the station logs?
A: Any necessary correction must be made by the person who made the error.

Q10: How should errors in the station logs be corrected?
A: Correction must be made by the person originating the entry by striking out the error and initialing the correction and date. Erasing is prohibited.

Q11: Under what conditions can messages be rebroadcast? [M]
A: Rebroadcast is permissible only with the express authority of the originating station.

Q12: What messages and signals may not be transmitted?
A: A licensed radio operator shall not transmit unnecessary, unidentified, or superfluous radio communications or signals. Communications containing obscene, indecent, or profane language or meaning are likewise prohibited, along with deceptive or false signals or communications and call letters not assigned by proper authority to the station in operation.

Q13: May an operator deliberately interfere with any radio communication or signal?
A: No.

Q14: What type of communication has top priority in the mobile service? [M, O]
A: Distress calls, distress messages, and distress traffic with an order of priority in the mobile service is as follows:
   (a) Distress calls, distress messages, and distress traffic
   (b) Communications preceded by an urgency signal.
   (c) Communications preceded by a safety signal.
   (d) Communications pertaining to direction finding.
   (e) Communications relative to navigation and safe movement of aircraft.
   (f) Communications relating to the navigation, movements, and needs of ships, and weather observation messages destined for an official meteorological service.
   (g) Government radiotelegrams: priority nations.
   (h) Government communications for which priority has been requested.

(i) Service communications relating to the working of radio communications previously exchanged.

(j) All other communications.

Q15: What are the grounds for suspension of operator licenses?
A: (a) Violation of any provision of an act, treaty, or convention or any regulation made by the Commission under such act, treaty, or convention.

(b) Failure to carry out a lawful order of one in charge of a ship or aircraft on which he/she is employed.

(c) Willfully damaging radio equipment or permitting it to be damaged.

(d) Transmitting prohibited signals or communications as outlined in Question 12.

(e) Willfully or maliciously interfering with any other radio transmissions.

(f) Aiding or abetting another to attempt to obtain a license by fraudulent means.

Q16: When may an operator divulge the contents of an intercepted message?
A: Messages pertaining to ships in distress or those transmitted for the use of the general public may be divulged. However, new rules have tightened the secrecy provisions of the Communications Act.

Q17: If a licensee is notified that he/she has violated an FCC rule or provision of the Communications Act of 1924, what should be done?
A: Within 10 days of the receipt of notice or such period as may be specified therein, the licensee must send a written reply in duplicate to the FCC office that originated the violation notice. If an answer or acknowledgment cannot be made within the 10-day period due to illness or other unavoidable circumstances, an answer must be made with satisfactory explanation for the delay at the earliest practicable date. The answer to each notice must be complete within itself, and abbreviation by reference to other communications or answers to other notices are not acceptable. In every instance, the answer must contain a statement of action to correct the condition or omission and preclude its recurrence. *Note*: If the notice relates to violations that might have resulted from the physical or electrical characteristics of transmitting apparatus and the new apparatus is to be installed, the reply must give date of order, manufacturer, and estimated date of delivery.

Q18: If a licensee receives a notice of suspension of his license, what must he/she do?
A: The operator must send the license to the FCC on or before the effective date of the order. Actually, the notice of suspension is not effective until received by the licensee. From the date of receipt, the licensee has 15 days to mail an application for a hearing on the suspension order. Upon such compliance, the suspension will be held in abeyance pending conclusion of the hearing.

Q19: What are the penalties provided for violating a provision of the Communications Act of 1934 or a rule of the FCC?
A: Violation of the Act, upon conviction, carries a fine of not more than $10,000, one year in prison, or both. The prison term can be increased to two years for second offenders. Violation of an FCC rule, if convicted, provides a fine of not more than $500 for each and every day during such offense, in addition to any other penalties.

Q20: What is meant by "harmful interference"?
A: Any emission, radiation, or induction that endangers the proper functioning of the radio navigation service or of other safety services or seriously degrades, obstructs, or

repeatedly interrupts a radio communication service operating in accordance with these regulations.

ELEMENT 1: SAMPLE QUESTIONS

Here are some sample questions, similar to those in the exam, that pertain to Element 1, Basic Law.

1. When may a license be renewed?
   (a) Any time within six months before or after expiration.
   (b) Within a year of expiration of current license.
   (c) At all times, if no violations have been made.
   (d) Only within one month of expiration date.
   (e) None of the above.

2. Urgency signals have second priority—what has first?
   (a) Overseas commercial messages.
   (b) Safety communications.
   (c) Distress calls and messages.
   (d) International bulletins.
   (e) D-F communications.

3. Who may make corrections in the station log?
   (a) Any licensed operator.
   (b) Only an officer of the station.
   (c) Any licensed operator on duty.
   (d) Any person competent and familiar with facts.
   (e) The person who made the initial entry.

4. What is the usual license term for radio operators?
   (a) Two years.
   (b) Eight years.
   (c) One year.
   (d) Five years.
   (e) Until 65 years of age.

5. Who may inspect radio stations in the U.S.?
   (a) The Federal Communications Commission.
   (b) The U.S. Dept. of Commerce.
   (c) The General Services Administration.
   (d) Internal Revenue Service.
   (e) Secretary of Interior.

6. When do the secrecy provisions of the law not apply?
   (a) Commercial bulletins.
   (b) Distress messages.

(c) Position reports.
(d) Private wire service messages.
(e) Weather bulletins.

7. An operator who loses his/her license must take what action?
   (a) Notify the FCC within 30 days.
   (b) Exhibit a copy of the application for the duplicate while continuing work.
   (c) Stop operating until his duplicate is received.
   (d) Notify the field office the next business day.
   (e) Continue operating until an FCC inspector arrives.

8. False signals of distress are:
   (a) Permissible with low power only.
   (b) Proper tests for rescue efficiency.
   (c) Prohibited by law.
   (d) Allowed when checking emergency equipment.
   (e) Permitted only from midnight to sunrise.

9. If you receive a notice of violation from the FCC, what must you do?
   (a) Reply within 24 hours to the nearest field office of the FCC.
   (b) Reply within 10 days to the FCC office that originated the notice.
   (c) Reply within three days to the main office of the FCC.
   (d) Respond immediately to the nearest Federal District Court.
   (e) None of the above.

10. A message may be rebroadcast:
    (a) If authorized by the operator in charge.
    (b) Provided the FCC engineer in charge of the district is notified.
    (c) With proper credit given to the source.
    (d) With the express authorization of the originating station.
    (e) If the originating station is promptly notified.

11. How are errors in station logs corrected?
    (a) Erase the error and enter the correction in ink.
    (b) Cross out the mistake, initial, and enter the name of correction.
    (c) Remove the error with correction fluid and type in the correction.
    (d) Line-out the error, initial the correction, and indicate the date made.
    (e) Have corrections notarized and notify the FCC district office.

12. An operator's license may not be suspended for which of the following offenses?
    (a) Allowing another to willfully destroy or damage radio equipment.
    (b) Transmitting false or deceptive signals.
    (c) Transmitting unnecessary communications.
    (d) Transmitting obscene, indecent, or profane language.
    (e) Refusal to carry out orders from the station manager.

13. If a notice of suspension is received, what must you do?
    (a) Request a hearing within 15 days.
    (b) Cease operating at once.
    (c) Return your operator's license to the FCC immediately.
    (d) Request a hearing within 30 days.
    (e) None of the above.

14. To obtain a commercial FCC license or permit, you must be:
    (a) Of average intelligence.
    (b) A citizen of the United States.
    (c) At least 21 years of age.
    (d) A citizen of a friendly country.
    (e) A licensed operator of a friendly country.

15. A transmitter in a public place:
    (a) May never be left unattended unless turned off.
    (b) Must be posted with a warning sign.
    (c) Must be locked when unattended.
    (d) Must be fenced in to keep children away.
    (e) Must have the output stage removed when unattended.

16. A person willfully violating a provision of the Communications Act of 1934 is subject to:
    (a) A fine of not more than $500.
    (b) A fine not to exceed $10,000.
    (c) A fine of $5000 and two years imprisonment.
    (d) A fine not exceeding $10,000 and one year imprisonment.
    (e) A fine not exceeding $5000 and one year imprisonment.

17. An operator violating a rule of the FCC can be subject to:
    (a) Two years imprisonment.
    (b) Five years imprisonment.
    (c) $500 fine for each day during which the violation occurs.
    (d) $1000 fine for each day the violation occurs.
    (e) None of the above.

18. Under what conditions may an operator divulge the contents of an intercepted message?
    (a) If the sender's permission is obtained.
    (b) By notification of the FCC within 24 hours.
    (c) When receiving a distress message or a message intended for general public use.
    (d) If the regular commercial rate is paid.
    (e) Not permitted at any time.

19. Deliberate interference with radio communications is permissible:
    (a) At SHF frequencies only.
    (b) In the 500 kHz channel, only at reduced power.

(c) When required to establish frequency checks.
(d) Under no conditions.
(e) When operating above 54 MHz (daytime only).

ELEMENT 2: SAMPLE QUESTIONS

The following questions are broken down into those pertinent to Category O and those pertinent to Category M.

CATEGORY O: GENERAL STATION OPERATING PROCEDURES

Q1: **What should an operator do when leaving a transmitter unattended?**
A: Transmitters must be locked or made inaccessible to unauthorized personnel.

Q2: **What are the meanings of clear, out, over, roger, words twice, repeat and break?** [M]
A: *clear* or *out*—The transmission is ended and no response is expected.
　　over—My transmission is ended and a response is expected.
　　roger—Your last transmission has been received and completely understood.
　　words twice—Each word will be given twice due to poor reception.
　　repeat—Say again.
　　break—This is the end of this part, another will follow shortly.

Q3: **How should a microphone be treated when used in noisy locations?**
A: Shield the microphone with cupped hands to avoid background noise pickup.

Q4: **What can happen to the received signal when an operator shouts into a microphone?**
A: This causes overmodulation, distorts the signal at the receiving station, making it difficult to understand. Interference with stations on adjacent frequencies also result.

Q5: **Why should radio transmitters be off when signals are not being transmitted?**
A: The transmitter can cause interference with other stations even when not modulated.

Q6: **Why should an operator use well-known words and phrases?**
A: Simple phrases and plain words are easy to understand; this reduces errors and avoids undue repetition, saving time.

Q7: **Why is the station's call sign transmitted?**
A: This provides positive identification of the sending station to avoid possible confusion.

Q8: **Where does an operator find specifications for obstruction marking and lighting (where required) for the antenna towers of a particular radio station?**
A: Simply examine the station authorization as issued by the FCC. (See part 17 FCC Rules and Regulations for specifications.)

Q9: **What should you do if you hear profanity being used at your station?**
A: Cut the speaker off, note the incident in the station log, and forward a report of the infraction to the FCC.

Q10: When may an operator use his/her station without regard to certain provisions of the station license?
A: During a period of emergency in which normal communications facilities are disrupted as a result of a hurricane, flood, or earthquake, or similar disaster. Notice must be sent to the FCC in Washington, D.C. *and* to the engineer in charge of each district of the station location as soon as possible following beginning of such emergency use. Emergency use of the station shall be discontinued as soon as substantially normal communication facilities are again available, along with immediate notification to the Commission and the engineer in charge when such special use of the station is terminated. To summarize,

(a) As soon as possible after beginning such emergency use of the station, send notice to the FCC in Washington, D.C., and the engineer in charge of the district in which the station is located.

(b) Emergency use of the station must be discontinued as soon as substantially normal communications facilities are restored.

(c) Notify the FCC in Washington, D.C., and the engineer in charge immediately when special use of a station is terminated.

(d) Under no circumstances may any station engage in emergency transmission on frequencies other than, or with power in excess of, that specified in the instrument or authorization or as otherwise expressly provided by the Commission or by law.

(e) Any emergency communication undertaken under this section must terminate upon order of the Commission.

Q11: Who bears the responsibility if an operator permits an unlicensed person to speak over his station?
A: The licensed operator in charge of the station, as he/she is responsible for the proper operation.

Q12: What is the *phonetic alphabet* in radiotelephone communications?
A: A list of 26 words, each starting with a different letter of the alphabet is used to avoid possible misunderstanding of similar sounding words or letters. For example the word *bad* is recognized by the phonetic spelling bravo, alpha, delta.

Q13: How does the licensed operator of a station normally exhibit operational authority?
A: Simply by posting a valid operator license or permit at the transmitter control point.

Q14: What precautions should be observed in testing a station on the air?
A: The operator should indicate testing by clearly giving the station call sign or name of the station. Tests must be brief; before starting, check the frequency to make sure that the test will not interfere with other communications already in progress.

CATEGORY M: MARITIME SERVICES OPERATING PROCEDURE

Q1: What is the importance of the frequency 2182 kHz?
A: The frequency 2182 kHz is an international distress frequency for radiotelephony. It is used for this purpose by ship, aircraft, and survival craft stations using frequencies in the authorized bands between 1605 and 4000kHz when requesting assistance from the maritime services. It is also the international general radiotelephone calling frequency for

the maritime mobile service, and it may be used as a carrier frequency for this purpose by ship stations and aircraft stations operating in the maritime mobile service.

Q2: **Describe completely what actions should be taken by a radio operator who hears (A) a distress message, and (B) a safety message.**

A: (A) Distress Message

1. Acknowledge receipt of the distress message.

a. Stations of the maritime mobile service that receive a distress message from a mobile station that is, beyond any possible doubt, in their vicinity must immediately acknowledge receipt. However, in areas where reliable communication with one or more coast stations is practicable, ship stations may defer this acknowledgment for a short interval so that a coast station can acknowledge receipt.

b. Stations of the maritime mobile service that receive a distress message from a mobile station that is *not*, beyond any possible doubt, in their vicinity must allow a short interval of time to elapse before acknowledging receipt of the message to permit stations nearer to the mobile station in distress to acknowledge receipt without interference.

2. Form of acknowledgment.

a. The acknowledgment of the receipt of a distress message is transmitted (when radiotelegraphy is used) in the following form: (1) The call sign of the station sending the distress message, sent three times; (2) the letters DE; (3) The call sign of the station acknowledging receipt, sent three times; (4) The group RRR; (5) The distress signal SOS.

b. The acknowledgment of receipt of a distress message is transmitted, when radiotelephony is used, in the following form: (1) The call sign or other identification of the station sending the distress message, spoken three times; (2) The words, "This is"; (3) The call sign or other identification of the station acknowledging receipt, spoken three times; (4) The word, "received"; (5) The distress signal "mayday."

3. Information furnished by the acknowledging station.

a. Every mobile station that acknowledges receipt of a distress message must, on the order of the master or person responsible for the ship, aircraft, or other vehicle carrying such mobile station, transmit as soon as possible the following information in the order shown: (1) Its name; (2) Its position; (3) The speed at which it is proceeding; and (4) The approximate time it will take to reach the mobile station in distress.

b. Before sending this message, the station must ensure that it will not interfere with the emissions of other stations better situated to render immediate assistance to the station in distress.

4. Transmission of distress message by station not in distress.

a. A mobile or a land station that learns that a mobile station is in distress must transmit a distress message in any of the following cases: (1) When the station in distress is not itself in a position to transmit the distress message; (2) When the master or person responsible for the ship, aircraft, or other vehicle not in distress or the person responsible for the land station considers that further help is necessary; (3) When, although not in a position to render assistance, it has heard a distress message that has not been acknowledged. When a mobile station transmits a message under these conditions, all necessary steps must be taken to notify the authorities who might be able to render assistance.

    b. The transmission of a distress message under conditions prescribed must be made on either or both of the international distress frequencies (500 kHz radiotelegraph; 2182 kHz radiotelephone) or on any other available frequency on which attention might be attracted.

    c. The transmission of the distress message must always be preceded by the call indicated below. The code should be preceded whenever possible by the radiotelegraph or radiotelephone alarm signal.

When *radiotelegraphy* is used, this call consists of:
- DD SOS SOS SOS DDD
- The letters DE
- The call sign of the transmitting station, sent three times.

When *radiotelephony* is used, this call consists of:
- The signal, "mayday relay," spoken three times;
- The words, "This is";
- The call sign or other identification of the transmitting station, spoken three times.

    d. When the radiotelegraph alarm signal is used, an interval of two minutes should be allowed (whenever this is considered necessary) before the transmission of the call.

(B) Safety Message

The safety message contains information concerning the safety of navigation or important meteorological warnings. All such messages should be reported to the ship's master, and the radio operator should not make any transmission likely to interfere with a safety message.

Q3: What information must be contained in a distress message? What procedure should be followed? What is a good choice of words to use in sending a distress message?

A: (A) Distress Signals
1. The international radiotelegraph distress signal consists of three dots, three dashes, three dots (...- - -...), symbolized herein as SOS, transmitted as a single signal in which the dashes are slightly prolonged so as to be distinguished clearly from the dots.
2. The international radiotelephone distress signal is the word "Mayday" (from the French expression *m'aidez*.)
3. These distress signals indicate that a mobile station is threatened with grave and imminent danger and requests immediate assistance.

(B) Distress Calls
   1. The distress call sent by radiotelegraphy consists of:
      a. The distress signal SOS, sent three times;
      b. The letters DE;
      c. The call sign of the mobile station in distress, sent three times.
   2. The distress call sent by radiotelephony consists of:
      a. The distress signal, "Mayday," spoken three times;
      b. The words, "This is";
      c. The call sign, or name if no call has been assigned, of the mobile station in distress, spoken three times.
   3. The distress call has absolute priority over all other transmissions. All stations that hear it must immediately cease any transmission capable of interfering with the distress traffic and continue to listen on the frequency used for the distress call. Do not address the call to a particular station. The receiving station should not acknowledge receipt before receiving the distress message that follows.
(C) Distress Messages
   1. The *radiotelegraph* distress message consists of:
      a. The distress signal, SOS;
      b. The name of the mobile station in distress;
      c. Particulars of its position;
      d. The nature of the distress;
      e. The kind of assistance desired;
      f. Any other information that might facilitate rescue.
   2. The *radiotelephone* distress message consists of:
      a. The distress signal "Mayday";
      b. The name of the mobile station in distress;
      c. Particulars of its position;
      d. The nature of the distress;
      e. The kind of assistance desired;
      f. Any other information that might facilitate rescue (for example, the length, color and type of vessel, and number of persons aboard).
   3. As a general rule, a ship signals its position in latitude and longitude (Greenwich) using figures for degrees and minutes and either north, or south and east or west. In radiotelegraphy the signal *dot, dash, dot, dash, dot, dash* (.-.-.-) is used for separation of the degrees and minutes. When practicable, the true bearing and distance in nautical miles from a known position is appropriate.
(D) Radiotelephone Distress Call and Message Transmission Procedure
   1. The radiotelephone distress procedure shall consist of:
      a. The radiotelephone alarm signal (if possible);
      b. The distress call;
      c. The distress message.
   2. The radiotelephone distress transmissions must be made slowly and distinctly, with each word clearly pronounced to facilitate transcription.

3. After the transmission by radiotelephony of its distress message, the mobile station might be requested to transmit suitable signals, followed by its call sign or name, to permit a direction-finding station to determine its position. This request may be repeated at frequent intervals if necessary.
4. The distress message, preceded by the distress call, must be repeated at intervals until an answer is received. This repetition must be preceded by the radiotelephone alarm signal whenever possible.
5. When the mobile station in distress receives no answer to its message transmitted on the distress frequency, the message can be repeated on any other available frequency on which attention might be attracted.

The following is an example of an accurate distress message: "Mayday, mayday, mayday. This is the Freighter Brown, 32 degrees 28 minutes North latitude, 48 degrees 12 minutes West longitude. Abandoning ship due to fire. 23 crewmen aboard. Launching four lifeboats. Ship will sink in 30 minutes. Over."

Q4: What are the requirements for keeping watch on 2182 kHz?
A: Each station on board a ship navigating the Great Lakes and licensed to transmit by telephony on one or more frequencies within the 1605- to 3500-kHz band must, during its hours of service for telephone, maintain an efficient watch for reception of emissions on the authorized carrier frequency 2182 kHz, whenever the station is not being used for transmission on that frequency or for communication on other frequencies.

Except for stations on board vessels required by law to be fitted with radiotelegraph equipment, each ship station (in addition to those ship stations specified in the above paragraph) licensed to transmit by telephony on one or more frequencies in the band 1605 to 3500 kHz must, during its hours of telephony service, maintain an efficient watch for the reception of emissions on the authorized carrier frequency of 2182 kHz whenever such station is not being used for transmission on that frequency or for communication on other frequencies. When the ship station is in Region 1 or 3, the watch must be maintained at least twice each hour for three minutes commencing on the hour and the half-hour.

Q5: Under what circumstances may a coast station contact a land station by radio?
A: For the purpose of facilitating the transmission or reception of safety communication to or from a ship or aircraft station.

Q6: What do distress, safety, and urgency signals indicate? What are the international urgency, distress, and safety signals? In the case of a mobile radio station in distress, what station is responsible for the control of distress message traffic?
A: The *distress signal*, "Mayday" or SOS, indicates that a mobile station is threatened by grave and imminent danger and requests immediate assistance.

The *safety signal*, "Security" or TTT, indicates that the station is about to transmit a message concerning the safety of navigation or giving important meteorological warnings.

The *urgency signal*, "Pan" or XXX, indicates that the calling station has a very urgent message to transmit concerning the safety of a ship, aircraft, or other vehicle, or the safety of a person.

The *international urgency signal in radiotelephony* is the word "pan" spoken three times and transmitted before the call. In *radiotelegraphy*, the urgency signal consists of three repetitions of the group XXX, sent with the individual letters of each group and the successive groups, clearly separated from each other.

The *international safety signal in radiotelephony* is the word "Security" spoken three times and transmitted before the call. In *radiotelegraphy*, the safety signal consists of three repetitions of the group TTT, sent with the individual letters of each group and the successive groups, clearly separated from each other.

The *international distress signal in radiotelephony* consists of the word "Mayday'',, spoken three times and transmitted before the call. In *radiotelegraphy*, the distress signal consists of the code SOS sent three times.

The control of distress traffic is the responsibility of the mobile station in distress or of the station which, pursuant to FCC Rule 83.242a, has sent the distress message. These stations may, however, delegate the control of the distress traffic to another station.

Q7: In regions of heavy traffic, why should an interval be left between radiotelephone calls? Why should a radio operator listen before transmitting on a shared channel? How long may a radio operator in the mobile service continue to attempt to contact a station that does not answer?

A: In regions of heavy traffic (many stations operating), the radio operator must leave an interval of time between radiotelephone calls to permit other stations to transmit on the same frequency without interference. This is required by FCC rules, as many stations are sharing a few allotted channels. A radio operator should listen before transmitting on a shared channel to make sure that no one else is transmitting on that same channel.

Calling a particular station should not continue for more than 30 seconds in each instance. If the called station does not respond, that station should not be called again until after an interval of at least two minutes. When such station does not answer to a call sent three times at two-minute intervals, the calling must stop and not be resumed for an interval of 15 minutes unless it is obvious that other communications in progress will not have harmful interference at the time. In the latter case, calls may be resumed after an interval of at least three minutes. However, the provisions of this paragraph do not apply in the case of an emergency involving safety.

Q8: Why are test transmissions sent? How often should they be sent? What is the proper way to send a test message? How often should the station's call sign be sent?

A: Test transmissions are sent to make sure that the equipment is in proper operating condition. They should be sent on a regular basis, once a day, before the normal day's communications are scheduled. Regular tests often reveal defects which, if corrected promptly, can prevent needless delays if a problem arises.

Ship stations must use every precaution to ensure that, when conducting operational transmitter tests, the emissions of the station will not cause harmful interference. Radiation must be reduced to the lowest practicable level, and if feasible, entirely suppressed. The proper way to send a test message is as follows:

1. The licensed radio operator or other person responsible for operation of the transmitting apparatus must ascertain by careful listening that the test emissions will not interfere with transmissions in progress; if they are likely to interfere with the working of a coast or aeronautical station in the vicinity of the ship station, the consent of that station or stations must be obtained before the test emissions occur.

2. The official call sign of the testing station, followed by the word "test," must be announced on the channel being used for the test as a warning that test emissions are about to be made on that frequency.

3. If, as a result of the announcement prescribed in subparagraph 2, any station transmits by voice the word "wait," testing must be suspended. When, after an appropriate interval of time, such announcement is repeated and no response is observed with careful listening, indicating that harmful interference will not occur, the operator should proceed as set forth in subparagraph 4.

4. The operator must announce the word, "testing," followed by the count "1,2,3,4, etc." or by test phrase or sentences not in conflict with normal operating signals. If the transmission is not voice capable, use appropriate test signals not in conflict with normal operating signals. The test signals in either case need have a duration not exceeding 10 seconds. At the conclusion of the test, announce the official sign of the testing station, the name of the ship on which the station is located, and the general location of the ship at the time the test is being made. This test transmission must not be repeated until a period of at least one minute has elapsed; on the frequency 2182 kHz or 156.8 MHz in a region of heavy traffic, a period of at least five minutes must elapse before the test transmission is repeated.

5. When testing is conducted on any frequency within the bands 2170 to 2194 kHz, 156.75 to 156.85 MHz, 480 to 510 kHz (survival craft transmitters only), or 8362 to 8366 kHz (survival craft transmitters only), no test transmissions may occur that are likely to actuate any automatic alarm receiver within range. Survival craft stations using telephony shall not be tested on the frequency 500 kHz during the 500-kHz silence periods. The test signal shall have a duration not exceeding ten seconds. The official call sign of the testing station shall be given at the conclusion of each test.

Q9: In the mobile service, why should radiotelephone messages be as brief as possible?
A: This permits all stations to transmit their communications without undue delay, and the courtesy works both ways.

Q10: What are the meanings of *clear, out, over, roger, words twice, repeat* and *break*?
A: *clear* or *out*—Conversation is ended and no response expected.
   over—My transmission is ended and I expect a response from you.
   roger—I have received all of your transmission and understood same clearly.
   repeat—Say again.
   break—Hold, I will continue the transmission.

Q11: Does the Geneva World Administrative Radio Treaty of 1979 give other countries the authority to inspect U.S. vessels?
A: Yes. The governments of appropriate administrations of countries that a mobile station visits might require the operator to present the license. The operator of the mobile station, or the person responsible for the station, facilitates this examination. The license must be kept in such a way that it can be produced upon request. As far as possible, the license, or a copy certified by the authority that issued it, should be permanently exhibited in the station.

Q12: Why are call signs sent? Why should they be sent clearly and distinctly?
A: Call signs are sent to enable other stations to identify all callers. They should be sent clearly and distinctly to avoid unnecessary repetitions.

Q13: How does the licensed operator of a ship station exhibit his authority to operate a station?
A: When a licensed operator is required to operate a station, the original license of each such operator, while employed or designated as radio operator of the station, must be posted in a conspicuous place at the principle location on board ship at which the station is operated; provided that in the case of stations of a portable nature, including marine-utility stations, or in the case where the operator holds a restricted radiotelephone operator permit the operator may in lieu of posting have on his person either his required operator license or a duly issued verification card (FCC Form 758-F), attesting to the existence of that license.

Q14: When may a coast station not charge for messages it is requested to handle?
A: No charge is made for the service of any public coast station unless effective tariffs applicable to such service are on file with the Commission. No charge is made by any station in the maritime mobile service of the United States for the transmission of distress messages and replies thereto in connection with situations involving the safety of life and property at sea. No charge is made by any station in the maritime mobile service of the United States for the transmissions, receipt, or delay of the information concerning dangers to navigation, originating on a ship of the United States or of a foreign country.

Q15: What is the difference between calling and working frequencies?
A: A calling frequency is one on which all stations listen for incoming calls or on which they transmit a call for another station. Once a reply has been received to the initial call, both stations transfer to a working frequency to continue their communication.

ELEMENT 2: SAMPLE QUESTIONS

Questions 1 through 20 pertain to Category 0, and 21 through 40 pertain to Category M.

1. What should an operator do when leaving a transmitter unattended?
    (a) Transmitter should be inaccessible to authorized persons.
    (b) Notify the night watchman.
    (c) Make a note of the time and date on the log.
    (d) Turn the keys over to the security officer.
    (e) Pull the main circuit breaker.

2. What problem can result from shouting into a microphone?
    (a) Overmodulation.
    (b) Miller effect.
    (c) Linear amplification.
    (d) Demodulation.
    (e) The amplifier fuse will blow.

3. How should a microphone be treated when used in a noisy location?
   (a) Reduce the audio gain.
   (b) Speak in a normal tone at about six inches.
   (c) Cover the microphone with a handkerchief.
   (d) Cup hands over the microphone to help exclude noise.
   (e) Speak softly into the microphone at close range.

4. The term *clear* means
   (a) I have received your last transmission fully.
   (b) This message ended, another will follow.
   (c) Message ended, no response expected.
   (d) Speak each word more distinctly.
   (e) My transmission ended, I expect response.

5. The word *break* indicates
   (a) End of this message, another will follow.
   (b) My transmission is ended, response expected.
   (c) My message ended, no response expected.
   (d) Last transmission received completely.
   (e) Standby for further instructions.

6. The word *roger* indicates
   (a) I have received all of your last transmission.
   (b) My transmission is ended; no response expected.
   (c) This completes my message; another will follow.
   (d) Please repeat each group twice.
   (e) None of the above.

7. Why should a transmitter be off when transmissions are not being made?
   (a) For economy reasons.
   (b) To avoid wear on equipment.
   (c) In order to avoid interference with other stations.
   (d) To prevent overheating the power supply.
   (e) To check operation of the main switch and regulators.

8. Parts of a single message may be separated by the following:
   (a) Stop
   (b) Repeat
   (c) Break
   (d) Over
   (e) Clear

9. Responsibility for the proper operation of the radio station falls on:
   (a) The station licensee.
   (b) The owner of the station.
   (c) The person using the microphone.

(d) The licensed operator in charge of the station.
(e) Any operator over 21 years of age.

10. During an emergency, the operator should:
    (a) Change frequency to avoid interference.
    (b) Reduce power to a predetermined level.
    (c) Discontinue operation at once.
    (d) Increase power above that authorized.
    (e) Stand by for further instructions before cutting the carrier.

11. How does the licensed operator show his authority to operate the station?
    (a) Posting his license in the station manager's office.
    (b) Posting his license at the transmitter control room.
    (c) Posting his license inside the antenna house.
    (d) By carrying a card attesting to same (FCC Form 758-F).
    (e) Any of the above.

12. What should the operator do if profanity is used at his/her station?
    (a) Send a copy of the incident to the FCC.
    (b) Enter the information in the station log.
    (c) Report the incident to the local authorities.
    (d) Turn off the speaker, enter a report in the station log, and submit the report to the FCC.
    (e) Notify the station owner and cut off the audio.

13. Where may specifications for obstruction marking and lighting of antenna towers be found?
    (a) In the radio station authorization.
    (b) Extracts from the Geneva Treaty.
    (c) Part 74 of the FCC rules and regulations.
    (d) Part 17 of the FCC rules and regulations.
    (e) None of the above.

14. Why is the station's call sign transmitted?
    (a) To provide positive identification of the sending station.
    (b) To reveal the location of the transmitter.
    (c) To permit determination of the output power.
    (d) Checking the frequency by the monitoring services.
    (e) To identify station ownership.

15. An operator testing the transmitter should:
    (a) Omit a statement of test.
    (b) Make the test as brief as possible.
    (c) Provide personal identification.
    (d) Not listen for a clear channel before the test.
    (e) Increase the power for the test only.

16. In radiotelephone communications, common words representing letters of the alphabet that are used to spell out words positively are called:
    (a) The communications methods.
    (b) The Morse code.
    (c) The Miller effect.
    (d) Alternate alphabet.
    (e) The phonetic alphabet.

17. If an unlicensed person speaks over the air, who bears the responsibility for his/her actions?
    (a) Only the individual speaking.
    (b) The general manager of the station.
    (c) The licensed operator in charge at the time.
    (d) The owner or owners of the station.
    (e) None of the above.

18. Why should an operator use well-known words and phrases?
    (a) To demonstrate familiarity with the use.
    (b) To eliminate distortion.
    (c) Avoids damaging the microphone internally.
    (d) Reduces biasing requirements.
    (e) None of the above.

19. If testing the radio transmitter, the operator should not:
    (a) Test for a brief period.
    (b) Interfere with normal communications.
    (c) Clearly indicate that a test is in progress.
    (d) Identify the station by the call sign.
    (e) None of the above.

20. The word *out*, used in radiotelephone communications, indicates:
    (a) Transmission complete, no response expected.
    (b) Transmission complete, response expected.
    (c) Ignore previous message and resume transmission.
    (d) All of your last transmission understood.
    (e) End of this message, another will follow.

21. What is the radiotelephone distress verbal signal word?
    (a) XXX
    (b) Pan
    (c) Mayday
    (d) Attention
    (e) Security

22. What is the importance of the frequency 2182 kHz?
    (a) It is the international distress frequency for radiotelephone.
    (b) It is the frequency for radio beacon purposes.
    (c) It is the international distress frequency for radiotelegraph.
    (d) This is the appropriate ship-shore working frequency.
    (e) It is the frequency for commercial messages between ships.

23. The control of distress traffic must be handled by:
    (a) A representative of the FCC.
    (b) Anyone willing to volunteer the distress signal.
    (c) The station originating the distress signal.
    (d) The nearest Coast Guard station.
    (e) Any government aircraft in the area.

24. What does the word *pan* indicate?
    (a) Urgency message.
    (b) Distress message.
    (c) Safety message.
    (d) Pan-American aircraft.
    (e) Weather message.

25. What does the word *security* indicate?
    (a) Navigational message.
    (b) Safety message.
    (c) Distress priority message.
    (d) Radio beacon signal.
    (e) Urgency message.

26. When is it not necessary to acknowledge receipt of a distress message at once?
    (a) If the ship is traveling in the opposite direction.
    (b) If the ship is too far away to be assisted.
    (c) To allow a closer station to acknowledge without interference.
    (d) If the ship is nearer a Coast Guard vessel.
    (e) When the ship in distress belongs to an unfriendly country.

27. What is the international general calling and distress frequency for radiotelephone in the maritime mobile service?
    (a) 500 kHz
    (b) 1650 kHz
    (c) 2,182 MHz
    (d) 88.5 MHz
    (e) 2,182 kHz

28. What are the "top three" priority messages in order of their priority?
    (a) Distress, safety, urgency
    (b) Distress, urgency, safety

(c) Safety, distress, urgency
(d) Distress, navigational, urgency
(e) Priority, distress, safety

29. When may a mobile station send a distress message for another mobile station in distress?
    (a) When the person not in distress considers further help needed.
    (b) When a station in distress is not in a position to transmit.
    (c) When it has heard a distress message not acknowledged.
    (d) All of the above apply.
    (e) None of the above apply.

30. The safety signal would have priority over:
    (a) D-F bearing communications.
    (b) Urgency messages.
    (c) Distress messages.
    (d) Communications preceded by an urgent signal.
    (e) None of the above.

31. When operating on a shared frequency, the radio operator must:
    (a) Never operate after local sunset.
    (b) Leave an interval between calls.
    (c) Limit transmission to five minutes.
    (d) Transmit on a fixed schedule only.
    (e) Increase power to override others.

32. What is the purpose of a test transmission?
    (a) To ensure proper operation of the equipment.
    (b) To avoid antenna icing conditions.
    (c) To locate Coast Guard stations in the area.
    (d) Provide a check on power supply regulation.
    (e) Acquire additional time on the air.

33. A station may not make charge for:
    (a) Distress messages.
    (b) International commercial messages.
    (c) News bulletins.
    (d) Personal messages if under 50 words.
    (e) Baseball scores.

34. What is a calling frequency?
    (a) One used only for special messages.
    (b) Frequency used after an initial call for communications.
    (c) Frequency for personal use only.
    (d) Frequency used for priority messages.
    (e) Frequency on which stations listen for incoming calls.

35. Calling a particular station should be limited to about:
    (a) 45 seconds
    (b) 15 seconds
    (c) 10 seconds
    (d) five seconds
    (e) one minute

36. When hearing the word *security* repeated three times:
    (a) Call all stations.
    (b) Increase power to attract other stations.
    (c) Continue listening until the message is completed.
    (d) Contact the Coast Guard for urgent information.
    (e) None of the above.

37. Why should all radiotelephone messages in the mobile service be as direct and to the point as possible?
    (a) So all stations may transmit their messages without delay.
    (b) To avoid a cross-talk problem.
    (c) To avoid overmodulation of the carrier.
    (d) To eliminate parasitic oscillations.
    (e) Harmonic suppression is improved.

38. What information must be contained in distress messages?
    (a) Position, nature of distress, kind of help needed.
    (b) Output power, call sign, number of operators.
    (c) Type of antenna, length of ship, and location.
    (d) Approximate distance from port, number of persons.
    (e) Speed and direction, assistance needed.

39. Under what circumstances may a coast station contact a land station by radio?
    (a) To aid transmission of safety communications to the ship.
    (b) When commercial messages are not getting through.
    (c) When the channel is not clear from ship to shore.
    (d) When power is not sufficient to contact the ship.
    (e) To report a violation of priorities.

40. How does the operator of a ship station exhibit his authority to operate a station?
    (a) Showing proficiency in Morse code.
    (b) Posting his license in plain view at the control point.
    (c) Exhibiting his school diploma.
    (d) Showing his Navy discharge papers.
    (e) Posting his latest proficiency certificate.

2
Basic Radiotelephone: Part I

As we begin our study of basic radiotelephone, an understanding of the fundamentals of electronics is a good first step. *Direct current* (or *dc*) flows in one direction and consists of a force or pressure known as a *voltage*. The amount of flow is the *current*, which is measured in amperes or milliamperes (thousandths of an ampere). *Power* is a multiple of the two: *volts* times *amperes* equals *watts*. In radio broadcasting, a term commonly used is kilowatts (kW) which is one thousand watts; ten kilowatts (10 kW) equals 10,000 watts.

ELECTRON THEORY

The physical structure of the atoms in a material determines whether or not it will conduct electric current. In a ring around the nucleus of each atom, there is a number of electrons. The fewer the number in this outer shell, the more easily other electrons can break away from each atom to become free electrons that can carry a current flow when a voltage is applied to the material. Thus, electric energy can be transferred from one point to another by the movement of free electrons, as in a metallic conductor.

The term *current direction* refers to a point of reference only and does not indicate the direction of movement for the electrical charges. Electron flow is from negative to positive because electrons are negative and unlike charges attract; therefore, it is much simpler to consider the actual movement of electricity to be in the same direction.

To determine whether an electric charge is positive (+) or negative (−), connect a voltmeter to the source. When connected correctly across the battery or source of voltage, the meter pointer should indicate the actual dc voltage on the meter scale, but if the meter leads are reversed, the meter pointer will swing left against the pin, below zero.

Aside from the positive (+) designations, the positive terminal is usually red or has a red wire or lead, while the negative normally uses black as an indicator, in addition to minus or (−).

PROPERTIES OF COMPONENTS

The various properties of resistance, power, impedance, capacitance, inductance, and reactance constitute the characteristics of different components.

RESISTANCE

Resistance of a device or material opposes the flow of an electric current. Resistance converts electrical energy into heat and is the only form of opposition to dc. (Opposition to ac or alternating current is called *impedance* and is described in more detail shortly.) The resistance of a conductor depends on the cross-sectional area, length, and material and is measured in ohms. If a pressure of 1 volt causes a current of 1 ampere to flow through a device, its resistance must be 1 ohm. That useful formula known as Ohm's Law uses the usual letter abbreviations for its factors—resistance in ohms (R), voltage in volts (E or V), and current in amperes (I):

$$E = IR$$

This formula will be used many times (along with others to be introduced) in FCC exams. Figure 2-1 illustrates the Ohm's Law "wheel," which demonstrates how to manipulate the equation to find the desired parameter.

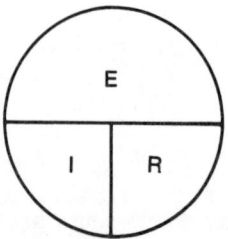

Fig. 2-1. Ohm's Law "wheel." Cover the desired quantity and the remaining "slices" of the pie show the expression you should use to calculate the parameter in question.

If resistances are connected in *series*, simply add the resistance of each to find the total. If R1 is 2 ohms, R2 is 3 ohms, and R3 is 4 ohms, the total resistance of the group of 2 + 3 + 4 is 9 ohms. To calculate the *parallel* connection of the same three resistors, the formula below applies to any number of resistors connected in parallel:

$$R_T = \frac{1}{\frac{1}{R1} + \frac{1}{R2} + \frac{1}{R3}}$$

so,

$$R_T = \cfrac{1}{\cfrac{1}{2} + \cfrac{1}{3} + \cfrac{1}{4}}$$

$$= \cfrac{1}{\cfrac{6}{12} + \cfrac{4}{12} + \cfrac{3}{12}}$$

$$= \cfrac{1}{\cfrac{13}{12}} = 0.923 \text{ ohms.}$$

If all resistors connected in parallel are the same value, simply divide that common value by the number of resistors so connected, or if three resistors are connected in parallel and each has a value of 3 ohms, 3 into 3 ohms equals 1 ohm. As a safeguard, when making quick decisions, the value of any parallel connection of resistors is always *less than the smallest of the group*. This can be an important fact to know on multiple-choice FCC exams! When only two resistors are connected in parallel, a simpler formula, known as Product/Sum can be used:

$$R_T = \frac{R1 \times R2}{R1 + R2}$$

High-power resistors with a rating of 4 watts or more, are usually wirewound with the ohmic values and tolerance printed on the body. However, the common carbon-composition resistors in ¼-, ½-, 1-, or 2-watt sizes use the standard color code for proper identification. To read the color code, begin with the ring closest to one end of the resistor. The color of this first ring represents the first significant figure. The second or next ring provides the second significant figure, while the third ring indicates the multiplier (either the number of zeros or decimal). The fourth ring tells you how close to that value the resistor should be, (the *tolerance*) in percentage. If the fourth band is silver, the actual value of the resistor should be within 10 percent of the indicated value. A gold band stipulates that the value must be a little closer—within 5 percent. If there is no fourth band at all, the tolerance is 20 percent. The gold band sometimes appears in the *third* position; it means something entirely different in position 3—a multiplier of 0.1. So if band one is yellow and band two is violet, that's 47 times 0.1 or 4.7 ohms. The silver band is also used at times as a multiplier in position three, but not as often, and it would change the 47 to 0.47 ohms by indicating a 0.01 multiplier. The complete color code appears in Table 2-1.

POWER

Power is the rate at which electrical energy is delivered and consumed and is measured in watts. The ordinary resistor opposes current flow and the resulting "friction" produces heat. If the resistor gets too hot, it will be damaged by the heat and crack, causing a change

Table 2-1. Resistor Color Code Chart

COLOR	1ST & 2ND BAND	3RD BAND (MULTIPLIER) OR	ADD ZEROS
Black	0	1	none
Brown	1	10	1
Red	2	100	2
Orange	3	1,000	3
Yellow	4	10,000	4
Green	5	100,000	5
Blue	6	1,000,000	6
Violet	7	10,000,000	7
Gray	8	100,000,000	8
White	9	1,000,000,000	9
Gold	N/A	0.1	
Silver	N/A	0.01	

4TH BAND (TOLERANCE VALUE): Gold ±5 percent
Silver ±10 percent
None ±20 percent

in value or even an open circuit. To figure the correct wattage size for a resistor, use the power formula.

$$P = IR$$

For example, if a 4.7-ohm resistor must carry a current of 0.5 amps or 500 mA, what size could be used without danger of overheating? Squaring the current 0.5 equals 0.25 times 4.7 or 1.175 watts, so a 2-watt resistor would do nicely. If only voltage and resistance values are known.

$$P = \frac{E^2}{R}$$

When voltage and current figures alone are handy.

$$P = IE$$

Relays

A relay is an electromechanical switching device that opens or closes one or more sets of contacts when its armature is actuated by a current flow through the relay coil. Exam problems are often given regarding relays. For example, if a relay coil resistance is 300 ohms, and the current through it is 0.25 amperes from a 115-volt source, what value resistor must be connected in series with the coil? By Ohm's Law, the voltage required across the relay coil is

$$E = IR$$
$$= .25 \times 300$$
$$= 75 \text{ volts}$$

so, a series resistor capable of dropping the 115-volt source to 75 volts is required. This resistor value is determined by

$$R = E/I$$
$$= 40/.25$$
$$= 160 \text{ ohms}$$

Naturally, the 0.25 ampere current flowing through the coil would also flow through the dropping resistor.

Relays usually have one or more sets of contacts that are *normally closed* or *normally open*. The normally closed (NC) contact means that the contact points are closed when no current is flowing through the coil. The normally open (NO) contacts are open when the coil has no current flowing through it. As soon as current is applied to the coil, the NC contacts open and the NO contacts close. Relay contacts are always shown in the de-energized (inoperative) position. The normal position is always without current applied to the coil unless otherwise noted.

IMPEDANCE

Impedance is the total opposition to an alternating current (ac) flow at a specific frequency and is a combination of the resistance and *reactance*. Impedance is represented by the letter Z and is measured in ohms. The formula for impedance is:

$$Z = \sqrt{R^2 + X^2}$$

The variable X is the reactance, which varies according to the frequency of the current and the inductive or capacitive value. When dealing with ac circuits, the Ohm's Law formulas are changed by substituting the impedance (Z) figure (in ohms) in place of the usual R figure normally used in dc circuits. So use

$$E = IZ$$

and derivatives thereof when dealing with ac circuits.

INDUCTANCE AND CAPACITANCE

Inductance, usually indicated by the letter L, is the property of a coil that causes an EMF (voltage) to appear in opposition to any *change* in current flow through that coil. This property also causes an EMF to be induced in adjacent coils and is effective only in circuits where a varying current is present. The measure of inductance is the *henry*, which represents a change in current of 1 ampere per second that induces an EMF of 1 volt. Because the henry is such a large unit of measure, the millihenry (mH) at one thousandth of a henry, or the microhenry (μH) at one millionth of a henry, are often used.

Inductance increases in direct proportion to the square of the number of turns, which means that a coil with twice the number of turns will have four times the inductance of the lesser coil. Inductance also varies with the cross-sectional area of the core and the permeability of the material used in the core. As the value of inductance increases, so does the coil's opposition to any change in current flowing through it. If two coils are wound in opposite directions and connected in series, they will oppose each other, and if they have the same number of turns, the total inductance of the pair is zero.

Remember that a shorted turn in a coil has a loading effect that acts in opposition to remaining windings of the coil, causing the total inductance to decrease. The result is a power loss through heating. This partially explains the problem that usually exists when a turn or two of a transformer winding becomes shorted due to an insulation breakdown. Minor heating is the result (which can eventually cause more turns to short), and the heat increases until the transformer burns out completely or forces a protective device to open the circuit.

The formula for inductive reactance in ohms is:

$$X_L = 2\pi FL$$

where F is the frequency in Hz and L is the inductance in henrys.

Capacitance is associated with a changing electric field instead of magnetic field and a capacitor is simply two conducting surfaces separated by an insulator known as the *dielectric*. The region between the two charged surfaces of the capacitor is an electrostatic field that blocks direct current but permits alternating current flow to a degree determined by the capacity and frequency of the current involved. The actual strength of the field is also dependent on the distance between the conducting surfaces and the dielectric constant (K) of the insulating material. Air has a K of 1, wax paper about 2, and mica or glass is about 5 or 6. If the value of the dielectric constant is increased, the force between opposite charges is decreased, but the value of capacity of the unit is greater.

The capacitor actually stores electrical energy as an excess of electrons on one plate, which causes an electric field in the dielectric because the plates are oppositely charged. Energy is stored in the dielectric, but the actual charge is on the inner surfaces of the plates or the outer surfaces of the dielectric. Being of opposite polarity, the charges are attracted to each other, but they can reach only the outer surfaces of the dielectric because they are unable to pass through it. When the capacitor is discharged, the charge on the plates is removed and the electric field in the dielectric collapses.

Capacitance or capacity is the measure of the ability of a capacitor to store a charge and depends on the voltage applied and the area of the plates. The capacitance is measured by the farad, but due to the very large size of this measure, the microfarad (μF), which is one millionth of a farad, and the picofarad (pF), one millionth of a microfarad, are commonly used instead.

The amount of charge in a capacitor is the product of the capacitance C and the applied voltage E:

$$Q = CE$$

where
- Q = charge in coulombs
- C = capacitance in farads
- E = applied voltage in volts

Capacitors are available in many sizes, shapes, and values, according to the specific need, and can be fixed or variable. The standard color code for EIA-MIL capacitors follows: Looking at a capacitor, with the arrows on the three top color dots pointing right, the first top dot (reading left to right) is black for MIL mica or white for EIA mica. The second dot is the first significant figure, the third dot the second significant figure, and the dot below the third dot denotes the decimal multiplier. The next bottom dot (middle) indicates tolerance percentage, and the extreme left bottom dot is the characteristic. The capacitor color code is listed in Table 2-2.

The energy stored in a capacitor is figured in *watt-seconds (joules)*, i.e. one watt of power for a time of one second.

$$W = \frac{E^2 C}{2}$$

where
- W is the energy in watt-seconds (1 J = 1 Ws)
- E is the applied voltage across the capacitor in volts
- C is capacitance in farads

Table 2-2. Capacitor Color Code Chart

Color	Fig. 1 & 2	Multiplier	Tolerance (Percent)	Voltage	Characteristic
Black	0	1	±20		A
Brown	1	10	± 1	100	B
Red	2	100	± 2	200	C
Orange	3	1,000	± 3	300	D
Yellow	4	10,000	GMV*	400	E
Green	5	100,000	± 5	500	F
Blue	6	1,000,000	± 6	600	G
Violet	7	10,000,000	±12.5	700	-
Gray	8	0.01	±30	800	I
White	9	0.1	±10	900	J
Gold	N/A	0.1	± 5	1,000	-
Silver	N/A	0.01	±10	2,0000	-

When no voltage is indicated, the EIA value is 500 volts. Voltage is shown by the left dot on the reverse side unless the capacitor is stamped with a value. The characteristic letter indicates the effect of temperature on the capacitance. "A" indicates a considerable change and "J" an extremely small change.

*GMV is Guaranteed Minimum Value

An electrolytic capacitor provides a large amount of capacity in a small space as a result of a thin-film manufacturing process. The space between two aluminum foil rolls is filled with a thick paste of aluminum borate, and the dc applied across the electrodes forms a thin film on the positive plate. The thin oxide film is actually the dielectric material between the positive plate and the electrolyte (which is part of the negative plate), because it is in electrical contact with the foil. By acting as the dielectric, the thin oxide film, being only a few millionths of an inch thick, provides very high capacity in a very small space. The polarity of an electrolytic capacitor is extremely important and must never be reversed, or damage to the capacitor will result. Such capacitors are used only in dc circuits or ac circuits of very low amplitude.

The value of capacitors connected in parallel are simply added for the total capacitance and the combined voltage rating is the same as the lowest of the group. When capacitors are connected in series, the only advantage is an increase in working voltage, but this gain is at a considerable sacrifice in capacitance (there are other considerations to be discussed later.) The formula for determining the total capacity of a series of capacitors is:

$$C_T = \frac{1}{\frac{1}{C1} + \frac{1}{C2} + \frac{1}{C3} \ldots}$$

The working voltages (WV) are added. Three 500-volt capacitors in series have an overall WV of 1500 V.

Capacitive *reactance* is expressed by the formula:

$$X_c = \frac{1}{2\pi FC}$$

where 2π equals 6.28, F is the frequency in hertz (Hz), and C is the capacitance in farads. X_C is the capacitive reactance in ohms. This expresses the opposition of the capacitor to an applied current. The opposition decreases as the frequency or capacitance is increased.

REACTANCE

When the frequency of an applied voltage increases, the reactance of the coil increases and the reactance of the capacitor decreases. If the frequency is lowered, the reactance of the coil or inductor goes down accordingly, but the reactance of the capacitor rises. Thus, the change in one component is opposed by the other, and in circuits where both are used, determine the total reactance figure by subtracting the smaller value from the larger. Thus, the formula for impedance states that:

$$Z = \sqrt{R^2 + (X_L - X_C)^2}$$

What is *resonance*? It is the condition where the X_L and X_C values are the same, in which case the total reactance of the pair is zero. Current flow in such a circuit is high because it is limited only by the pure resistance of the circuit. If no resistors are used in the circuit, there is nothing to oppose the current flow at resonance. This condition of equal reactances occurs at resonance regardless of whether the coil and capacitor are connected in series or in parallel. Determine the resonant frequency by:

$$F_r = \frac{1}{2\pi \sqrt{LC}}$$

where F_r is the resonant frequency in hertz, 2π equals 6.28, L is the inductance in henrys, and C is the capacity in farads.

VECTORS

To evaluate some quantities in electronics, it is necessary to use vectors. In a vector diagram, straight lines are drawn in the appropriate direction from a zero point for each value. The *length* of each line is proportional to the *magnitude* of the quantity involved. If two voltages are applied in the same direction, that is, negative of one to positive of the other (series connected), the total is obtained by adding the two voltages. But if 30 volts is applied in one direction and 50 volts in the opposite direction, the 50-volt potential would determine the combined direction with a net force of 50 less 30 or 20 volts, which is exactly what happens when two ac voltages are 180 degrees out of phase with each other. Between these extremes, the ac voltages may be out of phase to an extent less than 180 but more than 0 degrees and could be measured by drawing vectors accordingly.

CIRCUIT COMPONENTS AND PROPERTIES

You should be familiar with these properties and elements of electronic circuitry.

POWER FACTOR

Power factor is a rating determined by dividing the resistance of a circuit by its impedance, figured at the operating frequency. Multiply the result by 100 percent, the answer would be a percentage. This figure is the ratio between true power and apparent power. Therefore, the power factor is equal to the true power divided by the apparent power and the true power in watts is IE multiplied by power factor. Mathematically, it is the cosine of the phase angle between current (I) and voltage (E).

HIGH-PASS AND LOW-PASS FILTERS

When it is desirable to attenuate one frequency or group of frequencies, the use of a filter is indicated. If all frequencies above a specific cutoff point are to be passed without attenuation and those below that cutoff frequency must be attenuated, a high-pass filter would do the job. By the same token, the frequencies below a selected cutoff may pass while those above are attenuated; this is done with a low-pass filter.

Filters are also designed to pass or reject a selected band or group of frequencies as desired, known as *bandpass* or *bandstop* filters. The passbands of most filters are determined by the frequency at which attenuation rises 3 dB above the official passband value.

Vacuum Tubes

Although vacuum tubes are largely supplanted by solid-state devices, they still find applications in high-power radio transmitters. Of course, they are also found in older equipment that is technically obsolete but still within its capital service life.

According to the number of elements or parts, vacuum tubes are known as diodes (two elements), triodes (three elements), tetrodes (4 elements), and pentodes (five elements). All have a *cathode* to emit electrons when heated and an *anode* (or "plate") to attract those electrons. In order to attract the negative electrons, the plate must be positive because unlike electrical charges attract. The diode has only these two elements; therefore, it can only be used as a detector or rectifier. A diode does not amplify. A diode is merely capable of passing current when the plate is provided with a positive voltage to attract the electrons from the cathode. The plate repels those electrons if negatively polarized. This causes the electrons to return to the cathode, and no current can flow.

The triode has a third element known as the control grid (or simply "grid"). The control grid is a wire mesh or spiral between the cathode and the plate that regulates electron flow from cathode to plate. When the control grid is supplied with a negative potential called *negative bias*, some of the electrons are repelled and forced back to the cathode. Therefore, grid action regulates the actual amount of current that flows from cathode to plate. The more negative the grid, the more electrons are forced back to the cathode and the lower the current flow. The less negative bias applied to the control grid, the greater the cathode plate current will be. As the grid is biased negative with no signal or input to the circuit, you can see that a tiny ac signal coming into the grid will cause the normally negative grid bias to become less negative when the incoming ac signal is positive and more negative as it swings through the negative half cycle. So the small input signal to the control grid is regulating the much heavier current flow between cathode and plate and this is basically what is meant by amplification. The ordinary triode enables us to increase the magnitude of the input signal and, unlike the diode, provide an output that is many times the input.

Before looking at the tetrode (four-element tube), there is a major problem with the triode (that actually led to the development of the tetrode). Whenever two conducting surfaces (grid and plate) are separated by an insulator, a capacitance exists. This capacity is very small, but it becomes troublesome at radio frequencies, and the higher the frequency the greater its effect. The tetrode contains a fourth element, a screen grid, between the plate and control grid reduces control grid-plate capacitance to that smaller value. The screen grid is positive, though to a lesser degree than the plate; therefore, electrons from the cathode pass through the control grid, and most go on through the holes in the wire mesh screen grid to the plate. The few that are attracted to the screen grid cause a small dc current flow in the circuit.

However, with introduction of the screen grid, another problem arose: secondary emission. This results from the acceleration of the electrons by the positive screen grid. The fast-moving electrons strike the plate with sufficient force for many to bounce back to the screen grid. The undesirable effect of this action is more screen current and less plate current. Because most of the useful output signal is in the plate current, this presents a big disadvantage. The answer is the pentode, which has a fifth element called the suppressor grid between the plate and the screen grid. The suppressor normally operates at the same potential as the cathode. As electrons pass the screen grid, they are slowed considerably by the suppressor grid and strike the plate at too low a speed to bounce off. Even the few that manage to bounce from the plate will return as the suppressor isolates them from the screen grid attraction, and the secondary emission problem is thereby remedied.

Semiconductors

Solid-state or semiconductor diodes and transistors are available in many types and have replaced the vacuum tube type in most electronic equipment designed in recent years. Displaying more resistance than the usual conductor but far less resistance than an insulator, semiconductor materials such as germanium and silicon find widespread use today in diodes as well as transistors and integrated circuits. The pure semiconductor material is not useful for diodes or transistors until a small quantity of a suitable impurity is blended in to lower the resistance (called *doping*). The two basic materials when doped can form n-types (additional free electrons are created), or p-types (some electrons are taken away), leaving "holes" in the material. A hole is a place where an electron could be, but isn't. Because the holes always move in the opposite direction from electrons, they are considered to have a positive charge.

The solid-state diode consists of p-type material on one side and n-type material on the other; in other words, a pn junction forms the anode and cathode, respectively. The resistance of the junction is very low from cathode to anode but very high from anode to cathode, and, like its tube counterpart, it can be used to rectify, detect, or steer, but not to amplify. (Note: Some special diodes can be made to amplify at UHF or microwave frequencies, but these devices are not ordinary pn junction diodes).

The bipolar transistor, however, shows how a solid-state device can amplify just as well and even more efficiently than the common vacuum tube triode. The transistor has many advantages, including its small size, no cathode to heat, very short time lag, and it's less expensive to name a few. The transistor can be either an npn type or a pnp type. In either case, the connections are the base, emitter, and collector. In a schematic diagram, the emitter symbol for the pnp type always points in toward the base while the emitter points away from the base in the npn symbol. The memory aid is *points in* for *p*np. In both types, the center letter designates the polarity of the base with regard to the emitter.

During normal operation, the emitter-base junction is forward biased while the collector-base junction is reverse biased. The resistance is always low when the transistor is forward biased as the carriers move through the junction between emitter and base. When the junction is reverse biased, the resistance at the junction is extremely high as would normally be the case between collector and base.

When comparing the terminals of the transistor to the elements of the vacuum tube, the emitter compares roughly to the cathode, the base to the control grid, and the collector to the plate (see Fig. 2-2). If you apply a signal to the base of the transistor, the swing of the ac wave causes the base-emitter current to vary, and the carriers in the base increase to result in a greater change in collector current.

As the signal is applied across the forward-biased base-emitter terminals in a common-emitter amplifier, electrons flow from the emitter across the junction with the base and into the base according to the signal variation applied. When additional electrons flow

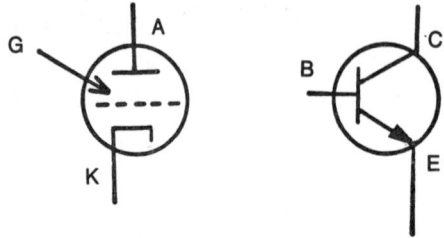

Tube	Transistor
Cathode (K)	Emitter (E)
Grid (G)	Base (B)
Anode (A)	Collector(C)

Fig. 2-2. Comparison of the parts of a vacuum tube to those of a transistor.

into the base in excess of the base-emitter needs, some proceed to the base-collector junction and are promptly attracted by the positive charge at that junction and thus continue into the collector. This free movement of carriers into the reverse-biased junction results in a current flow through the collector load resistor into the common emitter and provides an amplified potential across the high resistance of the collector load circuit.

Needless to say, additional study regarding the subject of electronics might be desirable in many cases, and many texts are available from the publisher of this book to fit the direction of greatest interest. However, I can only scratch the surface while proceeding toward our major goal of preparing for the exam. But if the inclination is there, the material is seemingly unlimited and you should enhance your technical know-how to any level you like.

Power Supplies

The demand for dc power exists regardless of the size or type of electronic equipment being used. Only the smaller transistorized devices use batteries of the dry-cell type. Mobile transmission equipment operates from the vehicle battery directly (e.g. most solid-state equipment) or indirectly through dynamotors, generators, inverters, or power supplies. Base stations draw from the 117 Vac line (broadcast stations typically use 220 or 440 Vac lines) and convert that power into direct current as needed. As a rule, this is done by using a rectifying device to convert the ac line current to pulsating dc. This pulsating current might be smoothed out by a filter section. The rectifier can be either a vacuum tube diode or a solid-state (semiconductor) diode that allows the current to flow in one direction only. Modern equipment, including high-power transmitters, uses solid-state rectifier diodes.

A rectifier converts ac to dc by allowing it to flow freely in one direction and stopping it completely in the other direction. The filter section may consist of one or more capacitors and a resistor or filter choke coil to smooth out the rectified voltage by removing the remaining ac ripple and increasing the average voltage. The larger the capacitor, the greater the filtering. There are many types of rectifiers: the simple half-wave, the full-wave, and the bridge type (full wave), the latter of which offers many advantages. These types, along with the various filtering sections and regulation facilities, are covered later in this book.

The "bleeder" resistor, by maintaining a minimum load on a power supply, improves the stability or regulation of the output and removes charges stored in the capacitors when the equipment is turned off. This charge could be hazardous, and the bleeder resistor, which is across the output of the supply, acts to dissipate this dangerous potential.

QUESTIONS AND ANSWERS

Q1: By what other expression can a "difference of potential" be described?
A: Voltage, EMF, IR drop, fall of potential, electromotive force, voltage drop, voltage difference, difference of charge.

Q2: By what other expression can an electric current flow be described?
A: *Electron flow*. If electrons pass a point at the rate of one coulomb per second, the current is one ampere.

Q3: Explain the relationship between the physical structure of the atom and electric current flow.
A: The atom has an inner nucleus around which electrons revolve in rings, with the outer ring determining the electrical (conductive) characteristics of the material. The loosely held electrons are free-moving, and this movement is current flow. Such material is a conductor, and the movement of free electrons in a general direction under the influence of voltage force constitutes an electric current flow through the conductor.

Q4: With respect to electrons, what is the difference between conductors and nonconductors?
A: A conductor has a large number of free electrons that can move from atom to atom, because they are not bound tightly to any one atom; therefore, there can be a current flow. Nearly all of the electrons in a nonconductor are tightly bound to their atoms and are not free to travel when voltage is applied.

Q5: What is the difference between electric power and electric energy? In what units are each expressed?
A: Electric power is the rate of doing electrical work or work per unit of time. Electric energy is the amount or capacity for doing work. Electric power is expressed in watts, electric energy in joules. One watt is equal to one joule per second.

Q6: If the diameter of a wire is doubled, how is the resistance of the wire affected?
A: The resistance varies inversely with the cross-sectional area. By doubling the diameter, the area is increased four times and resistance reduced to one fourth.

Q7: If a relay having a coil resistance of 400 ohms is designed to operate when 250 milliamperes flows through the coil, what value of resistance must be connected in series with the coil for operation from 115 volts dc?

A: Because the total resistance of the circuit is E divided by I, R equals 115 divided by .25, or 460 ohms. The coil is 400 ohms, so the series resistor would be 460 minus 400 or 60 ohms.

Q8: Draw a circuit with three resistors (50, 100, and 150 ohms) connected in a "pi" network to a 12V battery, as shown in Fig. 2-3A.
  (a) What is total current through each resistor?
  (b) What is the voltage across each?
  (c) What power is dissipated in each resistor and the total power used in the circuit?

A: (a) Solving for R1, I equals E divided by R1, or 12 divided by 50, which is 240 mA. Because R2 and R3 are in series across the battery, I equals E divided by R2 + R3, or 12 divided by 100 + 150, or 12 divided by 250, which is 48 mA.

(b) The voltage across R1 is the battery voltage, 12 volts. The voltage across R2 is I2 × R2, or .048 × 100, which equals 4.8 volts. The voltage across R3 is I3 × R3, or .048 × 150, which equals 7.2 volts. (Because R2 and R3 are in series with each other across the 12V battery, the voltage drop across the two must equal 12 volts.)

(c) Power dissipated in R1 equals E × I, or 12 × .240, which equals 2.88 watts. R2 equals 4.8 × .048 or .2304 watts. The power across R3 is 7.2 × .048, or .3456 watts. The total power dissipated in the circuit would be the sum of the three resistors R1 + R2 + R3, or 2.88 + .2304 + .3456 or 3.456 watts.

Q9: What is the relationship between wire size and resistance?

A: The resistance of wire varies in an inverse proportion to the cross-sectional area.

Q10: What is the meaning of "skin effect" in conductors carrying RF energy?

A: The term "skin effect" describes the tendency of RF currents to flow in that area of the conductor nearest the surface rather than throughout the entire cross-sectional area. This causes the effective resistance of the conductor to increase with the frequency.

Q11: Why is impedance matching between electrical devices important? Is it always desired? Can it always be attained in practice?

A: Impedance matching is important between electrical devices where a maximum efficiency of energy transfer is required. This is especially desirable in circuits handling appreciable power, but in the case of others, such as an amplifier in a PA system, a mismatch can contribute to amplifier stability. Perfect impedance matching is seldom attained.

Q12: A loudspeaker with an impedance of 3.2 ohms is operating in a plate circuit with an impedance of 3200 ohms. What is the impedance ratio of an output transformer needed to match the plate circuit to the speaker? What is the turns ratio?

A: The impedance ratio is 3200 divided by 3.2 or 1000 to 1. The turns ratio is the square root of the impedance ratio or the square root of 1000, which is 31.6 to 1.

Q13: Compare some properties of electrostatic and electromagnetic fields.

A: An unchanging electric field, such as the static charge between the plates of a charged capacitor after the charging voltage has been removed, is an example of an electrostatic field.

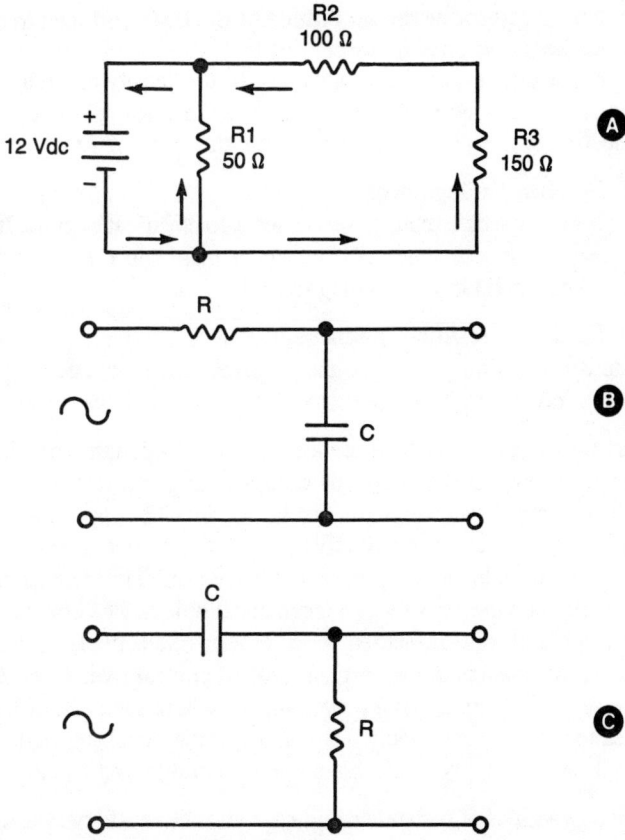

Fig. 2-3. A) Resistor "pi" network; B) RC integrator circuit; C) RC differentiator circuit.

The current in a wire is composed of moving electric charges or electrons. Each electron produces an electric field in the space around the wire. Because this electric field moves with the electrons, a magnetic field is produced that surrounds the wire, producing an electromagnetic field. If the current is turned off, the electromagnetic field collapses and soon disappears, while the electrostatic field remains in the capacitor even after the charging voltage is removed. The electrostatic field can cause induction while stationary, but the electromagnetic field must be in motion to cause induction.

Q14: In what way are the electrical properties of common circuit elements affected by electromagnetic fields? Are interstage connecting leads susceptible to these fields?

A: The electrical properties of most components are not usually affected by electromagnetic fields, although problems can be caused when electromagnetic fields exist near interstage leads. Unwanted voltages can be induced, resulting in hum, oscillation, or distortion. To eliminate such problems, interstage leads must be separated or shielded to prevent electromagnetic lines of force from leaving the wire or outside ones from entering.

Q15: Which factors determine the amplitude of the EMF induced in a conductor that is cutting magnetic lines of force?
A: Strength, rate, length, and angle. In other words, the density or strength of the magnetic field, plus the speed at which the conductor cuts the lines of force, the length of the conductor, and finally the angle at which it cuts the lines of force.

Q16: Define the term "reluctance."
A: Reluctance is to magnetic circuits as resistance is to electrical circuits. It is the opposition to the formation of magnetic lines of force in a magnetic circuit and is equal to the magnetomotive force divided by the magnetic flux.

Q17: Define the term "residual magnetism."
A: The magnetism remaining after the magnetizing force has been removed, as would be the case in the metal core of dc generator field coils when the generator is turned off.

Q18: In what way does an inductance affect the voltage-current phase relationship of a circuit? Why is the phase of a circuit important?
A: An inductance opposes a change in current but does not oppose a change in voltage. When an ac signal is applied to an inductive circuit, the output current will lag the output voltage by 90 degrees. Inductance, then, exhibits the same effect on current in an electric circuit as inertia does on the velocity of a mechanical object. The effects are advantageous in many respects and disadvantageous in others. Selectively using the phase-lag characteristics of inductance, allows a great deal of control over electrical circuits. One example is the series inductance of a power supply where the current-lag characteristics of a coil is used to maintain the supply's B+ at a constant level (through "inertia"), even though the load on the supply might change considerably and often.

Q19: Explain how values of resistance and capacitance in RC networks affect the time constant. How would the output waveform be affected by the frequency of the input in the RC network?
A: Because the time constant in seconds (T) equals the resistance in ohms times the capacity in farads (T equals RC), increasing either R or C would increase the time constant. The flow of current into or out of the capacitor is limited or slowed down by the series resistance; the greater the value of resistance, the more the flow of current is delayed. The greater the capacity, the longer the time it requires to charge. The time constant is the time in seconds needed for the voltage across the capacitor to reach 63 percent of the applied voltage when charging or a drop to 37 percent when discharging. T (in seconds) equals R (in megohms) times C (in microfarads) for simplification.

A simple sine wave would pass through an RC network (Fig. 2-3B) with no change of shape, but will experience a loss in amplitude and a phase shift, increasing with frequency. A complex wave would pass with a far greater attenuation of its high-frequency component, tending to change the shape of the output waveform. By reversing the take-off point of the RC network (Fig. 2-3C), by taking the output across the resistance instead of the capacitor, there is a differentiator or high-pass filter network producing the opposite characteristics. The low-frequency component of the complex wave is attenuated much more than the high; thus, the output consists almost entirely of frequencies above the designed cutoff point of the network.

Q20: Explain how the values of resistance and inductance in an RL network affect its time constant.

A: Resistance shortens the period necessary for the current to reach its final level after a voltage is applied, and the time constant is equal to the inductance divided by the resistance, T equals L divided by R, with T in seconds, L in henries, and R in ohms. The time constant is the time required in seconds for the current to assume 63 percent of its final value on charging or to drop to 37 percent of its original level upon discharge. Therefore, the greater the resistance, the lower the time constant, and the greater the value of inductance, the greater the time constant.

Q21: Explain the theory of molecular alignment as it affects the magnetic properties of materials.

A: The theory assumes that magnetic materials contain tiny magnets known as *magnetic dipoles*. When property aligned with all like poles pointing in the same direction, the material is completely magnetized with north and south poles appearing at opposite ends. In most materials, the tiny magnets do not line up in the same direction because of collisions and temperature vibrations going on within the atomic structure, which keeps the electrons in constant motion. Thus, metals like copper, aluminum, silver, and numerous others have little or no magnetic property. Other materials like iron retain their molecular alignment only as long as an electric current is applied. When this EMF is removed, only slight magnetism (known as residual) remains. Although iron is "soft" magnetically, certain hard magnetic materials like Alnico retain their magnetic power indefinitely with reasonable care. Excessive heat or mechanical shock in addition to opposing electromagnetic fields might weaken or demagnetize even such "permanent" magnets.

Q22: What factors influence the direction of the magnetic lines of force produced by an electromagnet?

A: The direction of the current flow and the way the coil is wound. By applying the "left-hand rule," which states the thumb points in the direction of the lines of force within the coil with the fingers wrapped around it in the same direction as the current is flowing, this indicates the north pole of the electromagnet. The lines of force from the north pole return to the south pole outside the coil to form a closed loop.

Q23: Explain how self and mutual inductance produce transformer action.

A: When ac flows in the primary winding of a transformer, the magnetic field around it collapses and builds up at the frequency of the applied sine wave, resulting in a counter EMF across the winding and adjacent turns. This action is called *self induction*. The turns of the second coil (or secondary), if close to the primary, are cut by the continuously expanding and collapsing magnetic field, inducing a current in it. The secondary will also induce a current in the primary, and the reaction between the coils or windings is mutual inductance, as transformer action occurs. If there is no load on the secondary, primary current will be small due to the opposing voltage (resulting from self induction) canceling out much of the source voltage. However, with the secondary loaded, the current flow tries to set up a magnetic field just as the current through the primary does, but it opposes it and thus reduces the self-induction bucking voltage of the primary winding, allowing its current to increase.

Q24: How does the capacitance of a capacitor vary with the area of the plates, the spacing between the plates, and the dielectric material between plates?
A: Capacitance varies in direct proportion to the area of the plates. The greater the area of each plate or the number of plates, the greater the capacitance.

The capacitance is *inversely* proportional to the spacing between the plates. *Doubling* the spacing between the plates will *halve* the capacitance.

The dielectric material between the plates varies the capacitance in direct proportion to the dielectric constant. Dielectric constant is a measure of the ability of any given material to conduct electric lines of force as compared to air. If the dielectric constant is 7, as in the case of mica, a capacitor having a value of .0005 μF with air would increase to .0035 μF by using mica as the dielectric between its plates.

Q25: What does "coefficient of coupling" mean?
A: Coefficient of coupling is a measure of how much of the flux from one coil cuts the turns of the other, when two inductances are positioned to interact with each other. When two coils are placed very close together and side by side, the coefficient of coupling is maximum—all the flux from one coil cuts all the turns of the other. In this case, the coefficient of coupling is said to be unity (1). When the coils are separated, or when one of the coils is turned on its axis, the coefficient of coupling decreases. Coefficient of coupling (K) is expressed as a decimal and is equal to the mutual inductance of the coils divided by the square root of the product of the two inductances.

Q26: Assuming the voltage on a capacitor is at or below the maximum allowable value, does the value of the capacitor have any relationship to the amount of charge it can store? What relationship does this storage of charges have to the total capacitance of two or more capacitors in series; in parallel?
A: The amount of charge a capacitor can store is equal to the capacitance C times the voltage across it E:

$$Q = C \times E$$

where
    Q = charge in coulombs
    C = capacitance in farads
    E = voltage across capacitor in volts

The coulomb is a measure of quantity and is equivalent to the accumulated change after one second if the current flow rate is one ampere per second.

The storage capacity of capacitors in series is reduced, and for two identical capacitors connected in series, the total capacitance is halved, but the working voltage is doubled. When connected in parallel, two identical capacitors have double the storage capacity as well as total capacitance.

Note that when the working voltage dc (i.e. WVdc) ratings are unequal, the situation for series-connected capacitors is more complex. The capacitor with the lower WVdc rating dominates, especially if the capacitances are unequal. In parallel-connected circuits, the WVdc rating of the pair is simply the WVdc rating of the two that is *lowest*.

Q27: How should electrolytic capacitors be connected in a circuit in relation to polarity? Which type of low-leakage capacitor is used most often in transmitters?

A: Electrolytic capacitors can be used only in dc circuits and must be connected with the positive terminal to the voltage point that is more positive than the voltage applied to the negative terminal. Simply stated, always observe polarity with electrolytic capacitors; otherwise they will be destroyed (i.e. could explode dangerously). Mica capacitors are used most often in transmitters where low leakage is a must, providing low values of capacity are required. If large values are needed, the oil-filled paper or electrolytic types are often used.

Q28: How much power would be used to operate a 120-volt bulb having an internal resistance of 100 ohms for a period of 24 hours on power supplied at 9 coulombs per kilowatt hour?

A: To find the power used by the bulb:

$$P = \frac{E^2}{R} = \frac{14,400}{100} = 144 \text{ watts}$$

144 watts × 24 hours = 3,456 watt-hours or 3.456 kilowatt-hours (kwh)

3.456 kwh × 9C = 31C

Q29: Name four materials that make good insulators at low frequencies but not at UHF or above.

A: Glass, fiber, rubber, and paper.

Q30: In an iron-core transformer, what is the relationship between the transformer turns ratio and primary-to-secondary current ratio; between turns ratio and primary-to-secondary voltage ratio? Assume no losses.

A: The primary-to-secondary *current* ratio is approximately an inverse relationship to the actual turns ratio. The primary-to-secondary *voltage* ratio is proportional to the turns ratio. Taking ideal examples and ignoring normal losses, if the secondary has twice as many turns as the primary, the secondary voltage will be twice that applied to the primary, but the current will be only half as much as that in the primary.

Q31: What prevents high currents from flowing in the primary of an unloaded transformer?

A: The inductance of the winding that presents a high inductive reactance to the flow of current.

Q32: How is power lost in an iron-core transformer? In an air-core transformer?

A: Power is dissipated through iron losses and copper losses in the iron-core transformer. The copper loss is caused by the actual resistance of the winding, the iron loss results from hysteresis and eddy currents in the core. The molecular friction in the iron core produces heat, which represents lost power, and is known as a *hysteresis loss*. The *eddy currents* are induced in the iron core by the alternating current flowing in the windings of the transformer and tend to heat the core, which again represents a loss. Transformer cores are constructed with thin sheets of metal insulated from each other to help reduce these eddy current losses. All the transformer losses appear as heat in the iron-core type.

In the air-core transformer, power is lost as a result of radiation, absorption and shield losses, skin effect, and bandwidth loading resistance. Because air-core transformers are normally used at radio frequencies, the copper loss increases due to skin effect and radiation loss goes up sharply as the lines of force are no longer confined by an iron core. This results in absorption loss as the lines of force spread out and induce currents in surrounding metal parts.

Q33: What is the value and tolerance of a resistor that is color coded (reading from left) yellow, violet, orange, silver? What if the silver band is replaced with gold? What if there is no fourth band?

A: The first resistor is 47,000 ohms (47k), plus or minus 10 percent. The second resistor (with gold band) is 47,000 ohms, plus or minus 5 percent. The third resistor with no fourth band has the value indicated by the first three bands (47k) and a tolerance of 20 percent.

Q34: What is the impedance or resonance of a parallel circuit that is composed of pure inductance and pure capacitance? Of a series circuit at resonance?

A: The impedance of the parallel circuit would be infinite, and the series could be zero. This is easy to remember, because the impedance of the two are opposite at resonance; parallel is high (very high) and series is low (very low). To avoid getting them confused, they are in alphabetical order: parallel *high*, series *low*. (The terms *infinite* and *zero* are meaningless in practical circuits because of inherent resistive losses, but these terms are used properly in considering ideal circuits for evaluation or study purposes.)

Q35: Explain the operation of a break-contact relay and of a make-contact relay.

A: A break-contact relay is one in which the contacts are closed when the relay coil is not energized, in other words, NC or "normally closed." The make-contact relay is one in which the contacts are open when the relay coil is not energized and closed when energized; this is referred to as NO or "normally open." A spring holds the movable contact arm in the not-energized position, and the energizing of the coil pulls this arm in the opposite position against the spring action. A relay in the energized position always reverses the contact position. NO contacts are closed, and NC contacts are open. Contact positions are *always* stated in the not-energized position unless specifically noted otherwise.

Q36: Draw a circuit diagram of a low-pass filter composed of a constant-K and an M-derived section.

A: See Fig. 2-4 for a two-section low-pass filter.

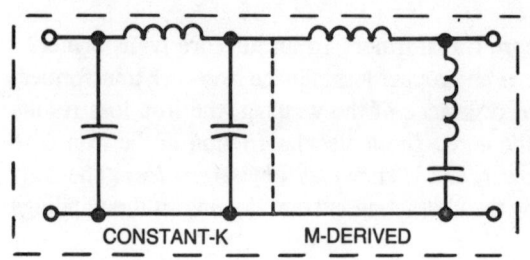

Fig. 2-4. Two-section low-pass filter.

Q37: List three precautions that should be taken in soldering electrical connections to assure a permanent junction.

A: Clean well and join mechanically; apply sufficient heat to the connection to make the solder flow freely onto the connection. Rosin-core solder should be used, *never* acid core.

Q38: Explain how to determine the sum of two equal vector quantities having the same reference point but whose directions are 90 degrees apart; 0 degrees apart; 180 degrees apart. How does this pertain to electrical currents or voltages?

A: A vector can be used to represent direction and magnitude of a quantity. In Fig. 2-5, two sine waves with differing phase relationships are represented by two vectors, E1 and E2. Thus, the solution of problems, such as adding out-of-phase currents or voltages, is simplified by the use of vectors. Vectors may be subtracted only when their directions are exactly opposite and added only when their directions are the same.

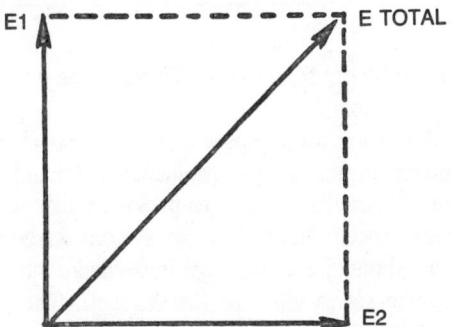

Fig. 2-5. Vector diagram representing two sine waves 90 degrees apart.

Q39: What would be the value, tolerance, and voltage rating of an EIA-coded capacitor whose first row colors are (from left) white, red, green and the second row is green, silver, red?

A: Reading from the white dot, which merely indicates that the EIA system is used, the red dot indicates the number 2, the green dot the number 5, and the red dot on the bottom row dictates the number of zeros, giving 2500 (picofarads). Continuing, the silver dot marks the tolerance of 10 percent, and the green dot indicates the temperature characteristic or variation due to change in temperature; green represents 0 to 70 parts per million per degree C.

Q40: Draw a circuit composed of a voltage source of 100 volts at 1000 Hz with a capacitor of 1 μF in series with the source and a T network composed of an inductor (2 mH), resistor (100 ohms) and inductor (4 mH). The load resistor following the network is 200 ohms.
  (a) What is the total current and the current through each circuit component?
  (b) What is the voltage across each component?
  (c) What is the apparent power consumed and what is the real or actual power consumed by the circuit? By the load resistor?

A: (a) The total current is 621 mA. The current through resistor C is 621 mA, through inductor L1 is 621 mA, through inductor L2 is 207 mA, through resistor R1 is 414 mA, and is load resistor R2 through 207 mA.

(b) Voltage across each component: C, 98.86 volts; L1, 7.82 volts; L2, 5.2 volts; R1, 41.4 volts; R2 (load resistor), 41.4 volts.

(c) The apparent power consumed is 62.1 watts and the actual power consumed by the circuit is 25.7 watts. The actual power used by load resistor R2 is merely 8.57 watts. See Fig. 2-6.

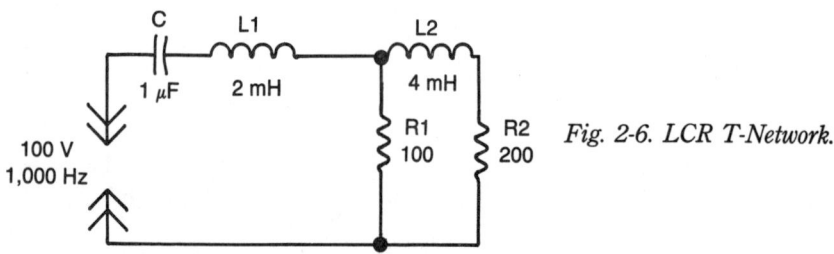

Fig. 2-6. LCR T-Network.

Q41: Why are filters used? Explain the purpose of *band-stop*, *high-pass* and *low-pass* filters.

A: Filters are used to pass all frequencies above or below a certain frequency, or to pass or reject certain bands of frequencies. A band-rejection (band-stop) filter is a combination of tuned circuits selected to provide a low-impedance short-circuit path for frequencies within a specific narrow range. All other frequencies (those not within the resonant band) see the shunt circuit as a high impedance. Such filters normally employ a parallel-resonant circuit in series with the line as well. The parallel-resonant circuit appears as a high impedance to the band that is to be rejected and as a low impedance to all other frequencies. Thus, all frequencies other than those within the resonant band are passed without opposition from input to output, while those within the resonant band are shunted to ground. A common application of the band-rejection filter is the familiar TV wavetrap, which is a pair of series-resonant circuits used on the lead-in of a transmission line to shunt out the signal from some interfering FM radio station.

A high-pass filter consists of a series capacitor and a shunt inductor. The capacitor is selected to pass all frequencies above a specific value, and the inductor is chosen to provide a low-impedance path for all frequencies below the desired value. Thus, high frequencies are passed and low frequencies are short-circuited. High-pass filters are commonly used in the lead-in wires of television antennas—the objective is to prevent signals from amateur or commercial transmitters operating below 54 MHz from interfering with the TV signals (which are above 54 MHz).

A low-pass filter is used to pass low frequencies and attenuate high frequencies. The low-pass filter is the opposite of the high-pass filter constructionally; that is, the low-pass filter consists of a series inductance and a shunt capacitance. A common use for the low-pass filter is in the transmission line of an amateur or commercial radio station. Use of such a filter minimizes the chances of radiating spurious energy at frequencies above the assigned frequency.

Q42: Summarize the physical characteristics and a common usage of each of the following electron tube types: diode, triode, tetrode, pentode, beam power, and remote cutoff.

A: Diode. The simple diode tube contains a heated cathode and a cold plate. The plate of the tube collects electrons emitted by the heated cathode when the plate is supplied with a high positive dc voltage with respect to the cathode.

In an electronic circuit, the two electrodes of a diode act in the manner of a flow valve in a water pipe. The behavior of a diode is observed after connecting the plate and cathode elements in series with a battery and a milliammeter. The cathode is brought up to normal temperature by applying voltage across the heater terminals. If the battery is connected so that the plate is positive with respect to the cathode, the meter will indicate a current flow. This phenomenon of the emission of electrons from hot bodies was first observed by Thomas Edison in 1883 is known as the *Edison Effect*. When the battery is connected in reserve polarity so that the plate is negative with respect to the cathode, the meter will indicate no plate-current flow. The principal characteristics of the diode are:

* When properly connected into a circuit and with proper voltages, electrons "boil off" the cathode and are attracted to the plate. Because electrons flow from cathode to plate and not from plate to cathode, the diode is a valve that allows current flow in one direction only.
* The diode has only two elements ("di" means *two* and "ode" stands for *electrode*)—a cathode (or filament), and a plate (or anode).
* The diode cannot amplify because it has no tube element between the cathode and plate with which to introduce signal voltages to control the electron flow.
* Diodes have a point of sharp transition between the conducting (current-gating) and nonconducting states. This nonlinear characteristic allows demodulation (detection) by rectifying the RF from modulated RF signals, leaving the audio for further processing.

Triode. The triode, or three-element tube, is similar in construction to the diode, except that a grid of fine-mesh wire is added between the cathode and plate. The addition of the grid gives the tube its most useful function—the ability to amplify. It is common practice to make the grid in the form of a spiral helix of circular or elliptical cross section with the cathode at the center. Other arrangements, however, can be used, provided the essential requirement of being able to control the flow of plate current with the potential applied to the grid is met.

The space between the meshes is sufficiently large not to block the flow of electrons from cathode to plate, yet it is still small enough and close enough to the cathode to effectively control the amount of electrons that are allowed to pass from the cathode to the plate.

While plate current in a diode depends only on plate voltage and cathode temperature, plate current in a triode depends on these factors as well as on voltage on the grid with respect to the cathode. A very small change in grid voltage causes a relatively large change in plate current.

The principal uses of triodes are as amplifiers. Triodes can be used to amplify audio signals as well as radio-frequency signals. When a triode is used to amplify signals of very high frequency, certain precautions must be observed to minimize the coupling that takes place between the elements inside the tube.

Tetrode. The relatively large values of interelectrode capacitances of the triode, particularly the plate-to-grid capacitance, impose a serious limitation of the tube as an amplifier at high frequencies. To reduce the plate-to-grid capacitance, a second grid, referenced to as a "screen" grid, is inserted between a grid and plate of the tube.

Because the screen grid is shunted by a screen bypass capacitor having a low reactance at the signal frequency, it acts as a shield between the plate and control grid. It effectively reduces the interelectrode capacitance coupling between the plate and the control grid circuits. The screen is supplied with a potential somewhat less positive than the plate. The positive voltage on the screen grid accelerates the electrons moving from the cathode to the plate. Some of these electrons strike the screen and produce a screen current. The larger portion of them, however, pass through the open-mesh screen grid to the plate.

Because of the presence of the screen grid, a variation in the plate voltage has little effect on the flow of plate current. The control grid, on the other hand, retains its control as in the triode. The tetrode has high plate resistance and a high amplification factor in comparison to the triode. The high amplification factor is brought about by the close proximity of the control grid to the cathode and the electrical isolation of the plate from the control grid. The transconductance of tetrodes is also relatively high when compared with that of triodes.

When the electrons flowing in a vacuum tube strike the plate with sufficient force, other loosely held electrons are knocked out of the plate material into the space near the plate. When the screen is at a higher positive potential than the plate the secondary electrons are attracted to the screen. The flow of these electrons to the screen is in the opposite direction to the normal flow from cathode to plate, so the plate current is decreased. This reduction in plate current continues until the potential of the plate approaches the screen grid potential. Further increases in plate voltage cause the secondary electrons to be pulled back to the plate and the plate current again increases. The tetrode's characteristic curve is thus considerably different from that of the triode.

The action in the region where plate current decreases as plate voltage increases is called *negative resistance*. This action is opposite to that encountered in a normal resistor. When the tetrode is used as an amplifier, plate voltage should not fall below the screen voltage. If plate voltage does fall below that of the screen, plate current will fail to follow the grid-signal waveform and the output-signal plate voltage variation will be clipped. This distortion can be eliminated by reducing the amplitude of the grid signal or increasing the supply voltage. However, the relatively large screen current and the effects of secondary emission from the plate limit the usefulness of the tetrode as to an RF voltage amplifier. Still, the tube has been successfully used in RF amplification applications both for transmitting and receiving for a number of years.

The most important single characteristic of the tetrode is the screen grid, which serves to overcome the disadvantageous interelectrode-capacitance problems of the triode.

Pentode. A pentode has, as the name implies, five elements. The secondary-emission problem of the tetrode is overcome with the addition of another grid element between the screen grid and the plate. This element, referred to as the suppressor grid, is typically at the same potential as the tube's cathode, so the effects of the screen's high potential are negated.

In the pentode, the suppressor grid serves to repel or suppress secondary electrons from the plate. It also serves to slow down the primary electrons from the cathode as they approach the suppressor. This action does not interfere with the flow of electrons from cathode to plate but serves to prevent any interchange of secondary electrons between screen and plate. The suppressor thus eliminates the negative-resistance effect that appears in a tetrode in the region where plate voltage falls below that of the screen. Thus, plate current rises smoothly from zero up to its saturation point as plate voltage is increased uniformly with grid voltage held constant. The amplification factor of pentodes is very high in comparison with triodes and tetrodes.

In the RF pentode, the chief purpose of the screen grid is to eliminate the effects of interelectrode-capacitance coupling between control grid and plate circuits. In the power pentode, at audio frequencies, the screen permits the output signal plate voltage variation to be relatively large without the degenerative action occurring as it does in the triode. Plate current is substantially independent of plate voltage in the power pentode, because the screen voltage is the principal factor influencing plate current. With the addition of the suppressor, the allowable output voltage variation is larger than that of the tetrode, and the distortion effects of the tetrode are eliminated. Thus, an audio-frequency power pentode has an allowable output voltage variation in which the plate voltage can fall a large amount below that of the screen on the positive half-cycle of input signal without clipping the plate signal current. And, the ratio of output power to grid-driving voltage is relatively large.

Pentodes have traditionally been extensively used in applications where high amplification factors or high frequencies are involved.

Beam Power. The beam power tube has the advantages of both the tetrode and the pentode. This tube is capable of handling relatively high levels of electrical power for application in the output stages of receivers and amplifiers and in different portions of transmitters. The power-handling capability stems from concentration of the plate-current electrons into beams, or sheets, of moving charges. In the usual type of electron tube, the plate-current electrons advance in a predetermined direction but without being confined into beams.

The external appearance of these tubes is like that of other standard receiving-type tetrodes or pentodes; they are slightly larger in dimension because they are called upon to handle somewhat more power, but they have no distinctive external identifying features.

The beam-forming plates influence the movement of the plate-current electrons from the time they pass the screen electrode until they strike the plate. The beam electrodes are connected internally to the cathode, and consequently they are at the same potential as the latter.

Because of this potential of the beam-forming plates, an effect equivalent to a space charge is developed in the space between the screen and the plate. The effect is as if a surface existed in the screen-plate space. This is identified as the "virtual cathode." The presence of this electric plane repels secondary electrons liberated by the plate and prevents them from moving to the screen.

In some tubes, the effect of a virtual cathode is achieved by the use of a third grid in the place of the beam-forming plates. The results are identical in both versions. Because the plate current becomes substantially independent of plate voltage at much lower values of plate potential than in the conventional pentode, the beam power tube can handle greater amounts of electrical power at lower values of plate voltage. In addition, the beam power tube produces less distortion than the ordinary pentode while accommodating an increased grid swing and plate-current change.

Remote Cutoff. The amplification constant, or *mu* of an electron tube has been described as being a function of the geometry of the tube—that is, of the shape and organization of the electrodes. Slight variations in its value can occur under different operating voltages, but for practical purposes it is considered to remain substantially constant. This accounts for the fact that each vacuum tube bears a single mu rating that is assumed to be fixed.

The amplification constant of a tube expresses the relationship between plate current cutoff and negative grid voltage when a fixed value of plate voltage is applied. High-mu tubes such as tetrodes and pentodes, especially the latter, reach plate-current cutoff at relatively low values of negative grid voltage. Low-mu tubes allow the application of much higher negative grid voltages before cutoff is reached.

Such plate-current/grid-voltage relationships and the fixed-mu constant stem from the kind of control-grid structure used in most of the tubes discussed previously—that is, the turns of the control grid are uniformly spaced throughout the length of the tube structure. Application of a voltage to the control grid results in the same effect on the plate-current electrons all along the control-grid wires. The fixed-mu state poses a problem when high-mu tubes are used in communications systems. Frequently, large amplitude signals are encountered and they must be controlled in the equipment to produce the desired intelligence with a minimum of distortion.

To minimize these effects, special kinds of tetrodes and pentodes are used. These are known as variable-mu, or *remote cutoff* tubes, and they differ from ordinary tubes in the construction of the control grids. In these tubes, the grid wires are unequally spaced. Turns are closer together at the top and bottom of the winding and wider apart at the center. This form of control grid construction produces a tube that does not have a constant mu; instead, mu changes with the value of grid voltage applied to the control grid.

At low values of bias, the grid operates in a normal manner. As the control grid is made more negative, the effect of the closely spaced grid wires becomes greater and the electron flow from the space charge in this region is cut off completely. The center of the grid structure also displays a greater effect but still allows electrons to advance to the screen and plate. The overall reduction in plate current, therefore, is gradual. Eventually, with sufficient negative voltage on the grid, all parts of the grid electrode winding act to cut off the plate current, but the negative grid voltage required to attain this is perhaps three to four times as much as for the conventional tube operated at like screen and plate voltages.

Remote-cutoff tubes are used in locations in communications equipment where high-bias voltages may be necessary to provide control of the signal level.

Q43: What is the principal advantage of a tetrode tube over a triode tube as a radio-frequency amplifier?

A: The tetrode tube has a screen grid to act as a shield between the control grid and plate. Thus, the interelectrode capacitance normally found in the triode is eliminated. This makes the tetrode useful in many amplifier circuits without the neutralization that is usually required for a triode.

Q44: Compare tetrode tubes to triodes tubes in reference to high plate current and interelectrode capacitance.

A: Because the screen grid in a tetrode normally operates at a lower positive potential than the plate, the plate current is not dependent on the plate voltage. Therefore, the tetrode is capable of handling higher plate current at lower voltages than the triode, other factors being equal, and higher gain is attained. The interelectrode capacitance is greatly reduced in the tetrode as a result of the screen grid shielding provided between the offering elements (control grid and plate).

Q45: Are there any advantages or disadvantages of filament-type vacuum tubes when compared with the indirectly heated types?

A: There are several advantages and disadvantages, with the indirectly heated tube apparently more desirable as more electrons are emitted from the specially coated cathode, and the ac filament variation does not affect electron emission, which prevents hum. In battery-operated (dc) circuits, the filament type is desirable for its instant-heating characteristic. Of course there would be no ac hum problems.

Q46: Draw a simple circuit of each of the following: a) single-stage FET audio preamplifier; b) push-pull "split secondary" transformer-coupled audio power amplifier; c) JFET phase splitter; d) two-stage bipolar transistor audio amplifier; e) noninverting follower op-amp audio preamplifier using a single dc power supply.

A: (a) Figure 2-7 shows a single-stage audio preamplifier based on a junction field effect transistor (JFET) connected in the common source configuration. Source bias is provided by resistor R2, which is bypassed by capacitor C2. This capacitor is used to keep the source

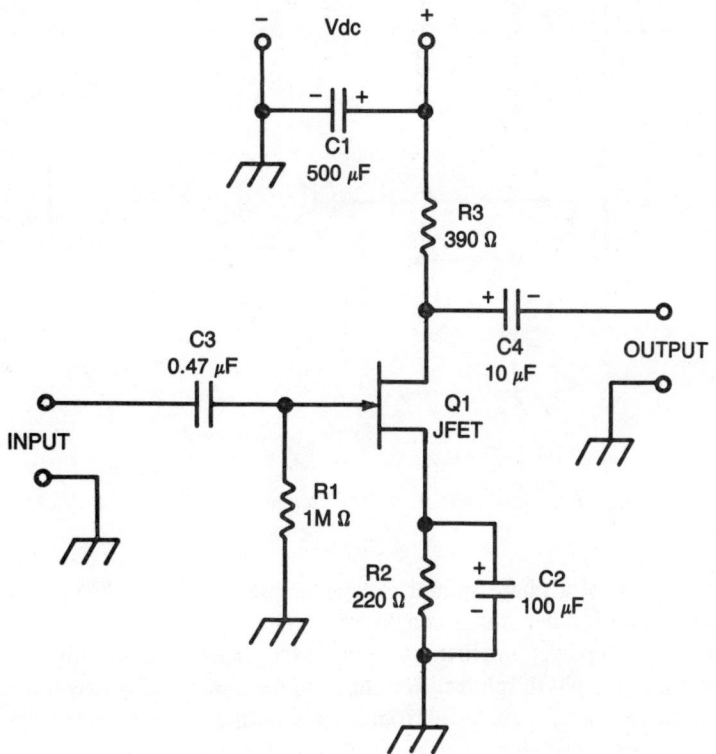

Fig. 2-7. JFET audio preamplifier circuit with source bias.

of the JFET at ground potential for ac signals while allowing it to be at a dc bias level established by the value of resistor R2 and the source-drain current of Q1. Input signal is applied to the gate, using resistor R1 to establish input impedance level and prevent C3 from becoming charged by the leakage current in the JFET gate. Output is taken across the drain-source path through capacitor C4.

(b) Figure 2-8 shows a push-pull audio power amplifier using identical npn bipolar power transistors. The necessary phase reversal required by the two transistors (to achieve push-pull operation) is supplied by a special interstage transformer that uses two opposite-phase secondary windings.

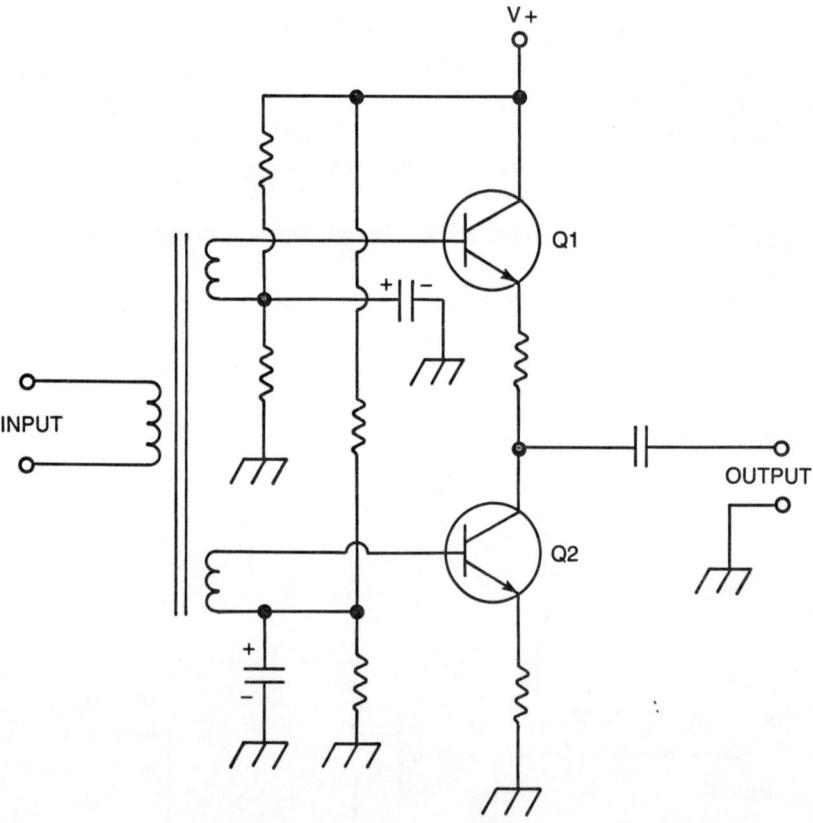

Fig. 2-8. Push-pull split-secondary "totem pole" power amplifier.

(c) The purpose of a phase splitter is to generate two signals that are 180 degrees out of phase with each other from a single-input signal. Such circuits are often used in transformerless audio power amplifiers to replace the transformer used in Fig. 2-8. Figure 2-9 shows a JFET phase splitter. The input signal is applied across the gate-source circuit, while the two outputs are taken from the source and drain, respectively. The drain signal is 180 degrees out of phase with the input signal, while the source signal is in phase with the input signal.

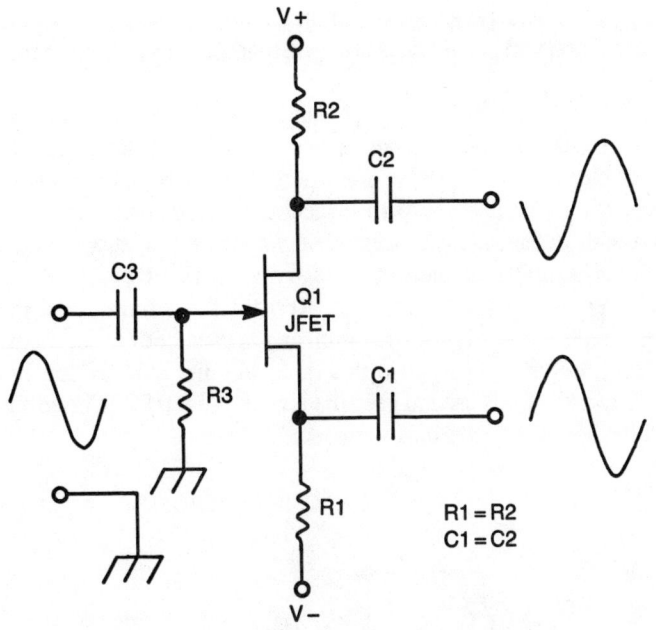

Fig. 2-9. JFET phase splitter.

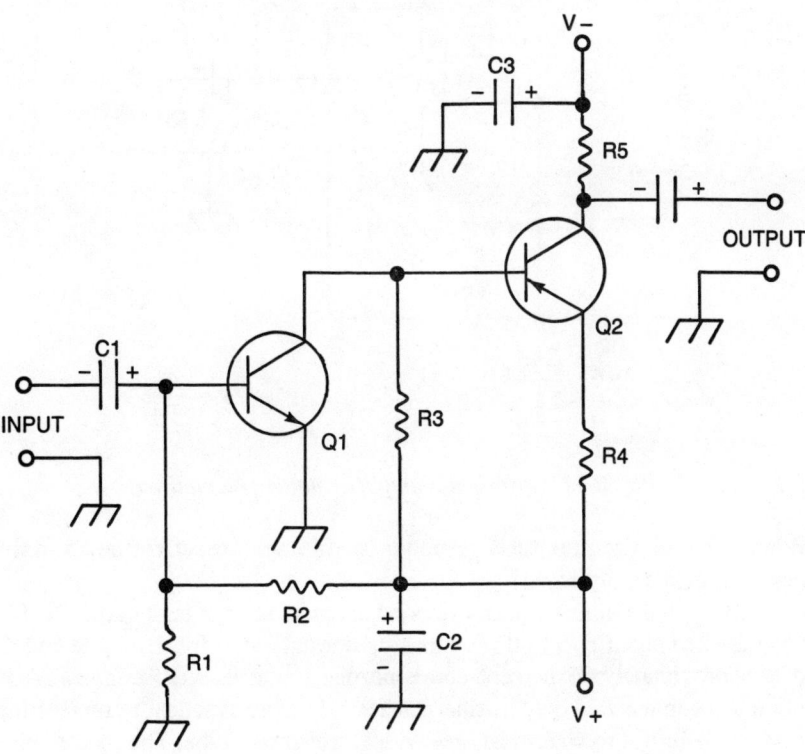

Fig. 2-10. Two-stage bipolar transistor audio preamplifier.

(d) Figure 2-10 shows a two-stage audio preamplifier based on bipolar transistors. In this particular case, npn and pnp devices are used together in a pair of common-emitter, direct-coupled, stages.

(e) The operational amplifier (op-amp) can be used in a wide variety of circuits and configurations. In fact, the op-amp is almost a universal amplifier device. In the circuit of Fig. 2-11, the circuit is a noninverting amplifier. Ordinarily, an operational uses two dc power supplies: "V+" is positive with respect to ground, while "V−" is negative with respect to ground. In Fig. 2-11, however, only a single V+ supply is used, and the V− terminal of the op-amp is grounded. A voltage divider (R3/R2) sets the noninverting input to a dc level that is (V+)/2 if R2 = R3. Resistor R1 is the normal input resistor for the operational amplifier in this configuration and is used both to set input impedance and prevent C1 from becoming charged with a dc level other than the level provided by the bias resistors. Capacitor C5 is used to keep the junction of R1/R2/R3 at ground potential for ac signals while preserving the bias network.

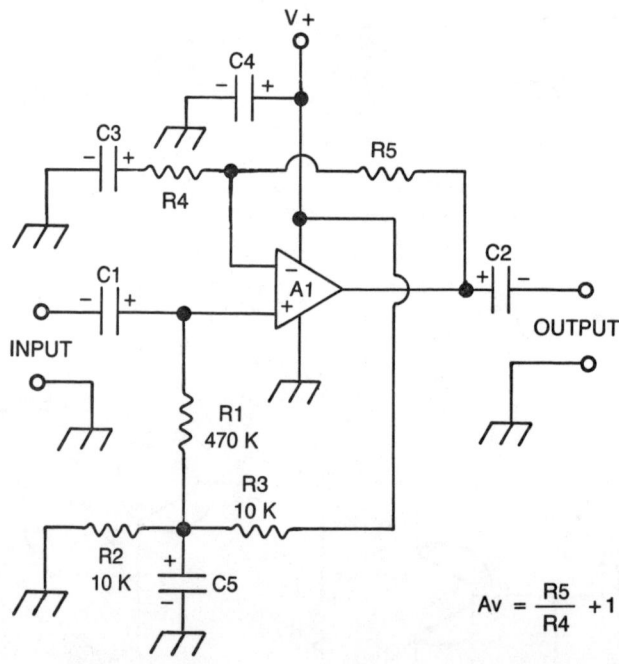

Fig. 2-11. *Operational amplifier audio preamplifier.*

Q47: What kind of vacuum tube responds to filament reactivation, and how is reactivation accomplished?

A: Thoriated tungsten filaments usually respond to reactivation at least once. The filament voltage is raised to about two to three times its normal value for 1 minute and then is reduced to approximately 25 percent above normal for at least 10 minutes. Actually, reactivation is recommended only in emergencies; it's more practical to replace the tube rather than risk failure at some critical time with a "repaired" tube. The figures and times for reactivation are quite flexible, so don't worry about splitting hairs if you use them.

Q48: Draw a rough graph of plate-current versus grid voltage (I_p vs. E_g) for various plate voltages on a typical triode vacuum tube.
  (a) How would output current vary with input voltage in Class A amplifier operation? Class AB operation? Class B operation? Class C operation?
  (b) Does the amplitude of the input signal determine the class of operation?
  (c) What is meant by *current-cutoff bias voltage*?
  (d) What is meant by plate current *saturation*?
  (e) What is the relationship between distortion in the output current waveform and:
   1. The class of operation?
   2. The portion of the transfer characteristic over which the signal is operating?
   3. Amplitude of the input signal?
  (f) What occurs in the grid circuit when the grid is "driven" positive? Would this have any effect on biasing?
  (g) In what way is the output current related to the output voltage?

A: Each curve in Fig. 2-12 is identified by a specific value of applied plate voltage and is therefore the resultant of the stated plate voltage and changes in the grid voltage. The curve is formed by noting the plate current as the control-grid voltage is increased in the negative direction, beginning at zero volts. These points are joined and form a curve. The change in plate current corresponding to a fixed change in bias voltage (grid voltage) is a function of the operating region on the plate-current curve.

  (a) In all classes of amplifier operation, the output current varies in direct proportion to input voltage as long as the input signal is swinging positive. In Class A operation, the plate current follows the input signal completely, increasing when the grid signal goes positive, decreasing when the grid signal goes negative. In Class B operation, the output current follows the grid voltage during positive half-cycles, but current is cut off completely during negative half-cycles. In Class C operation, because the tube is biased well below plate-current cutoff, plate current flows only during the latter two-thirds or so of its positive swing, so the tube is not drawing current at all during the greater portion of each input cycle. Class AB operation is quite similar to Class B, except that the tube is biased so that its operating point is slightly above cutoff; actually, the operating point is often midway between the center of the linear region and cutoff.

  (b) The amplitude of the input signal does not in itself determine the class of operation; however, the input-signal amplitude can alter the class. For example, when an amplifier stage is biased to operate Class AB and a very small input signal is applied to the grid, the tube will conduct during the complete input cycle and the tube can be said to be operating Class A. So long as the input signal keeps the tube operating within its linear region, the amplifier is operating Class A regardless of the biasing point.

  (c) The current-cutoff bias voltage is the amount of negative grid bias required to cause plate current to stop flowing at a particular value of applied plate voltage. As the grid voltage goes more and more negative, plate current decreases. At some point, called plate-current cutoff, the plate current will drop to zero. The value of the grid potential at this point is called *current cutoff bias voltage*.

  (d) Saturation occurs when plate current ceases to increase despite changes in grid voltage. As grid voltage approaches zero (from some negative value), plate current increases

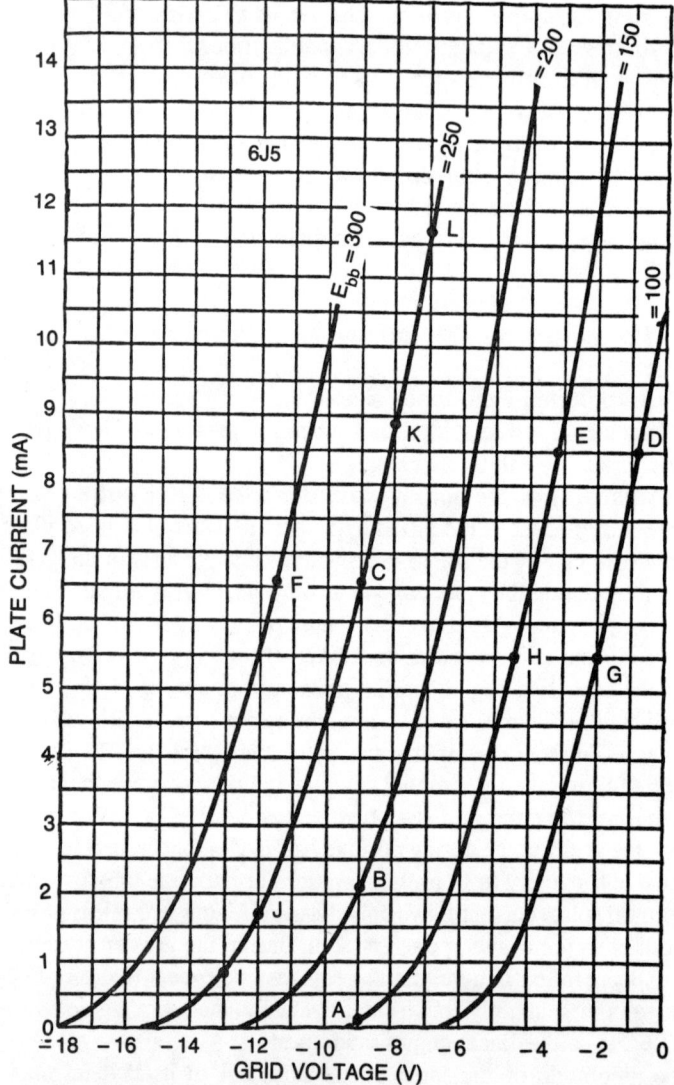

Fig. 2-12. Grid family of characteristic curves for the 6J5 triode.

in a linear manner. Close to zero grid voltage, though, the plate current increase does not follow the same constant rising characteristic as it did at the negative value.

(e) There is a direct relationship between distortion and class of operation. Class A operation is the only class that results in distortion-free operation, because it is the only class where the output waveform religiously follows the input waveform. Distortion occurs when the output waveform ceases to change in accordance with the changes of the input waveform.

The portion of the transfer characteristic being used is also a determining factor with respect to total distortion. If the transfer characteristic is linear and the tube is biased so that its no-signal current drain is plotted at the center of this linear region, then the output signal will be undistorted so long as the input voltage does not go beyond the limits established by plate-current cutoff and saturation. When a Class A amplifier is improperly biased—so that the no-signal operating point is above or below the center point on this transfer line—signal excesses will result in distortion.

The amplitude of the input signal is important in maintaining any given class of operation. A Class C amplifier, for example, requires a very high amplitude signal in order to bring the grid voltage up to the level required to make the amplifier conduct. If the driving signal is of insufficient amplitude, the tube may never be made to conduct at all. At best, it will be made to conduct for an insufficient period of the total signal's cycle.

(f) If the grid of an amplifier is driven positive, then the grid circuit will draw current. A current drain in the grid circuit is objectionable in most amplifier circuits because such drain represents a power loss and does not contribute to amplifier performance. The result of such power consumption is deterioration in amplifier operating efficiency. Similarly, a current drain in the grid circuit causes a change in biasing in most cases.

(g) An increase in output current causes a proportionate drop in plate voltage; thus, the output voltage of a grounded-cathode amplifier is precisely out of phase with the input voltage; and the plate-current waveform is equal and opposite to the plate-voltage waveform.

Q49: What is meant by *space charge*? By *secondary emission*?
A: The space charge is a cloud of electrons that forms around the cathode, that being negative, repels the electrons. Secondary emission is the effect caused by electrons striking the plate and knocking other electrons loose from the plate material.

Q50: What is meant by the *amplification factor* (mu) of a triode vacuum tube (amplifier)? Under what conditions would the amplifier gain approach the value of the mu?
A: Amplification factor is the change in plate voltage caused by a change in grid voltage with plate current constant. It is the maximum voltage gain of an amplifier.

When the highest possible plate load resistance is used, and when that load impedance is many times the plate resistance of the tube, amplifier gain will approach the mu of the tube.

Q51: What is meant by *plate resistance* of a vacuum tube? Upon what does its value depend?
A: Plate resistance is the opposition to current flow through a vacuum tube and actually is a measure of change in plate current caused by change in plate voltage with grid (E_g) bias constant.

$$r_p = \frac{\Delta E_p}{\Delta I_p} \bigg| E_g = \text{constant}$$

The value indicates the effectiveness of the plate voltage in producing a change in plate current and depends on the physical and electrical properties of the tube in question. The plate resistance of a triode is much less than that of a tetrode or pentode. Because the addition of screen and suppressor grids tends to isolate the control grid from the plate, changes in plate voltage have less effect on plate current and plate resistance increases.

Q52: What is meant by the voltage "gain" of a vacuum tube amplifier? How is this gain achieved?

A: Voltage gain is the ratio of output (plate) voltage to input (grid) voltage. Gain equals E-output divided by E-input. Voltage gain is achieved as a result of the amplifying ability of the tube and is related directly to the amplification factor.

Q53: Draw a rough graph of plate current vs plate supply voltage for three different bias voltages on a typical triode vacuum tube.

    (a) Explain, in a general way, how the value of the plate load resistance affects the portion of the curve over which the tube is operating. How is this related to distortion?

    (b) Operation over which portion of the curve produces the least distortion?

A: (a) The higher the plate voltage, the higher the value of negative grid bias required to cut off the tube; nonetheless, at some maximum plate-voltage value, there will be a determinable negative grid voltage that will cause the tube to cut off, or stop conducting. The lower right terminal of the load line of Fig. 2-13 is plotted at the point where applied plate voltage is maximum and current is zero. At the other extreme, there is a point at which plate current will be at its maximum—a point where further decreases in negative grid voltage will not result in additional plate-current flow, and where plate voltage will actually drop to zero. At this point, the factor that determines plate current is the plate load resistance. The maximum supply voltage divided by the plate load resistance will yield the total plate current at this extreme. The load line is plotted from this maximum-current at this extreme. The load line is plotted from this maximum-current point to the no-current, maximum-voltage point, as shown.

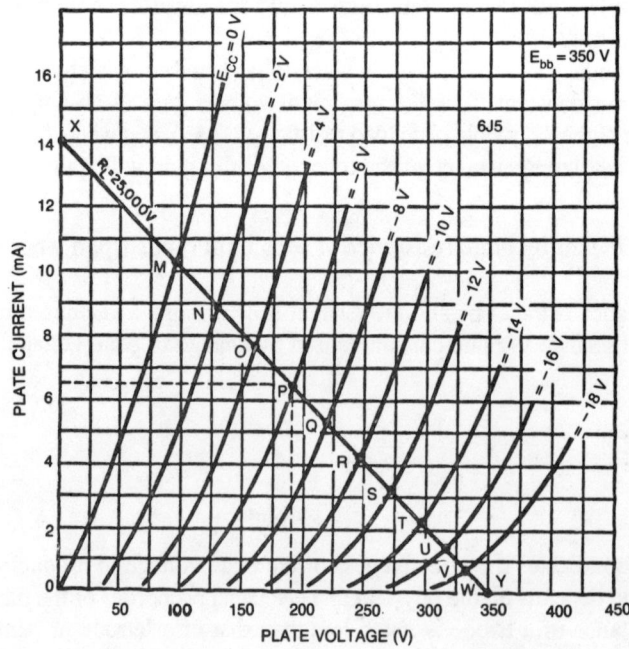

Fig. 2-13. Plate family of curves for 6J5 triode including load line.

(b) The least distortion is observed when the tube is operated over the lower part of the curve. The lower on the curve the load line is drawn, the more linear the operation of the amplifier will be and the more symmetrical the waveshape.

Q54: A triode "grounded-cathode" amplifier has a mu (amplification factor) of 30, a plate impedance of 5000 ohms, load impedance of 10,000 ohms, plate voltage of 300 volts, plate current of 10 mA and cathode resistor bias is used.
(a) What is the stage gain of this amplifier?
(b) What is the cutoff bias voltage E_{co}?
(c) Assuming the bias voltage is one-half the value of E_{co}, what value cathode resistor would be used to produce the required bias?
(d) What size capacitor should be used to sufficiently bypass the cathode resistor if the lowest approximate frequency desired is 500 Hz? (cps)

A: (a) The stage gain of this amplifier is 20. Stage gain is independent of plate current and voltage and is a function of the plate and load resistances and the individual tube's amplification factor, or mu. The equation for calculating stage gain is $(\mu R_L)/(R_L + R_p)$, where μ (Greek letter mu) is the amplification factor and the other symbols represent the resistance of the plate and load. With a mu of 30, a load resistance of 10k, and a plate impedance of 5k, the equation becomes

$$\frac{30 \times 10,000}{10,000 + 5,000} = \frac{300}{15} = 20$$

(b) The stage's cutoff bias is equal to the plate voltage divided by the amplification factor, or $E_{bb/\mu}$. The plate voltage is 300V and the amplification factor is 30; so the cutoff bias value is 10.

(c) The grid of the triode amplifier is biased at $-5V$ during the no-signal state. To maintain this 5V bias, the cathode resistor must keep the grid 5V negative with respect to the cathode. Given a plate current of 10 mA and a cathode-resistor drop of 5V, the value of the resistor can be determined from Ohm's Law, $R=E/I$. (The 10 mA of plate current flows through the cathode resistor, across the space charge, and out through the plate circuit). Because E is 5V and I is 0.01A (10 mA), the cathode resistor value is 500 ohms.

(d) The value of the cathode bypass capacitor must be such that its capacitive reactance is small compared with the resistance of the cathode resistor at 500 Hz. A practical value for audio applications would be in the range from 10 to 50 μF. Since there is no set rule for what constitutes "small" when compared with the cathode resistor value, some designers like to stay with an arbitrary percentage of the cathode resistance—such as 10 percent. Say that the capacitive reactance at 500 Hz must not exceed 50 ohms (10 percent of 500 ohms). Apply this formula: $X_c=1/(2\pi fC)$ using an arbitrary value of capacitance—say, 10 μF. By using the value of 10 μF and 500 Hz, the capacitive reactance works out to 31 ohms, well below the 10 percent established as the maximum.

Q55: Why is the efficiency of an amplifier operated class C higher than one operated class A or class B?

A: The efficiency of an amplifier depends on the time during each cycle that current flows. In the case of class C, this current is the least per cycle. Less power is dissipated and more output power is provided in class C operation than any other.

Q56: The following are excerpts from a tube manual rating a beam pentode. Explain the significance of each item.
  (a) Control grid-to-plate capacitance: 1.1 pF
  (b) Input capacitance: 2.2 pF
  (c) Output capacitance: 8.5 pF
  (d) Heater voltage: 6.3 V
  (e) Maximum dc plate supply voltage: 700 V
  (f) Maximum peak positive pulse voltage: 7,000 V
  (g) Maximum negative pulse plate voltage: 1,500 V
  (h) Maximum screen grid voltage: 175 V
  (i) Maximum peak negative control grid voltage: 200 V
  (j) Maximum plate dissipation: 20 W
  (k) Maximum screen grid dissipation: 3 W
  (l) Maximum dc cathode current: 200 mA
  (m) Maximum peak cathode current: 200 mA
  (n) Maximum control grid circuit resistance: 0.47 MΩ

A: (a) **Control-Grid-to-Plate Capacitance.** This important vacuum-tube specification gives valuable information with regard to the inherent coupling between the two key tube elements: the control grid and plate. The value of this interelectrode capacitance helps you to know the frequency limitations of the tube without neutralization. The lower the value here, the higher the signal frequency can be without neutralization. The specification is based on a measurement made with the screen and suppressor grids connected to the cathode element.

  (b) **Input Capacitance.** It would be hard to imagine a tube whose grid-to-plate capacitance was 1.1 pF having a total input capacitance of only 2.2 pF. This number is supposed to be the total sum of all the capacitances relative to the control grid; that is, it is the sum of these capacitances: control grid to plate, control grid to suppressor grid (and cathode, when no internal connection between the two is provided), and control grid to screen grid. The specified value is important because the values of capacitance on the control grid are additive to the capacitances in this part of the circuit, and will form part of any resonant circuit connected between these electrodes. A more realistic value for the total capacitance of a beam power pentode would be 10 to 20 pF.

  (c) **Output Capacitance.** The output capacitance is the sum of all capacitances associated with the plate, which include plate-to-grid, plate-to-suppressor, plate-to-screen, and plate-to-cathode. The importance of the specification lies in the fact that all these capacitance values are additive in nature and must be considered when a resonant network is part of the plate circuit.

  (d) **Heater Voltage.** The heater voltage is the voltage applied to the incandescent filament that heats the cathode. Since most tubes behave quite differently under different temperatures, it is important to adhere to this specified value when the tube is employed as an amplifier. Heater voltage ratings affect such important parameters as tube life, maximum plate current, and total output power.

  (e) **Max dc Plate Voltage.** As implied by the specification, this value is the highest permissible dc operating voltage the tube can tolerate under maximum-load conditions.

  (f) **Max Peak Positive Pulse Voltage.** In many applications, an amplifier tube will be subjected to pulse voltages far above the maximum safe value that can be applied under normal operating conditions. This specification permits the user to assess a tube's qualifications for such applications as TV horizontal output, where extremely high pulse voltages are applied to the plate for brief periods (microseconds). Exceeding the design values here could result in destruction of the tube, serious internal arcing, or erratic performance.

(g) **Max Negative Peak Pulse Plate Voltage.** In inductive plate loads, high negative voltages often develop for brief "flash" periods as the opposing EMF build across the load. This specification establishes the maximum flash-voltage value that can be tolerated without tube deterioration.

(h) **Maximum Screen-Grid Voltage.** Because the screen voltage (rather than plate voltage) is a determining factor with respect to the plate current capabilities of tetrodes and pentodes, this specification becomes one of particular importance. The value given represents the maximum working voltage that can be applied to the screen grid.

(i) **Max Peak Negative Control-Grid Voltage.** This specification describes the maximum safe value of negative voltage that can be tolerated by the tube without internal arcing. The value shown indicates that levels below 200 V can be handled without deleterious effects.

(j) **Max Plate Dissipation.** This specification describes the maximum power the plate can safely dissipate under a normal, sustained-load operation. The value may be exceeded where the tube is cooled by some external means (unless the specification is based on operation while tube is being cooled, as with many RF power amplifiers).

(k) **Max Screen-Grid Dissipation.** The screen, being positioned opposite to the control grid and in line with the cathode, is capable of dissipating power in the same manner as a plate, and many circuits involve use of the screen grid as a plate element.

(l) **Max dc Cathode Current.** The cathode supplies the current required in the plate circuit. The value specified here is the maximum current the cathode can deliver without premature burnout or weakening of the heater/cathode material. This current rating is *not* the maximum current the filament will draw from its low-voltage supply; this value would be considerably higher (perhaps more than an order of magnitude).

(m) **Max Peak Cathode Current.** Beyond the normal steady-drain value specified above, the cathode of many types of amplifiers is called upon to deliver very large surge currents. Understandably, the surge capability of a cathode is considerably higher than the normal dc constant-drain capability. The value given under this specification describes the maximum safe value of any high-current pulse. Exceeding the value even briefly could cause serious tube deterioration.

(n) **Max Control-Grid Circuit Resistance.** This specification describes the maximum grid-to-ground resistance of the tube's control grid and must not be exceeded. Higher values than that listed could result in cancellation of the grid bias, with a resulting excessive voltage applied to the grid, thereby destroying the tube or prohibiting normal operation.

Q57: Name at least three abnormal conditions that would tend to shorten the life of a vacuum tube. Also name one or more probable causes of each condition.

A: Excessive heater voltage, excessive power dissipation (plate), excessive dc operating voltage, operating the filaments from a dc supply, operating out of resonance, and improper bias are typical abnormal operating conditions that tend to degrade a tube's performance and result in premature failure. Causes are as follows.

(a) **Excessive Heater Voltage.** High line voltage is a common cause of this condition. An overheated heater boils off more electrons than the plate can handle; the condition results in rapid deterioration of the filament or heater.

(b) **Excessive Power Dissipation.** As mentioned in the answer to the preceding question, each tube designed for power applications has a rated dissipation that can be exceeded to the detriment of the tube. Excessive plate voltage on the tube results in a commensurate increase in plate current, the combination of which must be delivered to the load or dissipated by the tube. Excessive dissipation can also be caused by improper load impedance or an out-of-resonance condition in the output circuit.

(c) **Excessive DC Operating Voltage.** This condition might be caused by improper design of the tube's power supply, excessive line voltage, too low a value of plate load resistor, and—in mobile equipment—operation of the equipment when the voltage regulator in the automobile is adjusted for too high a voltage output. In the latter case, the excesses can be quite severe. The supply voltage should be about 12.6 volts (to a maximum of around 14.3 or so).

(d) **Filament Operation from a DC Supply.** Prolonged operation of a tube's filaments from a dc voltage supply will cause deterioration of one side of the heater before the other. Since one end of the heater winding is negative with respect to the other, it will release more electrons than the other end.

(e) **Operating Out of Resonance.** This condition is common with RF power amplifiers operated Class C. When the output circuit is in resonance, the plate current is at its minimum value. If any out-of-resonance condition occurs—such as any change in the plate load, change in the antenna, or a change in the setting of the capacitor in the final plate tank circuit—will cause the plate current to rise steeply, exceeding the maximum ratio value of the tube.

(f) **Improper Bias.** This is particularly true with Class C and Class B amplifiers, which are designed to operate only during portions of the input signal's complete cycle. When a Class C amplifier's operating point is moved closer to the tube's plate-current cutoff point (as happens with component value changes or other such causes), the tube is made to conduct for longer periods of each cycle. The maximum ratings for a particular class of operation are not necessarily applicable when the tube is operated under other classes.

Q58: Name at least three circuit factors (not including tube types and component values) in a one-stage amplifier circuit that should be considered at VHF but are not of particular concern at VLF.

A: Low-loss, high-quality components should be used. Length of wiring must be kept short to avoid capacitive and inductive feedback that result in circuit oscillation. Use of a common ground point for bypassing, and grounded-grid type amplifiers to reduce feedback problems. Neutralization may be necessary in many cases.

Q59: What factors tend to limit the operating frequency of a tube?

A: Two factors: interelectrode capacitance and (at VHF/UHF) electron transit time between cathode and anode.

Q60: Why are special tubes sometimes required at UHF and above?

A: As the operating frequency is increased, capacitive reactance between electrodes in a vacuum tube decreases. At frequencies higher than 100 MHz, the interelectrode capacitance of an ordinary vacuum tube bypasses radio frequencies very effectively. The electron transit time is about one-thousandth of a microsecond (0.001 μs). Although this might seem like an insignificant amount of time, it approaches and sometimes equals the time of a cycle within the tube, thus causing an undesirable phase shift.

A small number of ordinary vacuum tubes can be operated at frequencies higher than 100 MHz under certain critical operating conditions. The most suitable tubes of this type are triodes with low interelectrode capacitances, close spacing of the electrodes to reduce the transit time, a high amplification factor, and a fairly low plate resistance. Because some of these requirements are conflicting, tubes that strike a happy medium are generally selected. The operation of certain ordinary vacuum tubes at extremely high plate voltages is sometimes permitted in radar circuits to reduce the electron transit time.

The amount of interelectrode capacitance, the effect of the electron transit time, and other objectionable features of ordinary vacuum tubes are minimized considerably in the construction of special tubes for use at UHF. The UHF tubes have very small electrodes placed close together and often have no socket base. By reduction of all physical dimensions of a tube by the same scale, the interelectrode capacities are decreased without affecting the transconductance or the amplification factor. Transit time, likewise, is reduced, as is also the power-handling capacity of a tube of small dimensions.

In modern equipment, tubes are used at VHF/UHF only in high-power amplifiers. All small-signal applications use either silicon (Si) or gallium arsenide (GaAs) bipolar or field-effect transistors.

Q61: Describe the difference between positive (p-type) and negative (n-type) semiconductors with respect to:
   (a) The direction of current flow when an external EMF is applied.
   (b) The internal resistance when an external EMF is applied.

A: (a) An npn transistor is biased negatively at the emitter and positively at the collector (for a common base). The emitter injects free electrons into the p-type base region, where they are attracted across the collector junction by the positive potential. In a pnp transistor, the battery polarities would have to be reversed and the flow through the transistor would be by "holes" rather than electrons.

With the application of voltage to a transistor's input circuit, a current flows through the base-emitter junction. The action of the transistor is such that the same current density is caused to flow in the output circuit, which has a high internal resistance (considerably higher than the input circuit). Because the same current appears in both the input and output circuits, and the output circuit has a higher resistance than the input circuit, the result is a high equivalent power in the output circuit than the input circuit for any given input signal current.

Q62: What is the difference between forward and reverse biasing of transistors?

A: Forward bias provides a free flow of current through the transistor junction due to the movement of the majority carriers along the p- and n-type material. Forward biasing is illustrated for a pnp transistor in Fig. 2-14. Recall that holes are the majority carriers in p-type material.

Reverse bias retards or restricts current flow in a transistor. The small current flow is the result of the activity of the minority carriers. The minority carriers are few in number and consist of excess **electrons** in p material or excess **holes** in n material. The application

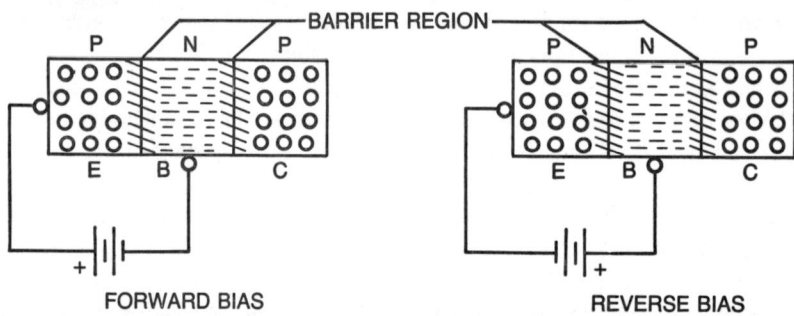

Fig. 2-14. *The effect of forward and reverse biasing of a pnp transistor.*

of an external EMF to reverse bias a junction actually widens the barrier region at that junction and stops the movement of majority carriers through it. (See Fig. 2-14.)

Q63: Show the connections of external batteries, resistance load and a signal source as would appear in a properly (fixed) biased, common-emitter transistor amplifier.

A: A properly biased common-emitter amplifier stage is shown in Fig. 2-15. It is common practice to show a signal source as a sine wave in a circle in series with the grid resistor (R_g). The stage could just as easily have been an npn transistor rather than a pnp. With an npn, the battery and electrolytic capacitors would be reversed in polarity. The current movement through a pnp is said to be by virtue of "holes" rather than electrons (A hole is the absence of an electron). In a common-emitter amplifier stage, the output signal is 180 degrees out of phase with the input signal; that is, as the input signal goes more positive the output signal goes more negative, and vice versa.

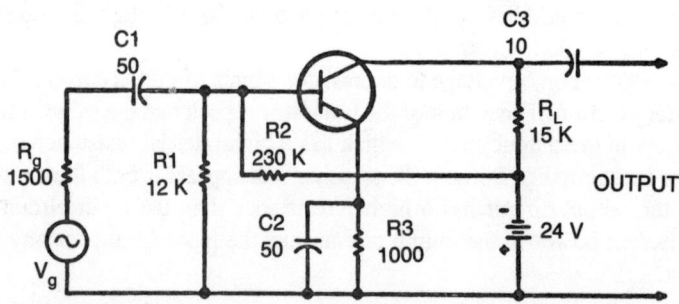

Fig. 2-15. *Common-emitter amplifier circuit using pnp transistor.*

Q64: The following are excerpts from a transistor handbook describing the characteristics of a pnp alloy-type transistor as used in a common-emitter circuit configuration. Explain the significance of each item.
Maximum and minimum ratings:
(a) Collector-to-base voltage (emitter open): −40 V maximum
(b) Collector-to-emitter voltage, (base-to-emitter volts is 0.5): −40 V maximum

(c) Emitter-to-base voltage: −5.0 V maximum
(d) Collector current: 10 mA maximum
Transistor dissipation:
(e) At an ambient temperature of 25°C, for operation in free air: 120 mW maximum
(f) At a case temperature of 25°C, for operation with a heatsink: 140 mW maximum
(g) Ambient temperature range, operating, the storage: −65° to +100°C.

A: (a) Collector-to-base voltage (emitter open) is the maximum voltage that can be applied between these elements without danger of breakdown of the junction. (No connection to emitter terminal.)

(b) Collector-to-emitter voltage (base-to-emitter reverse biased at −5 volts) refers to the maximum safe voltage that can be applied between the collector and emitter without danger of a breakdown.

(c) Emitter-to-base voltage is the maximum forward bias voltage that can safely be used to limit emitter-to-collector and base-to-emitter current.

(d) Collector current is the value that cannot be exceeded without possible permanent damage to the collector-emitter junction.

(e) Transistor dissipation at an ambient temperature of 25 degrees C for operation in free air is the maximum power that can safely be dissipated by the transistor without a heatsink.

(f) Transistor dissipation at a case temperature of 25 degrees C with a heatsink refers to the greatest thermal rating at which the transistor can be safely operated with a heatsink.

(g) Ambient temperature range for operating and storage is the temperature limit within which no electrical characteristic damage to the transistor will occur. Either simple storage or operation outside of these limits could be expected to result in degradation of the unit.

Q65: Draw a circuit diagram of a method of obtaining self-bias with one battery, without current feedback, in a common-emitter amplifier. Explain the voltage drops in the resistors.

A: See Fig. 2-16. Base-to-emitter negative bias is developed across R1, forming part of a voltage divider network. The network consists of R2 and R1, with electron flow going from the negative battery through R1, R2, and back to the positive battery terminal. Voltage drop is negative at the negative battery and positive at the transistor base end of R1. The R2 voltage drop is negative at the base end and positive at the battery terminal. The emitter is also connected to the positive battery, which makes the base negative with respect to the emitter. This form of bias is considered to be unstable for temperature variations.

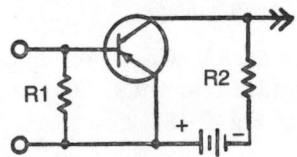

Fig. 2-16. Common-emitter amplifier circuit with self bias.

Q66: Draw a circuit diagram of a common-emitter amplifier with emitter bias. Explain its operation.

A: See Fig. 2-17, which places the emitter at ground potential with respect to the signal due to the capacitor across Vee, but at a positive dc level. Reverse biasing of the collector-emitter junction is accomplished by the negative terminal of Vee feeding the collector through R_L and the positive battery connection to the emitter. Forward biasing of the base-emitter junction is carried out by current flowing from negative Vee through current-limiting resistor R_b and on through the base-emitter junction back to positive battery.

Q67: Why is stabilization of a transistor amplifier usually necessary? How would a "thermistor" be used in this respect?

A: Stabilization of a transistor amplifier is normally required because the reverse-bias collector current or leakage current increases with temperature and changes the operating point of the transistor. Collector current may be stabilized by lowering the base current to compensate for the increase in leakage current. This can be done with a thermistor, which is a temperature-sensitive resistor with a negative temperature coefficient. Connecting the thermistor between the base and emitter provides the desired circuit stabilization, because the constant current through the base-biasing resistor is divided between the transistor base and the thermistor. As temperature increases, the resistance of the thermistor decreases, causing more current to flow through it and leaving less for the base-bias current. By the use of proper values, the decrease in base current may be sufficient to cause a decrease in collector current that is equal to the increase in collector leakage current. Thus, the collector current is constant under varying temperature conditions.

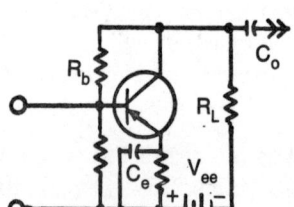

Fig. 2-17. Common-emitter amplifier with emitter bias.

Q68: The value of the alpha cutoff frequency of a transistor is primarily dependent upon what one factor? Does the value of alpha cutoff frequency normally have any relationship to the collector-to-base voltage?

A: Alpha cutoff frequency is primarily dependent on the physical thickness of the transistor base. The thinner the base, the higher the alpha cutoff frequency. Indirectly related to the collector-to-base voltage, the alpha cutoff frequency increases as the permissible collector-to-base voltage decreases, because the thinner the base, the lower the allowable base-to-collector voltage.

Q69: Draw the following power supply circuits with capacitor-input filters: Half-wave, full-wave center-tap, full-wave bridge, full-wave voltage-doubler.

A: Refer to Fig. 2-18.

Q70: What advantage does a bridge rectifier have over a conventional full-wave rectifier?

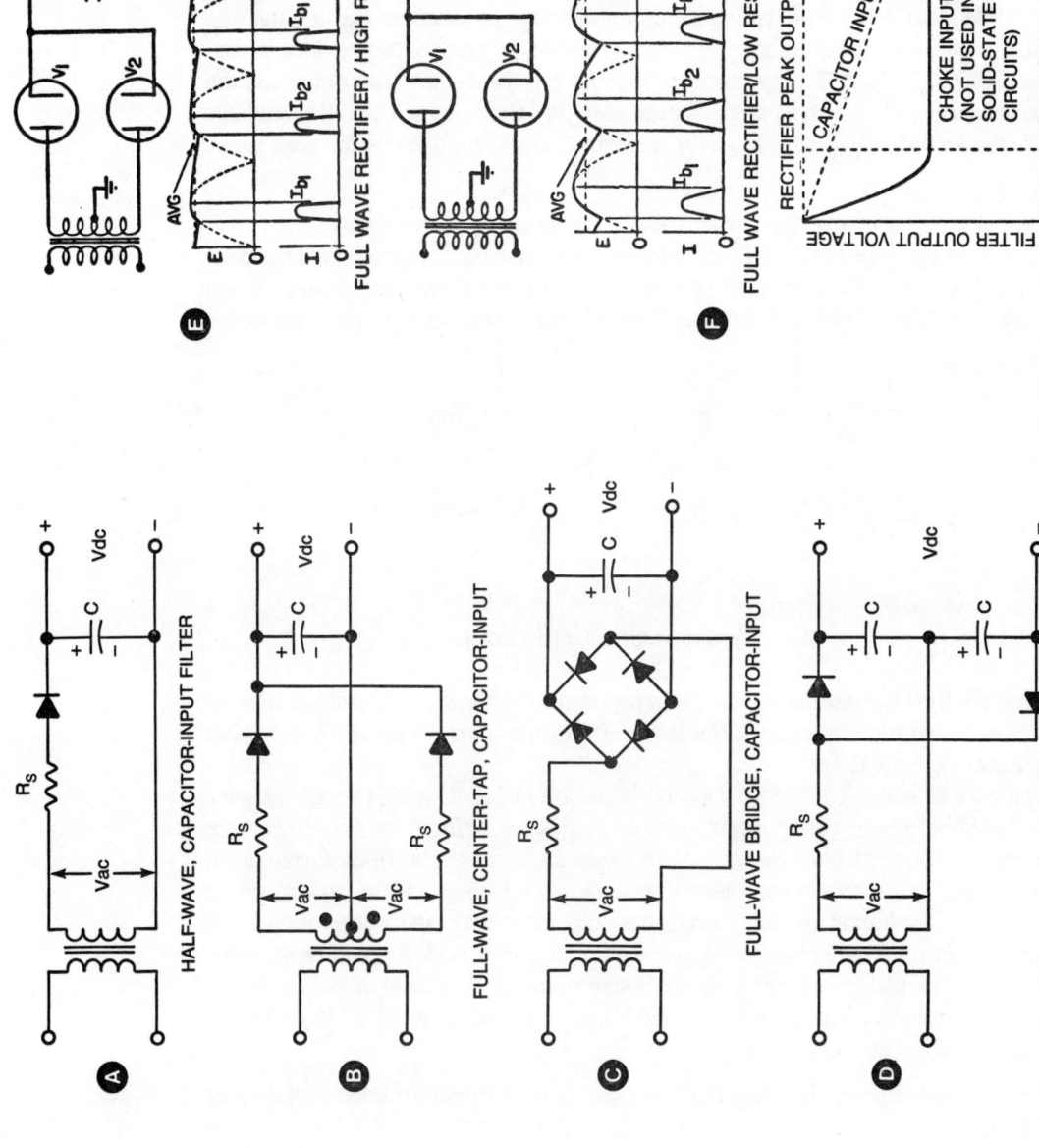

Fig. 2-18. Basic semiconductor and vacuum tube rectifier circuits.

A: The main advantage of the bridge rectifier over the conventional full-wave rectifier is that twice the voltage output from the same input is available. However, the current would be half as much in such cases, with the transformer limiting the VA or power output to the rectifier. The bridge rectifier does not require a center-tapped secondary, or in fact, even a transformer of any kind as is necessary in the other full-wave supplies.

Q71: What are swinging chokes? How are they used?
A: A swinging choke varies in inductance according to the actual load or inversely with the load current. The smaller the load, the greater the inductance required for adequate filtering. Aside from being more economical, voltage regulation under varying loads is greatly improved. An air gap in the iron core with the proper width provides partial saturation from the dc load. The greater the load, the greater the core saturation and the lower the inductance value. An ideal use for the swinging choke is the Class B modulator supply where the load changes from nearly zero to extremely high levels for peak audio inputs.

Q72: Show a method of obtaining two voltages from one power supply.
A: See Fig. 2-19 for a voltage divider circuit that provides the best regulation. The resistance of the dropping resistor is found by Ohm's Law ($R = E/I$) or the desired voltage drop divided by the sum of the current through the reduced voltage terminal plus the bleeder current.

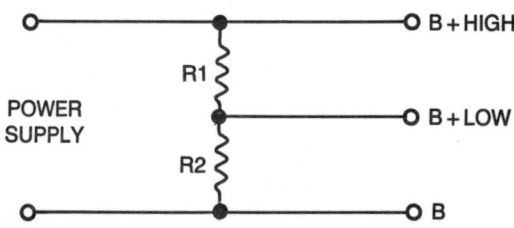

Fig. 2-19. Typical voltage-divider circuit.

Q73: What are the characteristics of a capacitor input filter system as compared with a choke input filter system? What is the effect on a filter choke of a large value of direct current flow?
A: Comparative characteristics of the capacitor input are higher dc output voltage, higher peak surge current, poorer voltage regulation. Capacitor input filters are not satisfactory for the mercury vapor rectifiers used in older equipment. Large values of dc current have no adverse effect on a filter choke if properly designed; otherwise, if the normal rating is exceeded, core saturation occurs, causing reduced inductance and overheating.

Choke-input filters offer better voltage regulation, lower peak surge current, more efficient use of tubes and transformers, lower voltage output (90 percent of the secondary rms), but they cannot be used without provision for current limiting and CEMF despiking in solid-state rectifier circuits.

Q74: What is the purpose of a "bleeder" resistor as used in conjunction with power supplies?

A: The bleeder resistor improves the regulation of the supply output by maintaining a constant load. It also offers a safety factor by discharging filter capacitors after shutdown.

Q75: Would varying the value of the bleeder resistor in a power supply have any effect on the ripple voltage?

A: Decreasing the value of the bleeder resistor would increase the output ripple in either capacitor input or choke input filters, while increasing the bleeder value would have little or no effect on the choke input type. However, increasing bleeder value would reduce the output ripple in the capacitor input arrangement.

Q76: What effect does the amount of current required by the load have upon the voltage regulation of the power supply? Why is voltage regulation an important factor?

A: As a rule, the greater the amount of current required, the poorer the voltage regulation. It is very important to hold output voltages constant under varying loads to prevent intermodulation of circuits, possible damage to components, and to maintain power output limits. The formula used to determine the percentage of regulation in power supplies is:

$$\text{Percentage of regulation} = \frac{(E_{NL} - E_{FL})}{E_{FL}} (100\%)$$

where E_{NL} = voltage, no-load (the voltage when no current is drawn) and E_{FL} = voltage, full load (when *maximum* current for which the circuit was designed is being delivered).

Q77: What is meant by the *peak inverse voltage* rating of a diode? How can it be computed for a full-wave power supply?

A: The peak inverse voltage (PIV) rating of a diode is the maximum peak voltage that can be applied reverse direction. The silicon diodes are very critical in this regard, because exceeding the ratings even momentarily will destroy the unit.

Full-wave peak inverse voltage is computed by multiplying the entire secondary (end-to-end) rms voltage by 1.414, or take the actual peak-to-peak reading end-to-end. When no transformer is used, multiply the line voltage (rms) by 1.414. The point most often overlooked in figuring PIV is that the voltage rating of a transformer on either side of center tap must be multiplied by 2.83 to come up with the correct figure. Add a 20-percent safety factor; remember that a 15-percent line surge is possible and your silicon diodes might not survive. In addition, the PIV rating of a specific diode could be out of tolerance.

Q78: Discuss the relative merits and limitations of the following types of rectifiers as used in power supplies.
    (a) Mercury-vapor diode
    (b) High-vacuum diode
    (c) Copper-oxide diode
    (d) Silicon diode
    (e) Selenium diode

A: (a) Mercury-vapor diodes have a low internal voltage drop of 10 to 15 volts, which provides a higher voltage output and improves voltage regulation because the loss in the tube is small. Disadvantages are numerous and hard to overlook: the need to preheat the filament before applying plate voltage; low peak inverse rating; a source of RF interference;

vertical operation required; and they cannot be used where power interrupts are brief and frequent; are now very expensive; they cannot be used with capacitive-input filter.

(b) High-vacuum diodes have a high peak inverse voltage rating. They can stand considerable abuse without damage, can be used in any position, and do not generate RF hash. They are less efficient due to the relatively high internal voltage drop, poor regulation, higher heater current requirement and considerable heat loss. High-vacuum rectifiers are almost entirely replaced by solid-state Si diodes.

(c) Copper-oxide diodes, an obsolete early semiconductor type, are more rugged than tubes but limited to low voltages and small currents. Characteristics vary greatly with temperature, and most have been replaced with silicon or germanium types.

(d) Silicon diodes represent top, overall efficiency with a very low internal voltage drop, compact size, inexpensive, good regulation and high current ratings. The only disadvantage of this diode is its sensitivity to PIV ratings. Even momentary transient voltages can cause a silicon diode to be destroyed if its ratings are exceeded.

(e) Selenium diodes are more efficient than copper-oxide types: They have a lower internal voltage drop and are not as sensitive to voltage transients as the silicon diode. Unlike the silicon, which has a lower internal voltage drop, the selenium forward resistance increases with age until replacement is mandatory. Selenium diodes are much larger than silicon units having the same rating. What's more, they are completely obsolete. Selenium diode equipment is usually retrofitted with silicon rectifier diodes.

3

Solid-State Electronics

The transistor has eclipsed the vacuum tube. For several decades, the tube reigned supreme. But in the late 1940s, physicists at Bell Laboratories invented the transistor. Fortunately for the electronics industry as a whole, Bell Labs wisely decided to license the device to other manufacturers and teach them the technology required to produce these little devices. Today it is difficult to find any new products that still use vacuum tubes. In the radio transmitter industry—the last holdout—the power level of transistor transmitter final amplifiers has been increasing to the point where it is conceivable to make a final amplifier in the 5000-watt class with transistor devices. This is the only area where vacuum tubes are still dominant. Transistor manufacturers have yet to find a way to make very high power solid-state final amplifiers at a competitive price with vacuum tubes.

The transistor was not actually *invented* until the late 1940s, but it had been postulated as theoretically possible for at least a decade prior to its invention. The metallurgy of the time, however, was not sufficiently advanced to make the purity of materials needed in transistor construction. One popular story that floats around the industry tells us that the junction field-effect transistor (JFET) was "invented" on paper as a physicist's *thought experiment* in the late 1930s. But again, metallurgy was the problem. It has been said that the JFET has certain properties that are superior to the bipolar (npn and pnp) transistor and would have been the standard transistor if the metallurgists could have produced the device. This would have given our present electronics industry a decade lead over its present state of development! Today, in several important areas of electronics, the FET has replaced the bipolar transistor as the dominant "device of choice."

The transistor is a descendant of the simple *pn junction diode* (also explained in Chapter 2). These diodes consist of two sections of semiconductor material, one being given negative charge carriers (*electrons*) and the other positive charge carriers (*holes*). The transistor is likened to a pair of pn junction diodes connected back-to-back.

The pn junction diode is found in nature. Certain minerals, such as the lead oxide called *galena*, operate as a pn junction diode. The early crystal radio sets, widely used in the years before vacuum-tube amplifiers made radio a *real* practicality, used chunks of galena to demodulate the radio signal. Artificial pn junctions became available in the early days of radio. These were made of copper oxide, or *selenium*. But it took the metallurgical advances under the pressure of World War II to make manufacturing pn junction diodes that operated into the radio spectrum an economic reality. The 1N34 and 1N60 germanium diodes and the 1N21 silicon diode were developed through this effort. The 1N21 was a VHF microwave device that was used extensively in radar receivers as the mixer diode. With the technology of World War II available, the physicists at Bell Labs were finally able to make the first transistor.

The transistor was a slow starter in the electronics world. When it was born, there were still severe manufacturing problems to overcome, and yield rates (the percentage of a batch that actually worked up to specifications) were low—quite low, in fact. The *1955 Allied Electronics Catalog* featured the CK722 and 2N107 transistors for about $16 (poorly performing audio transistors). (Remember, the dollar was worth a lot more then! To put it in perspective, $16 was ⅓ the week's take-home pay for minimum wage workers.) These leaky germanium transistors operated only at audio frequencies and would easily burn out if abused. And it took a lot less to abuse those devices than it does with modern devices. We were warned to use heatsink clips on the leads of the transistor when soldering, lest we melt the semiconductor material inside. Only a few RF-range transistors were available, and these were expensive. The first all-transistor portable radios were seen in 1954 to 1955 with a Motorola car radio model available in 1956. It wasn't until 1962, however, that a major American automobile manufacturer introduced the first production OEM all-transistor car radio (Delco Electronics Division of General Motors), followed by the rest of the makers the following year. The *all-American-five* home radio, a standard vacuum-tube design (which used the same tube line-up, no matter who made it or what style case was used), was still made until the mid-60s.

BASIC SEMICONDUCTOR THEORY

It is common practice to start any study of semiconductors with the basic model of the atom. Consider a model of a nucleus of positively-charged and neutral particles surrounded by a cloud of electrons orbiting in shells. The electrons carry a negative charge (of the same magnitude), but oppose polarity of the positively-charged particles in the nucleus.

An elementary nucleus consists of positively-charged particles called *protons*, and electrically-neutral particles called *neutrons*. The protons and neutrons have almost the same mass, but they are approximately 1850 times heavier than the orbiting electrons. The magnitude of charge, however, is the same.

The cloud of orbiting electrons contains as many electrons as there are protons in the nucleus. This arrangement permits *electroneutrality* (nature just loves order and balance), which means that the positive charges in the nucleus are exactly balanced; therefore, they are effectively cancelled, by the negative charges of the electrons.

The electrons orbit in *shells*, or fixed distances from the nucleus. These shells are the only distances at which the electrons are allowed to exist, and they represent the energy level of the electrons in that shell. When an electron takes on sufficient energy, it jumps to the higher shell—the one farther from the nucleus—that corresponds to that level. When an electron falls from an outer shell to an inner shell, it loses energy. To maintain the law of conservation of energy, the "lost" energy is actually given off in the form of light, infrared, or some other form of energy.

It appears that each shell is ideally filled with an exact number of electrons. When this number is reached, that shell is stable and any additional electrons begin to fill the next furthest shell from the nucleus. Chemical reactions, and the flow of electricity, depend mostly on the last shell in any given element. When the outermost shell is filled with its ideal number of electrons, a large amount of energy is needed to strip electrons loose for participation in the reaction known as current flow. These electrons are called *valence electrons*. How many are required to make the shell stable? For the shell closest to the nucleus, the stable configuration is two electrons. Hydrogen, the smallest atom, has just one electron in this shell, while helium fills the shell with two electrons. The hydrogen atom is chemically active because its outer—indeed, only—shell is filled with less than the ideal number of electrons. Helium, on the other hand, has two electrons in this shell, so it is ideally filled for stability. Helium is not chemically active. The second shell is ideally filled with eight electrons, while the third shell is filled with 18 electrons and so forth.

An ideally filled shell does not want to give up electrons. It takes a large amount of energy to make such a shell deliver electrons. If the outermost shell of an atom is *not* ideally filled, however, the outer electrons are not tightly bound. Consequently, only a small amount of energy is necessary to strip electrons from that atom. Consider carbon, which contains six electrons. Two of the electrons will completely fill the innermost shell, leaving four electrons for the second shell. Since this shell "wants" to have eight electrons, it will be loosely bound and will give up electrons easily. These four outer electrons give carbon its ability to permit the flow of electrical current and to participate in chemical reactions.

There are three classes of materials: *insulators, conductors* and *semiconductors*. Insulators are those materials that have atoms with ideally filled outer shells. They require a large electrical potential field to strip away *free electrons* for current flow. Consequently, they allow only a tiny leakage current to pass through them. Insulators are available for which *picoamperes* would be too large a unit to describe the level of current flow, even under potential differences of kilovolts! The outer shells of atoms in materials called conductors are loosely bound, which means they do not contain the ideal number required for filling. These atoms can have the outer electrons stripped away relatively easily, with only a small amount of energy needed. A little heat in some cases or a low voltage can easily remove an electron or two from these atoms.

Semiconductors are elements that are neither good conductors nor good insulators. Examples of semiconductors include germanium (Ge) and silicon (Si). Both of these elements are used in the manufacture of diodes, transistors, and other semiconductor products. Germanium has an atomic number of 32, meaning that it contains 32 protons balanced by 32 orbiting electrons. There are two electrons in the innermost shell, eight in the second shell, 18 in the third shell and only four in the fourth, or outermost shell.

Silicon has an atomic number of 14, meaning that its 14 protons in the nucleus are balanced by 14 electrons in the orbiting cloud. There are two electrons in the innermost shell and eight in the second shell. Both of these shells are ideally filled. The outer shell takes the remaining four electrons, which makes it less than ideally filled.

Both silicon and germanium have only four electrons in the outer shell; therefore, both have four valence electrons. As such, these elements are said to be *tetravalent*. These shells can be made stable with eight electrons. Tetravalent atoms try to mimic this stable configuration by forming *covalent bonds* between adjacent atoms in a crystalline structure. Figure 3-1 shows how this is done. Note that the actual situation is three-dimensional, but to keep this discussion simple, we demonstrate only a two-dimensional plane. Each atom has four valence electrons each of which is shared with one electron of an adjacent atom. The result is a simulation of the "stable octet." Note that each bond contains two electrons, one contributed by each of the atoms forming the bond.

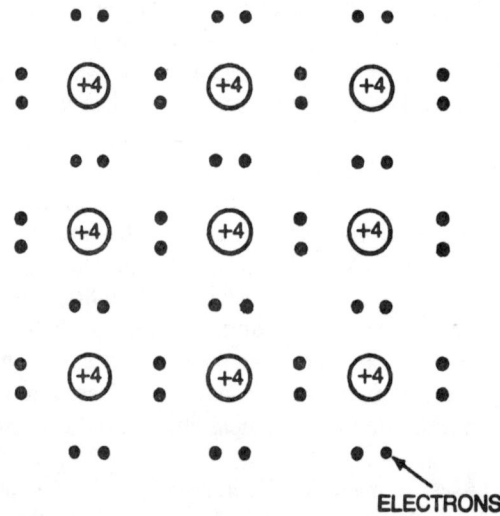

Fig. 3-1. Structure of the semiconductor crystal.

The bonding makes the element more stable, so there are few free electrons to participate in a current flow. There are a few free electrons floating around but not nearly as many as in conductors. Consider the situation if certain types of impurities are added to the pure semiconductor material. Suppose that the tetravalent semiconductor material (either germanium or silicon) is *doped* with a *penta*valent (five valence electrons) impurity (such as phosphorus, arsenic, or antimony). The rule requiring electroneutrality is not violated because each pentavalent atom is, in itself, electrically neutral; it contains as many electrons as there are protons. Figure 3-2 shows how a pentavalent atom forms covalent bonds in a tetravalent crystal (only one atom is shown with its four covalent atoms for the sake of simplicity). Because five valence electrons are in a pentavalent material, there is one *excess* electron for each pentavalent atom. This is because only four of the electrons in this atom can find "mates" in the tetravalent world it sees. The electrons that do not form covalent bonds with nearby tetravalent atoms are charge carriers in the flow of

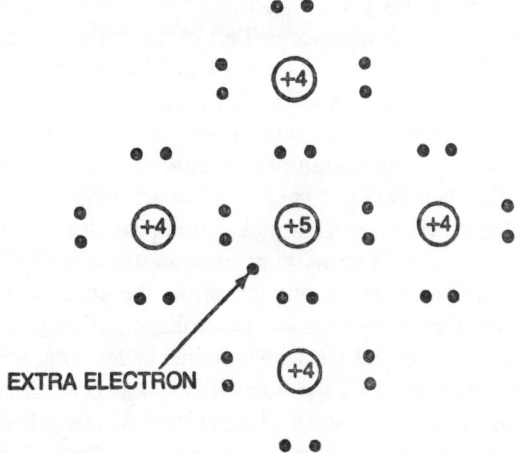

Fig. 3-2. Addition of pentavalent *atom creates an excess electron (n-type).*

electrical current. A material doped with a pentavalent impurity is an *n-type* semiconductor, and its *charge carriers* are *electrons*.

P-type semiconductor material is opposite the n-type: its charge carriers *appear* to be positively charged particles with the same size and mass of electrons. The p-type semiconductor is also doped with an impurity, but in this case a *tri*valent material is used in order to create an electron *deficiency*. The configuration of p-type semiconductor material is shown in Fig. 3-3. Each trivalent atom contains three valence electrons that form covalent bonds with the valence electrons of the tetravalent atoms; however, only three tetravalent atoms can form such bonds with the trivalent atoms. This leaves a *hole* in the crystal structure that seeks to be filled to become stable. The simplest definition of a *hole* is it is "a place where an electron should be, but isn't."

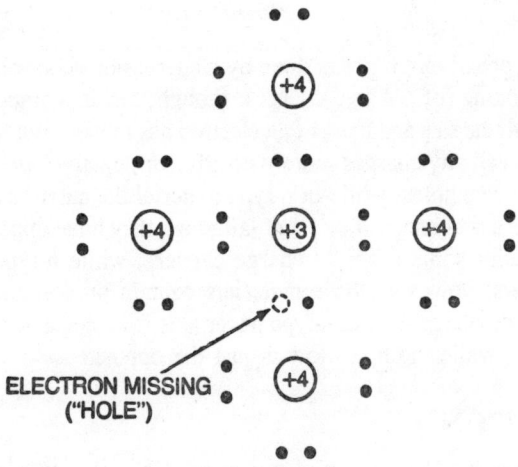

Fig. 3-3. Addition of a trivalent *atom creates an electron deficiency (p-type).*

The concept of holes usually gives students difficulty. It really should not, however, because it is essentially a very easy concept. Part of the problem is that many electronics textbooks refer to the holes exclusively in terms of positive charge carriers that *flow*. Holes don't flow; they only *appear* to flow. The electrons actually do the flowing. We treat the p-type semiconductor material *as if* holes—positively-charged particles—were actually flowing, but this is merely a convention of convenience. The situation is shown in Fig. 3-4. An electron can fill a hole in one of the bonds that are deficient, but only at the expense of creating a hole somewhere else. This makes it appear that the hole "flowed" from one atom to the other, but it was actually an electron that did all the moving. Consider Fig. 3-4. At the instant that voltage is applied across the doped semiconductor material (p-type), there is a hole located in atom A. The voltage field acts as a force that breaks loose an electron from atom B. This electron will drift under influence of the electric field until it is captured by atom A to fill the hole. A hole has been filled (obliterating it) at atom A, but only at the expense of creating a hole at atom B. The hole didn't move. Suppose that, a short time later, an electron is forced loose from atom C, and it drifts through the material until it fills the hole at atom B. This action creates a hole at atom C, but fills one at atom B. Again, there is an *apparent* movement of the hole from atom B to atom C. This action continues for some length of time, so there is an apparent hole drift over the path to atom F.

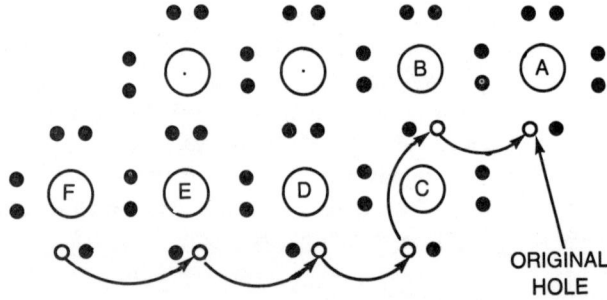

Fig. 3-4. "Hole flow" is electron flow viewed in reverse.

Once again, the actual moving was done by a succession of several electrons. To an external observer looking for positive charges, though, the appearance is that some sort of positive charge with the size and mass of an electron has moved. But this is in *appearance* only. We sometimes call any charges merely positive or negative *carriers*. In p-type material, the carriers are the holes, while in n-type material the carriers are electrons. Both of these carriers have a net charge of one unit, although they have opposite polarity. Also, p-type material contains some negative charge carriers, while n-type material contains some positive carriers; however, these materials contain predominantly more of their respective charges. The charge flow in n-type material is from negative to positive terminals of the power supply, while the hole flow is just the opposite.

THE PN JUNCTION

A pn junction consists of a section of p-type material interfaced with a section of n-type material. The basic model of an unbiased pn junction is shown in Fig. 3-5A. In the earliest, crude junction diodes, these sections were made separately and then joined together. Today, the practice is to make the pn junction from a monolithic piece of germanium or silicon, with the pentavalent and trivalent impurities diffused into the n-type and p-type sections, respectively. The germanium diode was the first to be commercially available outside of the radar business (circa World War II), but it has been largely eclipsed by the more thermally stable silicon diode.

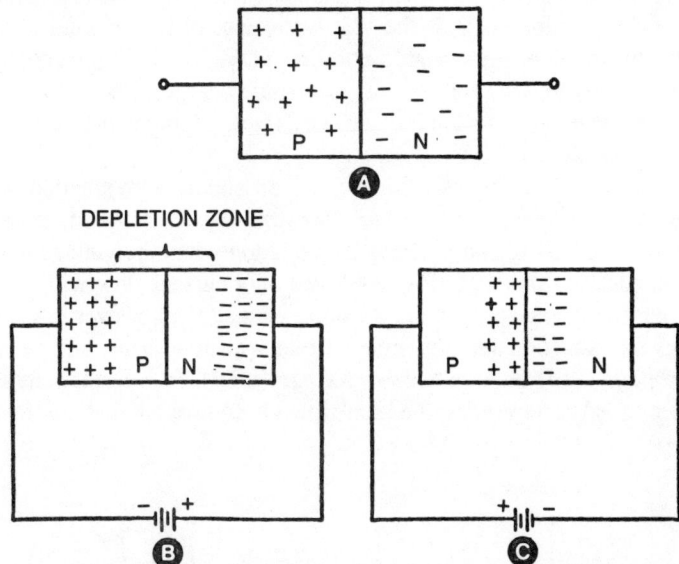

Fig. 3-5. (A) Unbiased pn junction; (B) reverse-biased pn junction; and (C) forward-biased pn junction.

The model for a *reverse-biased* pn junction is shown in Fig. 3-5B. In this case, the positive terminal of the battery is connected to the n-type material, and the negative terminal of the battery is connected to the p-type material. Recall that like charges repel each other and unlike charges attract each other. In the reverse bias situation, the negative charges of the n-type material are attracted to the positive terminal of the battery, and the positive charges of the p-type material are attracted to the negative terminal of the battery.

There is a zone near the junction in which all of the carriers have been removed, called a *depletion zone*. Very few carriers of either polarity exist inside the depletion zone. The depletion zone is severely deficient in carriers, so it has a very high electrical resistance. Little or no current will flow across the pn junction under this condition. Only a very small leakage current exists (leakage is also sometimes called *reverse current*).

The model for a *forward-biased* pn junction is shown in Fig. 3-5C. In this case, the polarity of the battery is reversed from the previous case—the positive terminal of the battery is connected to the p-type material, and the negative terminal is connected to the n-type material. Following the like-charges-repel rule, the positive carriers in the p-type are driven away from the battery and toward the junction. Similarly, the negative carriers are driven away from their battery terminal toward the junction. The depletion zone width reduces to zero, and charges can combine across the junction. This causes a current flow across the junction.

If a pn junction is reverse-biased, no current will flow. If a pn junction is forward-biased, current will flow. The diode current is given by:

$$I = I_s [e(V/nV_t) - 1] \qquad (3\text{-}1)$$

where I is the diode current, I_s is the reverse saturation current (leakage), e is the base of the natural logarithms, V is the voltage across the function, n is a constant (approximately) equal to 1 for Ge and 2 for Si, V_t is the volt equivalent of temperature and is defined as $V_t = KT/q$, where K is Boltzmann's constant (1.38×10^{-23} joules/°K), T is the temperature in degrees Kelvin, and q is the electronic charge (1.6×10^{-19} coulombs). The value of V_t evaluates to $V_t = T/11600$. If we assume room temperature to be 300°K, $V_t = 0.026$ volts, or 26 millivolts.

Figure 3-6 shows the current-vs-voltage characteristic for a pn junction diode. There are several regions of interest: *reverse bias, forward bias, saturation, breakdown* and *puncture*. The reverse bias region portrays the situation when the voltage applied to the diode causes no current flow, as in Fig. 3-5. The only current flow will be the reverse saturation current, or leakage current, I_s. At some voltage V_z, however, the diode breaks down in the reverse direction and a reverse current will flow. This voltage is called the *peak inverse voltage* (PIV) or *peak reverse voltage* (PRV). In a certain class of voltage regulator diodes called zener diodes, the breakdown is controlled, and the transition knee is sharper.

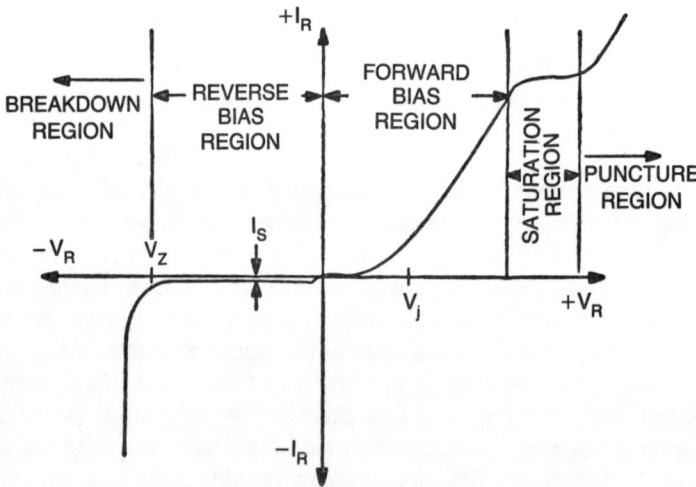

Fig. 3-6. I-vs-V characteristic curve for pn junction diode.

The forward current in a pn junction is not linear at low forward bias levels. The current is a nonlinear function of the applied voltage at potential levels between zero and some positive *junction voltage*, V_j. At voltages greater than V_j, however, the transfer curve straightens out and becomes linear. At some high value of forward bias, no additional current will flow for increases in applied voltage. This is the saturation region. Finally, at some high forward voltage, the junction breaks down (*punch through* region or *puncture* region).

Figure 3-7 shows the symbol commonly used for pn junction diodes used in electronics. The arrow indicates the direction of convention current flow, which is exactly the opposite of electron flow. The cathode end of the diode is the n-type material, while the anode end is the p-type material. When a voltage is applied that makes the anode more positive than the cathode, the diode is forward biased and a current flows. But when the anode is made negative with respect to the cathode, the diode is reverse biased and no current flows.

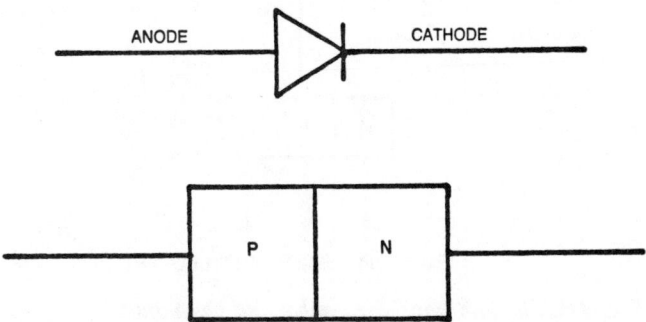

Fig. 3-7. Symbol for the pn junction diode and block diagram.

BASIC TRANSISTOR THEORY

The transistor is generally easier to apply and more efficient in several ways than tubes. Transistors are physically smaller and far more adaptable to printed circuit mounting than are tubes. They require no filament power, so they are inherently more efficient than tubes that do the same job—yet require filament power. The operating voltages for transistors are quite low. Transistors are usually happy with the type of voltages available from simple batteries, making them ideal for application in mobile or portable equipment.

The transistor is sometimes incorrectly modeled as a pair of pn junction diodes connected back-to-back. Figures 3-8 and 3-9 show the two basic types of bipolar transistor in model form. Both transistors can be modeled as having three sections. In both cases, a section of one-type semiconductor material (n or p) is sandwiched between two sections of the opposite material. The npn transistor is shown in Fig. 3-8, along with its circuit symbol. Note that the arrow on the emitter terminal is *pointing out* for the npn transistor. In the npn transistor, the center section (base) is made of p-type semiconductor material, and is sandwiched between n-type end sections (emitter and collector, respectively).

The pnp transistor is exactly the opposite of the npn. In this case, shown in Fig. 3-9, the base section is made of n-type material and is sandwiched between p-type emitter and collector sections. The symbol for a pnp transistor is also shown in Fig. 3-9. A little memory aid for remembering which symbol is to be used for pnp is *points in*.

Figure 3-10 shows an npn transistor biased for normal amplifier operation. The emitter-base junction is forward biased, and the collector is made positive with respect to the emitter. Note that biasing for a pnp equivalent amplifier is exactly the same, except that the battery polarities are reversed.

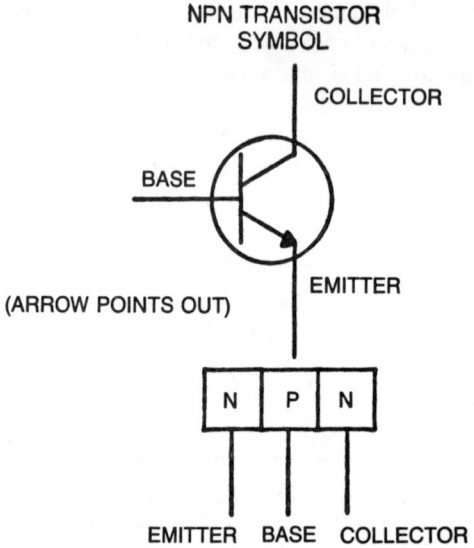

Fig. 3-8. The npn transistor symbol and block diagram.

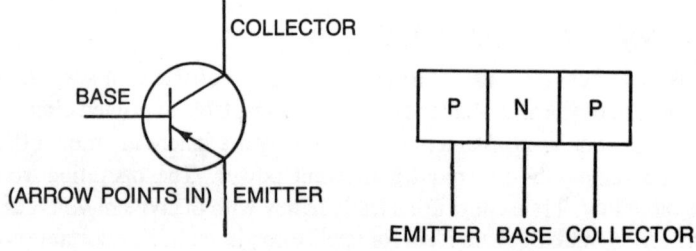

Fig. 3-9. The pnp transistor symbol and block diagram.

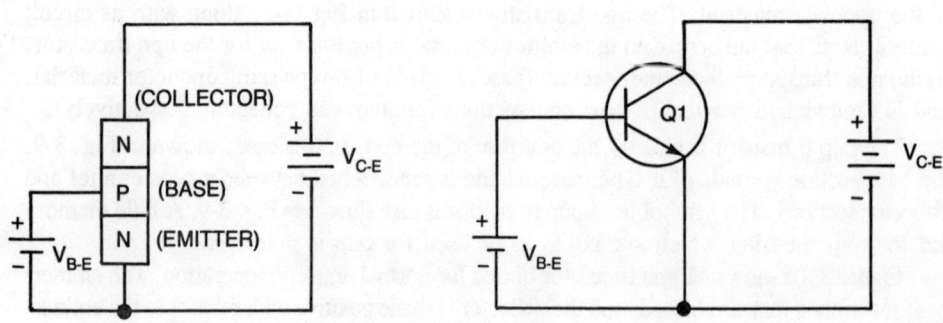

Fig. 3-10. The npn transistor showing biasing scheme for proper operation.

The circuit action in a transistor is a little more complex than our more simple descriptions. In general, though, we can claim that the charge carriers attracted into the base region from the emitter pass through to the collector. Only a small percentage of the current passes out of the base terminal of the transistor to the external circuit; on the order of 95 to 99 percent of the emitter current flows also in the collector circuit. The remaining 1 to 5 percent passes out of the transistor through the base circuit.

AMPLIFICATION

Amplification is the control of a larger voltage or current by a smaller voltage or current. In the case of the transistor circuits presented here, a small base current is capable of controlling a larger collector-emitter current.

The voltage appearing across the base-emitter junction is given by an expression that is derived from Equation 3-1. In this case, however, the expression for current is solved to give V, which this time is given as V_{be}:

$$V_{be} = \frac{KT}{q} \ln \left(\frac{I_c}{I_s} \right) \qquad (3\text{-}2)$$

where V_{be} is the base-emitter voltage, K is Boltzmann's constant 1.38×10^{-23} J/C, T is the temperature in degrees Kelvin, q is the electronic charge (1.6×10^{-19} coulombs), I_c is the transistor collector current, I_s is the reverse saturation current (approximately 10^{-10} amperes), and I_n denotes the use of natural (base-e) logarithms.

The ordinary base-emitter voltage for a properly forward-biased transistor falls within a narrow range that is dependent upon the semiconductor material used. Germanium transistors, for example, usually exhibit a junction potential of 0.2 to 0.3 volts, while silicon transistors exhibit a potential of 0.6 to 0.7 volts.

What happens when a signal source is applied to the input of the transistor amplifier? Figure 3-11 shows the circuit with an ac signal source connected in series with the collector-emitter circuit. As in all of our discussions, unless stated otherwise, the signal source is a sinusoidal waveform. It will add to or subtract from the ordinary base bias current. On positive excursions of the input sine wave, the collector-emitter current increases because

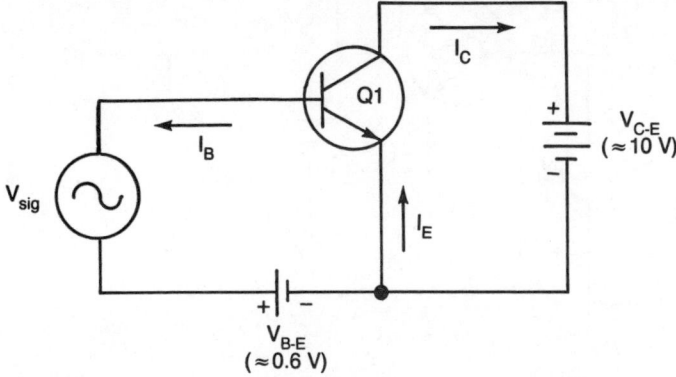

Fig. 3-11. Forward-biased transistor with signal superimposed on base current.

the current produced by the signal source adds to the base current. Similarly, on negative excursions of the input signal, the collector-emitter current decreases because the input signal subtracts from the bias current. Hence, the transistor is a basic current amplifier because I_b is always less than I_c; in some cases it is *very* much less.

Let's discuss a potential problem before it becomes firmly rooted in somebody's mind. In practice, the transistor can also be used as a voltage amplifier as well as a current amplifier, but many students have the *current* amplifier idea so firmly embedded into their minds that they seem unable to grasp the idea of voltage amplification. First, recall Ohm's Law, which relates voltage to the product of the current and resistance in circuit ($E = IR$). Consider the circuit shown in Fig. 3-12. Resistor R1 in Fig. 3-12 is connected in series with the collector of the transistor and the power supply. The power supply voltage is V_{CC}, while the collector-to-emitter voltage is V_C. The difference between V_C and V_{CC} is the voltage drop across collector load resistor R1, which is $I_C R_1$. The output voltage is the collector voltage, $V_C \times V_{CC} - (I_C R_1)$. Again, assume that the signal voltage is a sinusoid. As V_{SIG} increases, then so does collector current I_C. The collector voltage must then drop because collector current increases the voltage drop across the collector load resistor ($I_C R_1$), leaving less of the supply voltage for V_C. Similarly, when the signal voltage decreases, the collector current decreases, which causes the collector voltage to rise. In this case, the voltage drop across the collector load resistor is less as a result of the lower collector current, and there is more of the battery voltage available to make an output signal. The input signal voltage is less than the swing in the dc collector output voltage, so the transistor is operating as a voltage amplifier.

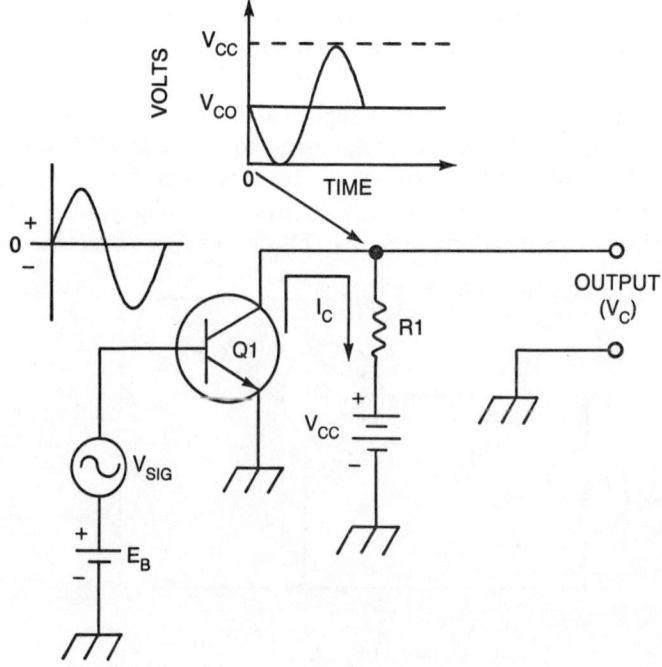

Fig. 3-12. Transistor as a voltage amplifier.

Thus far, the explanations apply directly to the npn transistor, but the same principles can also be applied to the pnp device. Simply reverse both the polarities in the circuit, and your thinking, and the same action occurs. At one time, most bipolar transistors were pnp devices, but today the npn predominates.

TRANSISTOR GAIN

There have been several popular ways to denote transistor gain. The two most popular are designated *alpha* and *beta*. The alpha gain can be defined as the ratio of the collector and emitter currents:

$$\alpha = I_C/I_E$$

where I_C is the collector current, and I_B is the base current, expressed in the same units as the collector current.

♦ **EXAMPLE**

Find the alpha gain when the collector current is 8 mA and the emitter current is 8.2 mA.

$$\alpha = I_C/I_E$$
$$= (8 \text{ mA})/(8.2 \text{ mA})$$
$$= 0.98$$

The value of the alpha gain is less than unity (1), with typical values between 0.7 and 0.99.

The parameters just used are static measurements, but a more realistic view is the ac alpha, which is defined as the ratio of a small change in collector current caused by a small change in emitter current.

$$\alpha = \Delta I_c / \Delta I_e$$

♦ **EXAMPLE**

A collector current that changes from 23 mA to 23.5 mA is caused by a change in base current of 1.12 mA to 1.64 mA. Find the alpha of this transistor.

$$\alpha = \Delta I_c / \Delta I_e$$
$$= (23.5 - 23.0)/(1.64 - 1.12)$$
$$= (0.5)/(0.52) = 0.96$$

The gain designation preferred for the common-emitter transistor amplifier, which is the circuit most frequently used, is the *beta gain*, defined as the ratio of the collector to the base current:

$$\beta = I_C/I_E$$

or for ac beta:
$$\beta = \Delta I_c / \Delta I_b$$

In some cases, beta is designated as H_{fe} for the dc beta and h_{fe} for the ac beta.

◆ **EXAMPLE**

Find the beta gain of a transistor in which a base current change of 100 µA causes a collector current to change from 66 to 79 mA.

$$\begin{aligned} h_{fe} &= \Delta I_c / \Delta I_b \\ &= (79 - 66)/(0.1) \\ &= (13)/(0.1) \\ &= 130 \end{aligned}$$

Alpha and beta are preferred for different classes of transistor amplifier circuit, but they are mathematically related. Recall that both involve the collector current, so they can be expected to have a relationship to each other. The expressions are:

$$\alpha = \frac{\beta}{1 + \beta}$$

$$\beta = \frac{\alpha}{1 - \alpha}$$

◆ **EXAMPLE**

Find the beta gain of a transistor that is known to have an alpha gain of 0.94.

$$\begin{aligned} \beta &= \alpha/(1 - \alpha) \\ &= \alpha/(1 - \alpha) \\ &= (0.94)/(1 - 0.94) \\ &= (0.94)/(0.06) = 15.7 \end{aligned}$$

◆ **EXAMPLE**

Find the alpha gain of a transistor that is known to have a beta gain of 100.

$$\begin{aligned} \alpha &= \beta/(1 + \beta) \\ &= (100)/(1 + 100) \\ &= (100)/(101) = 0.99 \end{aligned}$$

Examples such as those above have a way of showing up on FCC exams, so learn to work them.

FREQUENCY RESPONSE OF TRANSISTORS

The frequency response of transistor devices can be measured using any of three different methods: *alpha cutoff, beta cutoff* and *gain-bandwidth product*. The alpha cutoff frequency, F_{ab}, is defined as the frequency at which the ac alpha gain, h_{fb}, falls off to a value that is 3 dB lower than the alpha gain at some low frequency, such as 100 Hz; i.e., $h_{fb} = (0.707)h_{fbo}$, where h_{fbo} is the current gain at 1000 Hz.

The beta cutoff frequency is defined in a similar manner and is the frequency at which the beta gain of the transistor falls off to a value that is 3 dB down from its 1000 Hz value; i.e., $h_{fea} = (0.707) h_{feo}$. In general, the alpha cutoff is higher than the beta cutoff, and the beta cutoff frequency is a more ideal representation of the performance of a transistor.

The most useful transistor frequency response expression seems to be the *gain-bandwidth* product, f_T. This frequency is defined as the frequency at which the gain falls off to unity (beta gain is used). The gain-bandwidth product is valid for transistors operated in the common-emitter configuration. We can express f_T as:

$$f_T = h_{fe} \times f_o$$

where f_T is the gain-bandwidth product, h_{fe} is the ac beta, and f_o is the frequency at which the gain is measured. We find the value of f_T listed in transistor specification sheets as the frequency at which h_{fe} drops to unity. In many cases, we can find the gain-bandwidth product of a transistor if we know beta cutoff frequency f_{ae}:

$$f_T = f_{ae} \times h_{feo}$$

The f_T is often approximately equal to or slightly less than the alpha cutoff frequency.

In ordinary amplifier applications of the circuit shown in Fig. 3-12, it is normal to select a value of bias current that will cause a collector current to drop V_C to a point approximately equal to one-half of V_{CC}. This will allow an input signal with an amplitude that will exactly double the value of V_C on one peak of the sine wave and drop to exactly zero on the peak of the other half-cycle. A bias voltage must be provided to make the transistor operate at this *quiescent point*.

It is not economical to use separate dc power supplies for both the collector-emitter and base-emitter voltages, especially when the emitter is common to both circuits. A resistor *bias network* can instead derive the base-emitter potential from the collector-emitter power supply.

The simplest form of bias network is shown in Fig. 3-13. In this circuit, bias resistor R_B is connected between the base of the transistor and the $+V_{CC}$ power supply ($-V_{EE}$ if the transistor is pnp). The collector load resistor is connected from the collector to the same power supply. The collector current flows from the collector through the collector load resistor to the V_{CC} power supply. Similarly, base current I_B flows from the base terminal of the transistor, through base resistor R_B, and to the V_{CC} power supply. The value of the base current is selected to set the collector-emitter voltage to approximately one-half of V_{CC}.

The circuit shown in Fig. 3-13 provides the highest possible voltage gain, but it has terrible thermal stability. Temperature affects the base-emitter voltage of the transistor (Equation 3-2). In this circuit, the thermal effects are maximal. We select some needed, or convenient, level of collector current. Then we set the bias current according to Ohm's Law, considering the collector supply voltage, V_{CC}, and the junction voltage required for the transistor (0.6 volts in silicon transistors).

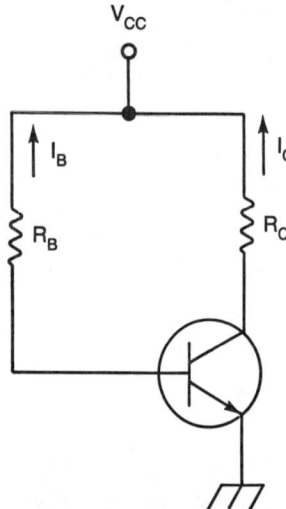

Fig. 3-13. Transistor bias scheme for operating from a single dc power supply.

The thermal stability of the circuit can be improved by creating a small amount of negative feedback in the emitter circuit. This is done by connecting a small resistor in series with the emitter (R_E in Fig. 3-14). The resistor works by using the emitter current to create a voltage drop across the emitter resistor. The emitter current is directly related to the collector current, so it too will change with temperature. When the temperature is stable, the voltage drops across the resistor, hence the voltage applied to the emitter terminal of the transistor is constant. But when the temperature changes, this voltage changes

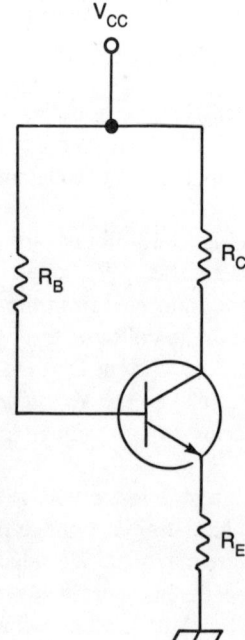

Fig. 3-14. Another bias scheme.

also. The bias on the transistor affects the collector current and is defined as the difference between the base and emitter potentials. Because the base voltage is fixed, the change in emitter voltage will effectively change the transistor bias and alter the collector current. Consider the case when the transistor temperature increases. According to Equation 3-2, the collector current will increase at this time, and that will cause an increase in the voltage drop across the emitter resistor, R_E. There is then a higher voltage on the emitter that tends to reduce the base-emitter voltage bias of the transistor, cancelling the change in collector current occasioned by the temperature rise. A similar change occurs in the opposite direction when the temperature decreases causing a decrease in the emitter current.

A measure of stability can also be obtained by using the bias configuration of Fig. 3-15. The emitter resistor is used (although in many cases it is deleted), and the base bias resistor is returned to the collector of the transistor instead of directly to the V_{CC} power supply. At frequencies up to 100 kHz (at higher frequencies, certain reactances enter into the picture) certain generalizations concerning the input impedance, output impedance, current gain and voltage gain can be made. For example, the low-frequency, small-signal output impedance is approximately equal to the collector resistor, R_C. The input impedance is approximately the product of emitter resistance R_E and the ac beta of the transistor. For example, assume that a 560-ohm emitter resistor is used in a circuit that has a beta of 90. What is the input impedance? By the rule-of-thumb just given, it would be (560 ohms) (90), or 50,400. The current gain of the stage is approximately the beta rating.

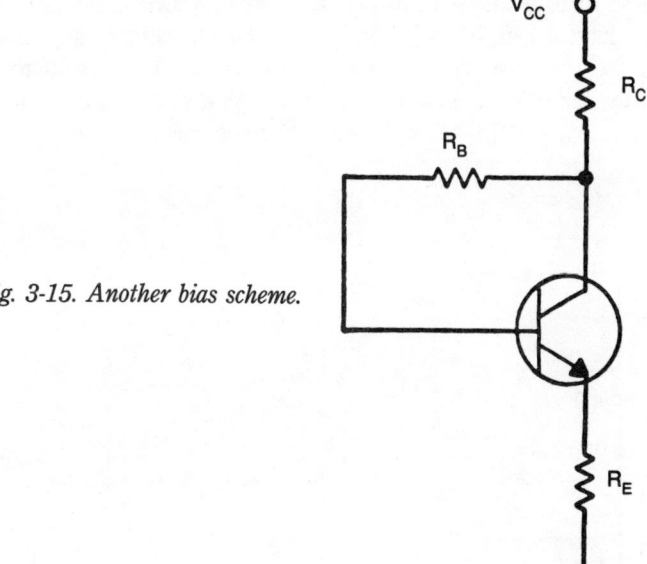

Fig. 3-15. Another bias scheme.

The voltage gain of the stage shown in Fig. 3-15 is the ratio of the collector resistor (R_C) to the emitter resistor (R_E), which is then multiplied by the ac beta:

$$A_V = \frac{R_C}{R_E} \times h_{fe}$$

where A_v is the voltage gain, R_C is the collector resistance in ohms, R_E is the emitter resistance in ohms, and h_{fe} is the ac beta.

In practical circuits you must make certain assumptions about the relative values of the emitter and collector resistances. Most small-signal circuits—not in power amplifiers—have an emitter resistor in the 100-ohm to 2000-ohm range with most being in the under-1000-ohm range. Also, some trade-offs must be made with respect to the relative values of these resistors. If the emitter resistor is a small percentage of the collector resistor (0.05 or less), the thermal stability of the circuit suffers, but the voltage gain is improved. Similarly, if the emitter resistor is made a large percentage of the collector resistor, the thermal stability is improved, but the voltage gain is lowered. The ratio of resistances must be adjusted for the most desirable trade-off between stability and gain. In general, the emitter resistor will have a value that is between 10 percent and 20 percent of the collector resistance.

The circuit of Fig. 3-15 achieves its stability from the emitter resistor and the configuration of the base resistor connection. Consider the action of the collector voltage on the base current as temperature changes cause the collector current to change. For the moment, assume that the emitter resistance is zero.

Perhaps the best bias circuit is the fixed-bias arrangement of Fig. 3-16. In this circuit, a resistor voltage divider fixes the base voltage at some specific point. Again, at low frequencies, the output impedance is approximately equal to the collector resistance, but there are some differences in the other parameters. The input impedance is approximately equal to the base resistor that is connected to ground, which is the resistance of R2 in Fig. 3-16. The current gain of the stage is given by the ratio of the grounded resistor in the base and the emitter resistance, or $R2/R_E$. The voltage gain is approximately equal to the ratio of the collector and emitter resistors. As a rule-of-thumb, the value of R2 should be approximately 10 times the value of the emitter resistor.

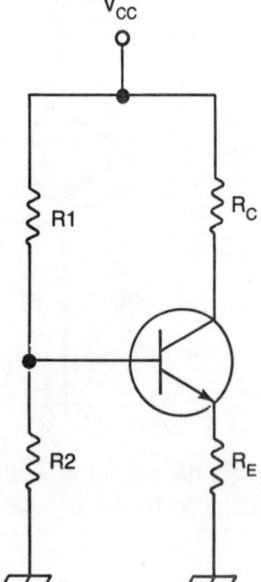

Fig. 3-16. Another bias scheme.

AMPLIFIER BASICS

Certain concepts are common to all amplifiers, and still others are common to large classes of amplifiers using all forms of active devices as the amplifying elements. The classification of amplifiers as to the action element is well known. It is easy, for example, to see why some amplifiers are vacuum-tube amplifiers and others are transistor amplifiers. Transistor amplifiers are even broken down into bipolar (npn and pnp) and field-effect (JFET and MOSFET) types. There are other methods of classifying amplifier circuits that are not so dependent on the technology of the active device. Several of these classification schemes are

* by common element
* by conduction angle
* by transfer function
* by feedback versus nonfeedback
* as to frequency dc, low-frequency ac, wideband, or RF.

This chapter considers some of the basics of amplifier circuits. Keep in mind that any of these could be constructed using transistors (of either basic type) or tubes. They can also be made with linear integrated circuits.

CLASSIFICATION BY COMMON ELEMENT

This method of classifying amplifier circuits revolves around noting which element (cathode/grid/plate, base/emitter/collector, or source/drain/gate) is common to both input and output circuits. Although technically incorrect, this is sometimes referred to as the *grounded* element, such as grounded-emitter amplifiers. We tend to use *common* and *grounded* interchangeably, so bear with us if you are a purist.

Consider this topic from the point of view of the transistor. You can extend the concept to tubes and FETs by keeping in mind the analogy of elements shown in Table 3-1. So, when the text states grounded base or common base, you can extend the same concept to vacuum tubes (grounded grid) and FETs (grounded gate) by substituting the analogous element.

Common Emitter

Figure 3-17 shows the different entries into this class. The circuit shown in Fig. 3-17A is the common emitter circuit. It gets its name from the fact that the emitter terminal of the transistor is common to both input and output circuits. The input signal is applied to

Table 3-1. Analogous Parts of Three Amplifier Devices

Bipolar Transistor	Vacuum Tube	Field-Effect Transistor
Base Emitter Collector	Grid Cathode Anode (plate)	Gate Source Drain

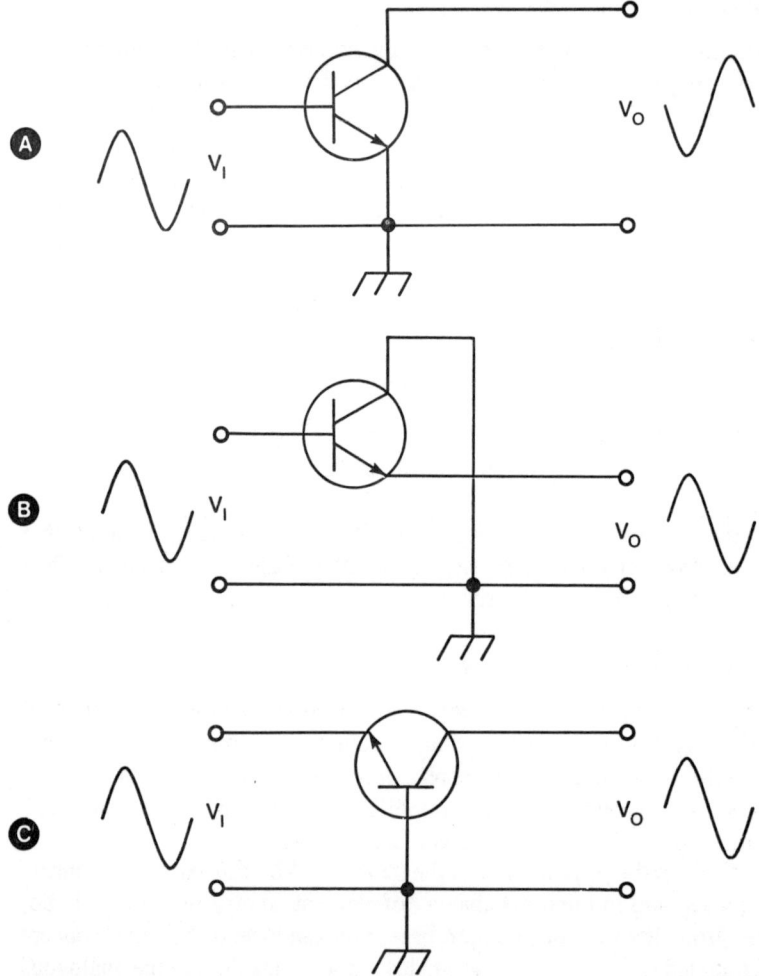

Fig. 3-17. (A) Common-emitter amplifier; (B) common collector amplifier; and (C) common base amplifier.

the transistor between the base and emitter terminals, while the output signal is taken from across the collector and emitter terminals.

The common emitter circuit offers high-current amplification (the beta rating of the transistor). But it also offers a substantial amount of voltage gain. Recall that the transistor is also a voltage amplifier, especially when a series resistor is placed between the collector terminal and the collector power supply. The values of gain for current and voltage are vastly different. The current gain is h_{fe}, but the voltage gain depends upon other factors. The next chapter shows how the voltage gain depends upon the R_L/R_E ratio in some cases and the product of the ratio and the beta in other cases.

The input impedance of the common emitter amplifier is medium ranged, or in the 1 kilohm range. The output impedance, though, is typically high, up to 50 kilohms. Typical values will be determined by the specific type of circuit, but there are some approximations that can be made. For most common emitter circuits, Z_{in} is equal to the product of the emitter resistor (R_E) and the h_{fe} of the transistor. The output impedance is essentially the value of the collector load resistor and will range from 5 to 50 kilohms.

The output signal in the common emitter circuit is 180 degrees out of phase with the input signal. This means that the CE amplifier is an *inverter*. The output signal is negative-going for a positive-going input signal and vice versa. The common-emitter transistor amplifier—and its tube/FET analogs—is probably the most often used configuration, and it is certainly the most often cited in textbooks.

Common Collector

The configuration is shown in Fig. 3-17B. In this circuit, the collector terminal of the transistor is common to both the input circuit and the output circuit. This circuit is also called the *emitter follower* circuit. The common collector circuit offers little or no voltage gain. Most of the time the voltage gain is actually less than unity. But the current gain is considerably better, i.e. $h_{fe} + 1$.

There is no phase inversion between output and input in the emitter follower circuit. The output voltage is in phase with the input signal voltage.

The input impedance of this circuit tends to be high, sometimes greater than 100 kilohms at frequencies less than 100 kHz. But the output impedance is very low, because it is limited to the value of the emitter resistor, which can be as low as 30 ohms. This situation leads us to one of the primary applications of the emitter follower: *impedance transformation*. The circuit is often used to connect a high-impedance source to an amplifier with a low-impedance amplifier.

The emitter follower is also used as a *buffer* amplifier, which is an intermediate stage between two circuits to provide some isolation between the two circuits. One primary example of this application is in the output circuit of oscillators. Many oscillators will *pull* or *chirp*, i.e. change frequency, if the load impedance they drive changes. Yet some of the very circuits used with oscillators (as in a superheterodyne receiver) naturally provide a changing impedance situation. The oscillator proves a lot more stable if a buffer amplifier is provided between it and the load. The emitter follower fills the bill because there is no need to provide any additional voltage amplification.

Common Base Circuits

Common base amplifiers use the base terminal of the transistor in both the input and output sections of the amplifier. The input signal is applied between the emitter and the base (see Fig. 3-17C), and the output is taken between the collector and the base.

The voltage gain of the common base circuit is high, on the order of 100 or more; however, the current gain is low, usually less than unity. The input impedance is also low, usually less than 1000 ohms, because it is limited by the emitter resistor. On the other hand, the output impedance is quite high. Again, there is no phase inversion between input and output circuits.

The principal use of the common base circuit is in high-frequency (HF) and VHF/UHF RF amplifiers in receivers. The circuit requires no neutralization at these frequencies, so it is superior to the common emitter circuit in that respect. Neutralization prevents oscillator action due to interelectrode capacitances providing a feedback signal that is in

Table 3-2. Characteristics for the Three Basic Amplifier Configurations

	Common Emitter	Common Collector	Common Base
Voltage Gain	High	Low ($\approx$ 1 or less)	High (> 100)
Current Gain	High	High ($\beta + 1$)	Low (<1)
Z_{in}	Medium ($\approx$ 1 K)	High (>100 K)	Low
Z_{out}	Medium to High (50 K)	Low (< 100 Ω)	High (>500 K)
Phase Inversion?	Yes	No	No

phase with the input signal. Table 3-2 summarizes the properties of the common emitter, common base, and common collector transistor amplifier circuits.

CLASSIFICATION BY CONDUCTION ANGLE

When one speaks of amplifier *classes*, it almost always means classification by *conduction angle*, or the portion of the input cycle over which output current flows. There are three traditional classes, labeled A, B, and C, and a combination class, AB. A new class is called class D.

* Class A: In this class, collector (plate/drain) current flows over the entire 360 degrees of the input cycle. This class is the least efficient of the three, but it is capable of linear (low distortion) operation using just one transistor.
* Class B: In this class, collector current flows over 180 degrees of the input cycle. This class is more efficient than class A, but requires two or more transistors in order to make the amplifier linear.
* Class C: This class is the most efficient of the classes and has collector current flowing over less than 180 degrees of the input cycle. A typical conduction angle for a class C amplifier is 90 to 150 degrees, with 120 degrees being most common.

The efficiency of an amplifier is dependent in part upon the class of operation. Figure 3-18 shows a chart of efficiency versus conduction angle. In this case, efficiency is defined as the ratio of the collector dc input power to the signal output power, expressed as a percentage

$$n = \frac{V_c I_c}{P_o} \times 100\% \qquad (3\text{-}3)$$

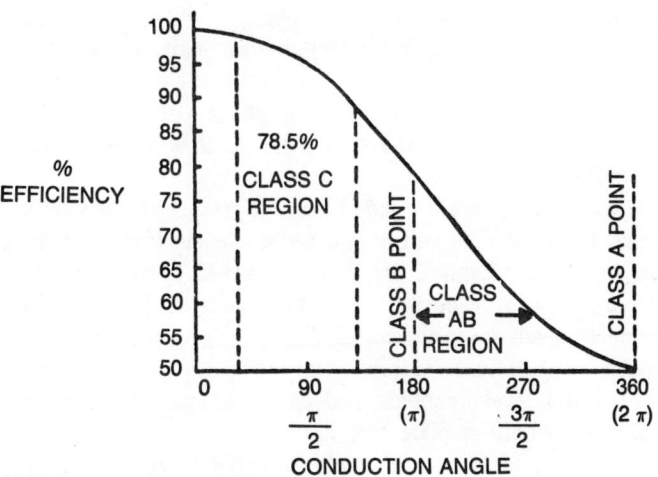

Fig. 3-18. Efficiency-vs-operating conduction angle for the four classes.

when n is the efficiency expressed as a percentage, V_c is the collector voltage (rms), I_C is the collector current (rms), and P_O is the output power in watts.

♦ **EXAMPLE**

Calculate the efficiency if an amplifier delivers 250 mW into a load when the collector current is 80 mA and the collector potential is 12V.

$$\begin{aligned} n &= (P_O)(100\%)/(I_c V_c) \\ &= (0.25)(100\%)/(12 \times 0.08) \\ &= (0.25)(100\%)/(0.96) = 26 \text{ percent} \end{aligned}$$

The class A amplifier does not operate at a great efficiency. In fact, most operate at about 25 percent efficiency. These amplifiers, then, would consume large amounts of dc power in order to generate any appreciable amount of signal output power. As a result, class A amplifiers are used almost exclusively as voltage amplifiers, which deliver little actual power but do build up the voltage level considerably. One exception to the rule is in situations where power is plentiful and component count is costly. In automobile and home broadcast receivers, for example, the designer worries little about the power supply but should be terribly concerned with the overall parts count in the radio. As a result, many have used class A power amplifiers despite the low efficiency. Almost all American-made auto radios from 1962 to 1980 used a single-ended, transistor, class A, audio power amplifier in the output stage.

The lost power—the difference between the dc power consumed and the signal output power delivered to an external load—is given off as heat in the transistor. The collector temperature of a transistor is considerably higher when operated in class A to deliver any given amount of output power. Also, at zero signal, when there is no input signal, the collector dissipation is maximum. In the class B and class C designs, on the other hand, the collector dissipation drops to zero when the input signal is zero.

The class B amplifier develops collector power only when the input signal is not zero. This is because the collector current flows over only one-half, or 180 degrees, of the input cycle. But it is impossible to make a linear amplifier using just one transistor in class B. It takes *two* transistors, driven 180 degrees out of phase with each other so that they will operate on alternate halves of the input signal. This theme is developed further in Chapter 7 (on power amplifiers).

The class C amplifier cannot be made linear, regardless of the number of transistors used. As a result, the class C amplifier cannot be used in places where it is important to preserve the input waveform. There are no class C hi-fi amplifiers, for example. Similarly, there are no radio frequency power amplifiers in class C if single-sideband or amplitude-modulated signals are to be boosted. The class C amplifier is, however, frequently used in radio power amplifiers if the transmitter is for CW (continuous wave), frequency modulation (FM) or high-level amplitude modulation (AM), in which the modulating signal is applied to the final amplifier plate circuit.

RF amplifiers operated in class C might be expected to produce a high harmonic output—not good for radio transmitters! But the high harmonic content of the raw pulse signal is eliminated by the action of RF tank circuits. Pulsed tank circuits oscillate at their natural resonant frequency. If the tank circuit is adjusted to have the same frequency as the plate/collector pulses, the output will be converted back to a sine wave by the flywheel effect. Class C efficiency is very high, typically on the order of 70 to 80 percent.

The class AB amplifier circuits are an attempt, not without success, to obtain some of the benefits of both classes. When two class AB transistors are used in push-pull, the circuit proves to be more linear than the class B and more efficient than the class A.

CLASSIFICATION BY TRANSFER FUNCTION

The transfer function of any electronic circuit can be expressed as the ratio of the output to the input. For a voltage amplifier, for example, the transfer function is V_O/V_{IN}, which results in the *voltage gain* of the circuit (symbolized by A_V).

There are four general subclasses of transfer function: *voltage, current, transconductance* and *transresistance*. Most readers are already familiar with the first two—voltage and current. A voltage amplifier has a transfer function that expresses the ratio of the output voltage to the input signal voltage:

$$A_V = V_O/V_{IN}$$

The units of the voltage gain A_V are dimensionless because the volts terms in both numerator and denominator cancel each other.

The current amplifier is given a transfer function that expresses the current gain, A_i, as a function of the output current and input signal current:

$$A_i = I_o/I_{in}$$

These two forms of amplifier are very well known and are understood by most readers. But what about transconductance and transresistance amplifiers? These are not quite as well known, except among engineering and technology students. Some hobbyists and amateur radio operators, for example, are totally unaware of these amplifier types, even though they see them repeatedly.

The *transresistance amplifier* has a transfer function that relates an output voltage to an input current. The units of *resistance* apply to the transfer function because in Ohm's Law, V/I is a resistance. The transfer function, then, is:

$$R_m = V_o/I_{in}$$

The resistance of the transfer equation is measured in the ordinary units of the ohm.

The *transconductance amplifier* is a little more familiar to most, because most ordinary vacuum tubes and field-effect transistors are described in terms of transconductance. The unit of transconductance is merely the reciprocal of resistance, so a transfer function can be of the form I/E. The unit used to describe transconductance is the *siemens* (S). Any transconductance amplifier can describe an output current caused by an input voltage:

$$g_m = I_o/V_{in}$$

♦ EXAMPLE

Find the transconductance in microsiemens of a field-effect transistor amplifier in which a 500 mV change in signal voltage will cause a 1 mA change in output current.

$$\begin{align} g_m &= I_o/V_{in} \\ &= (0.001A)/(0.500V) \\ &= 0.002 \text{ S} \\ &= 2000 \text{ } \mu S \end{align}$$

CLASSIFICATION BY FEEDBACK VERSUS NONFEEDBACK

Thus far, there has been no serious mention of the concept of feedback in amplifiers. But feedback can make a mediocre amplifier act like a much higher quality circuit. It is possible to cancel some distortion in the amplifier and make it a lot more stable by using a little *negative feedback*. This section deals briefly with the concept of negative feedback.

The basic block diagram for a feedback amplifier is shown in Fig. 3-19. The main amplifier will be a transistor, IC, or vacuum-tube stage with gain A. The output signal

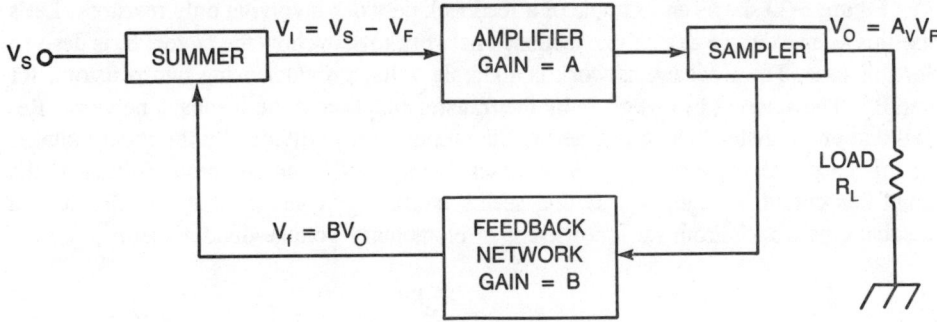

Fig. 3-19. Block diagram of a feedback amplifier.

voltage V_O is sampled, and this sample is passed back to the amplifier input via a feedback network. The gain of the feedback network is given by the Greek letter *beta* (β)—not to be confused with transistor gain. In most cases, the gain of the feedback network will be negative, as the network is made of all passive components (resistors, capacitors, indicators, etc.) and no amplification is provided. Of course, it is possible to make a feedback amplifier in which there is a main amplifier stage and a feedback amplifier stage.

The feedback signal is summed with the source signal V_S to form an input signal, $V_I = V_F - V_S$. The output signal is the product of this input signal and the voltage gain of the stage. This section deals with a voltage amplifier system, but it could just as easily be any of the other types of amplifiers by substituting the correct parameters.

The gain of the amplifier with feedback is given by the expression:

$$A_V = \frac{A}{1 + A\beta}$$

where A_V is the gain of the amplifier with feedback, A is the gain of the amplifier without feedback, and β is the transfer gain of the feedback network.

♦ **EXAMPLE**

Calculate the voltage gain of an amplifier when forward voltage gain A is 1000 and the feedback factor is 0.01.

$$\begin{aligned} A_v &= A/(1 + A\beta) \\ &= (1000)/[1 + (1000)(0.01)] \\ &= (1000)/[1 + (10)] \\ &= (1000)/(11) = 90.9 = 91 \end{aligned}$$

The use of feedback will improve the distortion situation of the amplifier. The *total harmonic distortion* (THD) will be reduced in the feedback amplifier, as in high-fidelity amplifiers. Audio amplifiers intended for high-fidelity applications have rather extensive feedback networks. The degree of feedback is often expressed in decibels:

$$N_{dB} = 20 \log_{10}(A_v/A) = 20 \log_{10}(1/(1 + A\beta))$$

Figure 3-20 shows an example of a feedback network involving only resistors. Let's use this as an example of a simple feedback network to show how the factor, β, is derived for this case. The feedback network is a simple voltage divider using two resistors, R1 and R2. The value of beta is given by the transfer function of the feedback network. Recall that any transfer function is merely the output voltage divided by the input voltage. In this case, the *output voltage* is feedback voltage, V_F, and the *input voltage* is the amplifier output voltage, V_O. If this seems confusing, consider that the circuit *is* a feedback network! From our knowledge of elementary voltage-divider theory,

$$V_F = \frac{V_O R1}{R1 + R2}$$

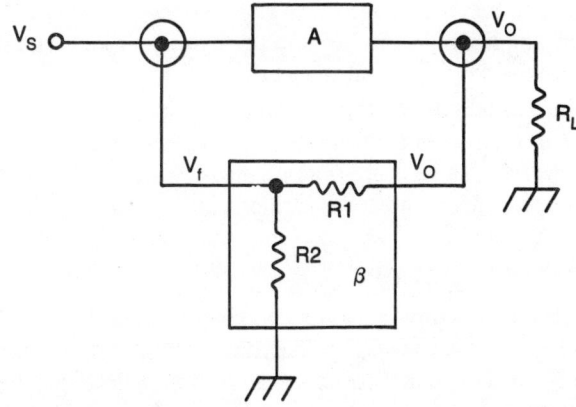

Fig. 3-20. Feedback amplifier with resistor feedback network.

The transfer function is obtained by dividing each side of the equation by V_o:

$$\frac{V_F}{V_O} = \frac{R1}{R1 + R2}$$

Because beta is defined as V_F/V_O,

$$\beta = \frac{R1}{R1 + R2}$$

This equation and the previous example show that you can set the gain of the amplifier by merely manipulating the resistors in the feedback network.

CLASSIFICATION BY FREQUENCY RESPONSE

Amplifiers are also classified as to the approximate frequency response. This method is almost anecdotal, and there is considerable overlap of the various classes. One class is a *dc amplifier*. This type of amplifier passes all ac frequencies (up to a practical limit) and dc levels. The fact that it is a dc amplifier does not mean that it won't pass ac, but it *will* pass dc signals as well as ac signals.

An *audio amplifier* is one that generally passes ac signals in the audio frequency range of roughly 30 Hz to 20,000 Hz. A communications audio amplifier may well have a much more limited bandwidth, such as 300 Hz to 3000 Hz. Hi-fi amplifiers, on the other hand, often have upper-end frequency responses into the 10 kHz range yet are still called audio amplifiers.

A *wideband amplifier* is exactly what its name implies: it presents a wide bandwidth. You might find a wideband amplifier presenting a bandwidth of dozens or hundreds of kilohertz. An amplifier that has a response up to 100 or 1000 kHz is surely a wideband amplifier. You could also call an amplifier with a 100 MHz bandwidth "wideband," but it is the practice of some to call amplifiers with bandwidths in the megahertz range *video amplifiers*. Because of current *m*onolithic *m*icrowave *i*ntegrated *c*ircuit (MMIC) technology, the term "wideband amplifier" is also used to refer to amplifiers with relatively flat response from dc (or near dc) to several gigahertz.

RF amplifiers are usually tuned to some specific frequency in the RF range. These amplifiers are used in radio communications (receivers and transmitters) to select only the frequency of interest. These amplifiers might be wideband but are centered about the frequency of interest. Most, however, are essentially narrowband devices.

AMPLIFIER COUPLING METHODS

It is usually necessary to provide some coupling between stages in cascade. When you want to pass a signal from one stage to the next, you must make sure that the proper conditions are met. For example, in some power amplifier applications, the coupling circuit must also match impedances between the stages (maximum power transfer occurs when the impedances of the source and load are equal). In other cases, we are not interested so much in matching impedances but are critically interested in keeping dc voltages from the input stage from interfering with the operation of the stage. In transistor amplifiers, for example, the collector voltage might be 10 V to 40 V while the base-emitter voltage of the next stage is in the 0 V to 5 V range. Clearly, when these stages are directly connected, trouble will result. But there is, however, a certain type of coupling circuit in which the transistor elements are connected together: *direct coupling*.

Direct Coupling

When direct coupling is used, the collector of the input stage is connected to the base of the next stage in cascade. It is necessary to design these stages such that the base voltage of Q2 (Fig. 3-21) is the same as the collector voltage of Q1.

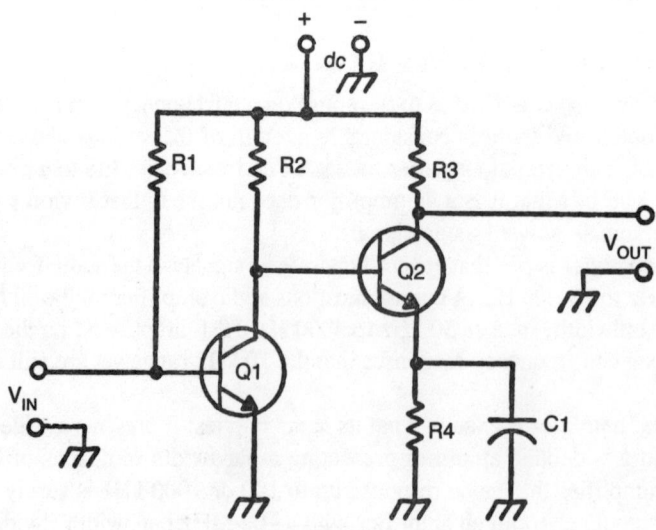

Fig. 3-21. Direct-coupled amplifier.

One advantage of the direct-coupled amplifier is that it can pass signals at all frequencies down to and including dc. These amplifiers are then sometimes called dc amplifiers, not after the name "direct coupled," but because they pass dc signals.

There could theoretically be almost any number of stages connected in this manner, but in most cases the amplifier has only 2 to 10 stages. Otherwise, you run into the supply voltage limit and there is no longer any swing available for the signal.

Transformer Coupling

Figure 3-22 shows an example of transformer coupling. This particular circuit shows two transformers. T1 is the interstage transformer and T2 is an output transformer. The principal reason for this type of coupling is impedance matching, so it is used mostly in power amplifier applications. On any transformer, the ratio of the impedances is set by the turns ratio of the windings:

$$(Z_P/Z_S)^{1/2} = (N_P/N_S)$$

where Z_P is the impedance seen by the primary (the reflected impedance across the transformer), Z_S is the impedance of the load connected to the secondary winding, N_P is the number of turns in the primary, and N_S is the number of turns in the secondary.

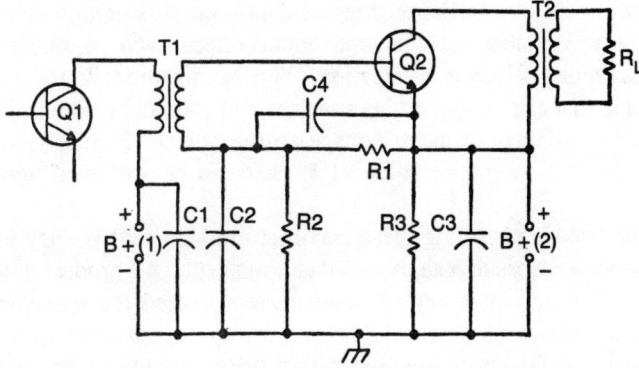

Fig. 3-22. Transformer-coupled amplifier.

Capacitor Coupling

Figure 3-23 shows an example of resistor/capacitor coupling. The idea here is to pass the signal from the input stage (Q1) to the output stage (Q2) without letting the dc voltage used to operate the input stage bias the output stage. The drain voltage of the input JFET can be as high as 25 V to 40 V, while it is almost a sure bet that the voltage applied to the gate of Q2 should be not only low, but negative. Both the level and the polarity, then, are wrong. Capacitor C1 is used to keep the dc that is present on the drain of Q1 from being applied to the gate of Q2.

TRANSISTOR AMPLIFIERS

Before you can understand transistors properly, you need to know some of the rules regarding designing practical transistor amplifiers. The actual process of design is given

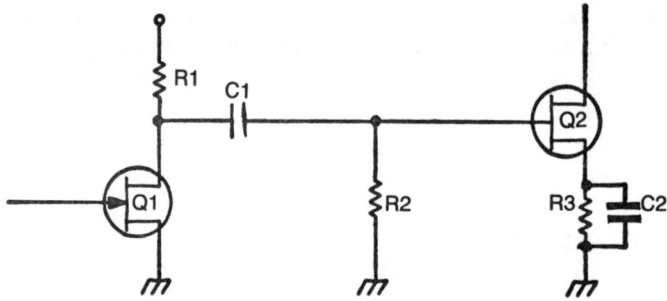

Fig. 3-23. Capacitor coupling.

in some greater complexity in other books, because they are dealing with generalized cases that can be applied to a wide range of situations. But we are discussing here the standard class A amplifier, in which the output is a voltage that is a linear reproduction (except for amplitude) of the input signal, and the frequency is less than 100 kHz.

Transistor Ratings

The manufacturer's specifications sheet or data book lists certain dc specifications for each transistor type number, including maximum collector voltage, maximum collector current, and maximum collector dissipation. The *maximum collector voltage* is the maximum potential that can be applied between the collector and emitter terminals of the transistor without causing permanent damage to the device. Similarly, the *maximum collector current* is the maximum value of I_c that can be sustained without causing damage.

There is little trouble regarding these maximum ratings until you try to operate the transistor at some values close to both maximums. Frequently, the product of the maximum collector voltage and the maximum collector current exceed the maximum allowable collector dissipation. While this would still be well within the maximum current and power ratings, it could exceed the maximum power dissipation. Therefore, be careful to check the maximum ratings under all circumstances.

Load Lines

It is desirable to vary the base current of the transistor in a manner that will cause the collector current to vary also in a linear manner. The transistor manufacturers make available data called a *family of curves* that relates the I_c-versus-V_c levels for assorted values of base current (Fig. 3-24). Each curve in the family represents the collector current caused by any given collector voltage for specified levels of base current. In this case, base bias levels of 100 µA, 200 µA, 300 µA, 400 µA and 500 µA have been selected.

The load line is drawn onto the family of curves by noting the points at which the collector current is zero and maximum. The load line is drawn between these two end points. The collector current is zero (neglecting the leakage current that invariably exists) when the transistor is cut off. In most amplifiers, a collector load resistance is in series with the dc collector power supply. Under conditions when the collector current is zero, the voltage drop across this resistor is also zero, so the collector-emitter voltage of the transistor is equal to the supply potential V_{cc}. The current is limited by the collector load resistor, so it assumes a value equal to V_{cc}/R_c under conditions where the transistor is saturated. Figure 3-24 shows a 1000-ohm load line (when the collector resistor is 1 kilohm). When the collector current is zero, the output voltage V_c rises to the full 28 V delivered by the power supply. And when the transistor is saturated, the output voltage is zero, and the collector current is maximum. All of the voltage from the power supply is dropped across the collector resistor, so the collector current is 28 mA (28 V/1000 ohms).

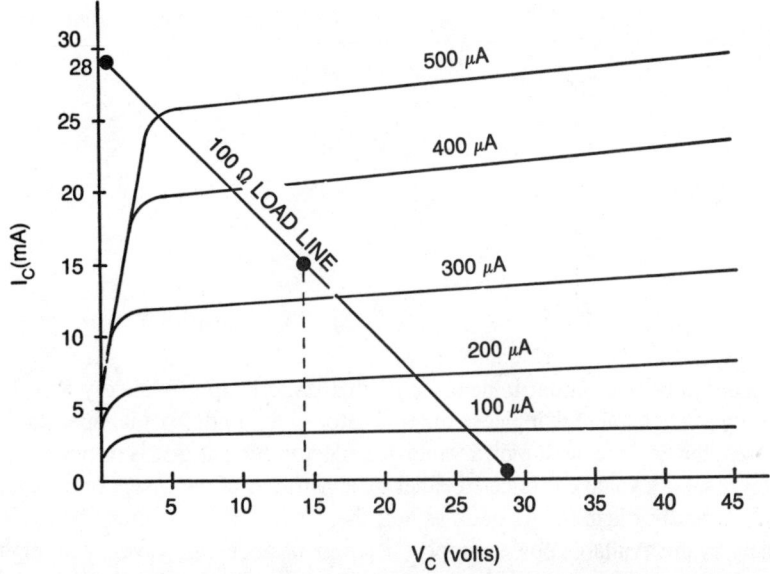

Fig. 3-24. Transistor operating curves with load line.

The *quiescent point* is the point defined when the input signal is zero. In most transistor, class A amplifiers, the quiescent point is the condition in which the collector voltage, V_c, is equal to one-half of V_{cc}, or in our present case 14 Vdc. In this example, the quiescent collect current is 14 mA. When an ac signal is applied to the base of the transistor, it modulates the base current up and down about the quiescent point, causing the collector current to vary. Voltage amplification is obtained because the varying collector current will cause the percentage of V_{cc} that is dropped across R_c to vary accordingly. One goal of designers is to choose values for these parameters that will achieve the correct results and then bias the transistor properly.

Perhaps the best way to demonstrate some of the principles is by example. Figure 3-25 shows the first of our simple transistor amplifiers (it is the simple bias network shown previously). In this case, there is an emitter resistor. This resistor improves stability under temperature variations and increases the input impedance of the amplifier.

The value of the collector resistor is used to set the quiescent point to bias the transistor. There are several different criteria for the value of the collector resistor. In some cases, for example, you might want to match some impedance and therefore select the collector

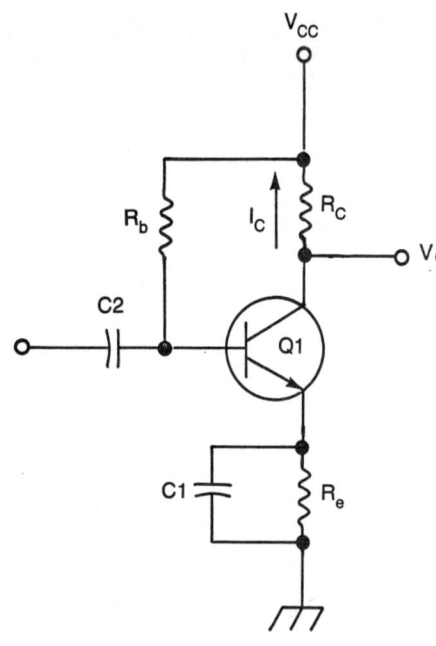

Fig. 3-25. Simple transistor amplifier.

resistor accordingly (the output impedance of this stage is approximately R_c). In other cases, merely select a value that places the transistor in the middle of its operating range. In most cases, the designer will pick a value of collector current that is deemed desirable, and the power supply voltage is selected either by constraints of the design or the equipment in which the amplifier is used. As often as not, the value of the collector voltage is determined solely by the available power supply. In some instances, however, you might want to achieve some specific output voltage swing and select the power supply voltage for this value.

The emitter resistor value is always a trade-off or compromise between thermal stability and gain. When emitter resistor R_e is made larger, the thermal stability is improved, but this improvement in stability is only at the expense of lost voltage gain. Recall the gain of this amplifier is approximated by R_c/R_e, so increasing R_e reduces the ratio. In most cases, the value of R_e is several hundred ohms (at least in the 100-ohm to 2000-ohm range). The value of R_e is obviously of critical importance. It has become almost standard practice to make the value of this resistor 1/10 to 1/5 of the value selected for R_c. Under very few circumstances will the value of the emitter resistor be greater than 1/5 of R_c. One of these situations is in phase-inverter circuits, but they are a special case that combines the properties of common-emitter and common-collector amplifiers at the expense of circuit gain. In the transistor whose family of curves is shown in Fig. 3-25, the emitter resistor should be between 100 ohms and 200 ohms.

The base resistor R_b is selected so the base current under zero signal conditions is approximately equal to the quiescent current. In the example of Fig. 3-25, the quiescent base current is 250 µA and the quiescent collector current is 14 mA. The emitter voltage is equal to $(0.014A)(R_e)$, so if an emitter resistor of 100 ohms is selected, the emitter voltage is 1.4 V. There is a 0.6 V base-emitter voltage for silicon transistors, so the total base voltage will be $V_e + V_{be}$, or 1.4 V + 0.6 V = 2.0 Vdc. The dc power supply is 28 Vdc, so the base resistor must drop a total of 28 V − 2.0 V, or 26 V, when delivering the quiescent base current of 250 µA. The value of R_b, then, by Ohm's Law is:

$$R_b = (V_{CC} - V_b)/I_{bq}$$
$$= (28 \text{ V} - 2\text{V})/(0.00025 \text{ A})$$
$$= (26 \text{ V})/(0.00025 \text{ A}) = 104{,}000 \text{ ohms}$$

In most cases, you would not try to find the value resistors called for in the calculations but would attempt first to make the circuit operate with the nearest standard value, in this case 100 kilohms.

Another version of the circuit shown in Fig. 3-25 connects the base resistor directly to the collector terminal, rather than to the power supply potential. This strategy gives better thermal stability because of some degenerative feedback present. The method for selecting the value of the base resistor is exactly the same as in the previous case, except that the voltage used will be the collector voltage, V_c, instead of V_{CC}. This is approximately one-half of the power-supply voltage in most applications.

The emitter bypass capacitor is used to keep the emitter at a low impedance to ground for ac frequencies. The emitter bypass capacitor should have a reactance that is one-tenth of the resistance of R_e at the lowest frequency of operation. The example used an emitter resistor of 100 ohms. Suppose that this amplifier was a speech amplifier in a two-way radio. The audio frequency response has a low-end limit of 300 Hz in most such equipment, so a capacitor that will present a reactance of 100/10, or 10 ohms, at 300 Hz must be selected. From the standard equation for capacitive reactance, this value is 53 µF or more. Fortunately, the voltage at this point is low—approximately 4 volts under maximum collector current conditions—so almost any small-size electrolytic capacitor will work.

The circuit of Fig. 3-25 suffers from several defects, so it is not quite as popular as the circuit of Fig. 3-26. In this more complex circuit, the base voltage is set by the resistor voltage divider in the base circuit. The collector and emitter circuits are the same as in the other case. Once again, the output impedance is equal to the collector load resistor. The input impedance, however, is limited by the value of R2. The design procedure is simplified below.

* Select an appropriate transistor type. It is absolutely essential that you know the type of transistor used, have a data sheet on that transistor, and are reasonably confident that the selected transistor will do the job.
* Select appropriate values of I_c, I_b and V_c. The value of I_c can almost be arbitrary, and the collector voltage is set by the power supply constraints.
* Select a value for the voltage on the emitter. In most class A amplifiers, the emitter voltage is roughly one-tenth of V_c. Check the selections to make sure that the emitter voltage is compatible with the normal range of emitter resistances when the selected value of collector current flows.
* Select a value for R_c. This resistor will be $(V_{CC} - V_e - V_c)/I_c$.
* Select a value for R_b (a combination of R1 and R2). A rule of thumb is used here, namely select a value that is 10 to 20 times R_e. Most use $15R_e$ for R_b.

* Calculate V_{BB} from V_b. A 0.6-volt junction potential is across b-e, and a voltage is on the emitter (V_e). The value of V_b is then $V_e + 0.6$. Then,

$$V_{bb} = V_b + I_b R_b$$

* Select R1 and R2 from the following expressions:

$$R1 = R_b(V_{cc}/V_{bb})$$
$$R2 = (R_b R1)/(R1 - R_b)$$

The parameters R_b and V_b are not real, but are used analytically to arrive at the correct values for R1 and R2 when the resistor network is connected directly to the V_{CC} power supply. This method is not foolproof, but it is *approximate*. It will yield positive results for most applications.

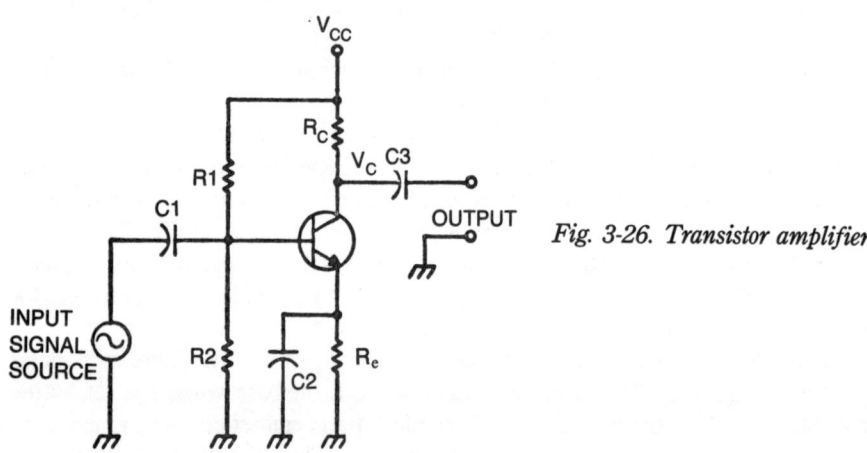

Fig. 3-26. Transistor amplifier.

FIELD-EFFECT TRANSISTORS

The field-effect transistor was a matter of speculation in the late 1930s. A paper published in a scientific journal at that time gave details of the construction of the junction field-effect transistor, and that transistor would have worked if the metallurgy had been developed by that time to provide the semiconductor materials needed. Following World War II, it was the bipolar transistor that received research attention. If the JFET had actually been built in 1939, then the semiconductor age would have had a 10-year headstart on itself, and most transistors today probably would be field-effect transistors instead of bipolar transistors.

FETs have a very high input impedance and are used for transconductance amplifiers. They tend to have performance properties that are far more reminiscent of the pentode vacuum tube than bipolar transistors.

There are two basic types of field-effect transistor: junction field-effect transistors (JFETs) and metal-oxide semiconductors FETs (MOSFETs). A basic JFET is shown in Figs. 3-27 and 3-28. The JFET consists of a semiconductor channel and a gate structure.

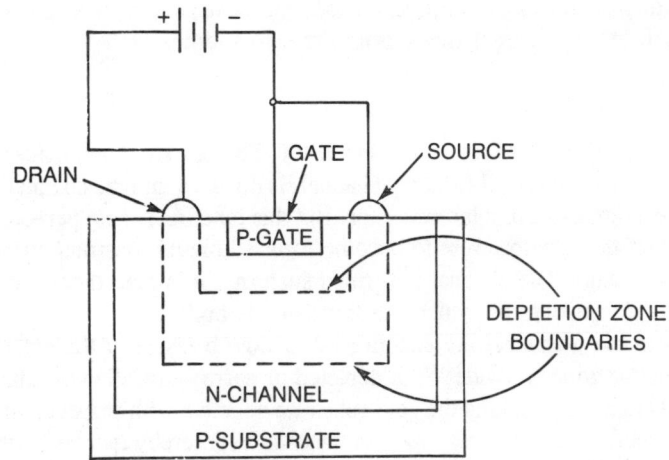

Fig. 3-27. Unbiased junction field-effect transistor structure.

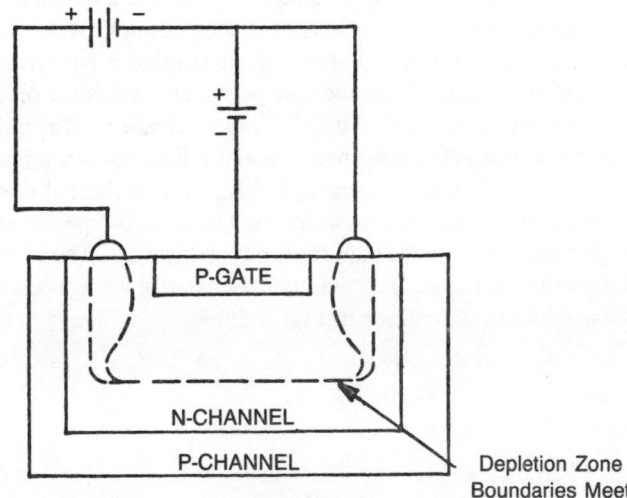

Fig. 3-28. JFET structure.

Two basic types of JFET are used: p-channel and n-channel, based on the type of material used to form the channel. The gate structure is always made of the *opposite* type of material.

In normal operation, a voltage is applied across the channel in the polarity shown. (Some early JFETs, incidentally, were capable of operation when the drain and source ends of the channel were reversed.) The channel has a low resistance, so current flows. When a potential is applied to the gate, however, it makes a *depletion zone* out of the channel where no electrical charge carriers (electrons or holes) can exist. This region has a high electrical resistance. At zero volts gate potential, this depletion zone is at a minimum, so the channel resistance is the lowest. But as the potential is increased, the depletion zone widens, thereby narrowing the portion of the channel in which charge carriers can flow.

At some specific potential called the *pinchoff voltage*, the depletion zone completely chokes off the channel, and no current flows from drain to source.

JFETs

The circuit in Fig. 3-27 is for an n-channel JFET. The gate structure, made of a different form of semiconductor material from the channel, is diffused into the channel in a manner resembling the emitter of a bipolar transistor. But this junction is kept perpetually reverse biased; in fact, if the junction were to become forward biased, destruction of the device would probably result. The channel is formed such that it is placed between the p-type gate and a p-type substrate on which the transistor is built.

Electrons in the n-channel are encouraged to flow because of the external battery. Here, the depletion zone is "wide" (not depleted or narrowed), allowing current to flow in the external circuit. By making the gate-substrate negative with respect to the channels, electrons are repelled (like charges repel) in the channel, thereby increasing the depletion zone. When the depletion zones from gate and substrate meet in the channel, pinchoff occurs. This is shown in Fig. 3-28.

Consider the channel resistance for a moment. When the gate voltage is zero, the depletion zones are at their least, so the channel resistance is low. When a high negative potential is applied to the gate and substrate, however, the depletion zones increase, causing the channel resistance to increase. When the gate potential reaches the pinchoff voltage, the channel resistance is extremely high. The JFET, then, can be used as a switch because it possesses a high resistance under one condition and a low resistance under the other.

The example in Fig. 3-27 was an n-channel JFET. The p-channel device is exactly the same, except for polarity. The channel structure is made of p-type semiconductor material, while the gate and substrate is made from n-type material. The circuits for the p-channel device are merely the n-channel circuits with the external voltage polarities reversed. Symbols for the two devices are shown in Fig. 3-29.

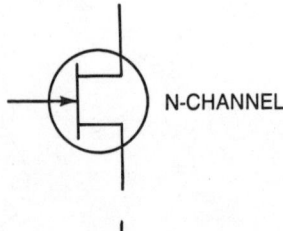

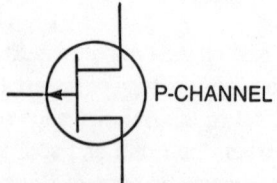

Fig. 3-29. Symbols for p-channel and n-channel JFETs.

MOSFETs

MOSFET devices are a little different. These devices are capable of much higher input impedances than JFETs because there is no ohmic connection between the gate and the channel. In the JFET, input (gate) impedance is limited by the reverse leakage current across the reverse-biased gate-channel junction. There are two basic types of MOSFETs: *depletion* and *enhancement*. These devices operate in somewhat different modes and are therefore discussed separately.

Both types are available in p-channel and n-channel versions. The possible MOSFETs are:

* p-channel depletion
* n-channel depletion
* p-channel enhancement
* n-channel enhancement

MOSFET devices are also sometimes called *insulated gate field-effect transistors* (IGFET), a name that more nearly describes their construction.

Depletion MOSFETs

The depletion type MOSFET is shown schematically in Fig. 3-30. As in the JFET, the channel is either of the two types of semiconductor material (in this case n-type), and the substrate is made from the opposite material. It is always desirable to keep the substrate-channel pn junction reverse biased (or at least zero) to prevent an excessive current flow. The gate is merely a metallic contact insulated from the channel material by a very thin layer of metal-oxide insulating material (the origin of the name "insulated gate"). The depletion-type MOSFET is a normally-on device, meaning that the channel resistance is low when the gate voltage is zero.

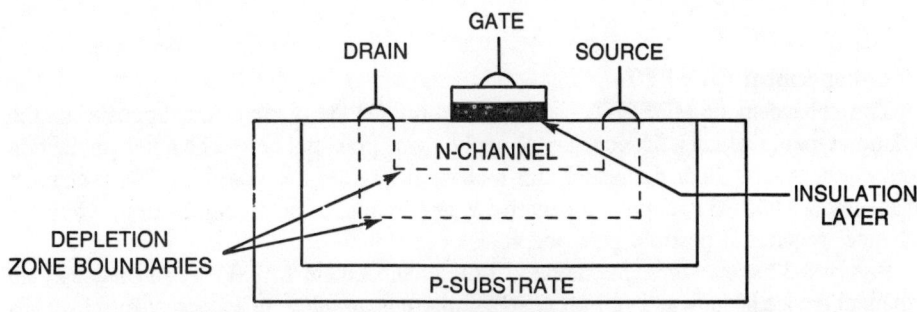

Fig. 3-30. MOSFET structure.

When an electrical potential is applied to the gate, however, an electric field is created in the channel. If the voltage is positive with respect to the channel, then electrons tend to draw closer to the gate structure. But if the voltage is negative, the field will repel the electrons in the channel, creating a depletion zone. The higher the gate voltage, the greater the depletion zone. When the depletion zone is finally large enough to pinch off the channel, current flow ceases. This, in effect, controls the channel resistance with a gate voltage.

Field-effect transistors are said to be *transconductance amplifiers*. The definition of such an amplifier is that a changing voltage at the input controls a change in current at the output. Because the channel resistance is controlled by the gate voltage, the output current is also under control of the gate, assuming that the drain-source voltage is constant.

The symbol for the depletion MOSFET is shown in Fig. 3-31. As in the case of the JFET devices, the arrow pointing inward means n-channel, and an arrow pointing outward means p-channel. It is interesting to note, incidentally, that the manufacture of p-channel depletion transistors is somewhat more difficult than the manufacture of n-channel devices. As a result, the n-channel depletion MOSFET predominates.

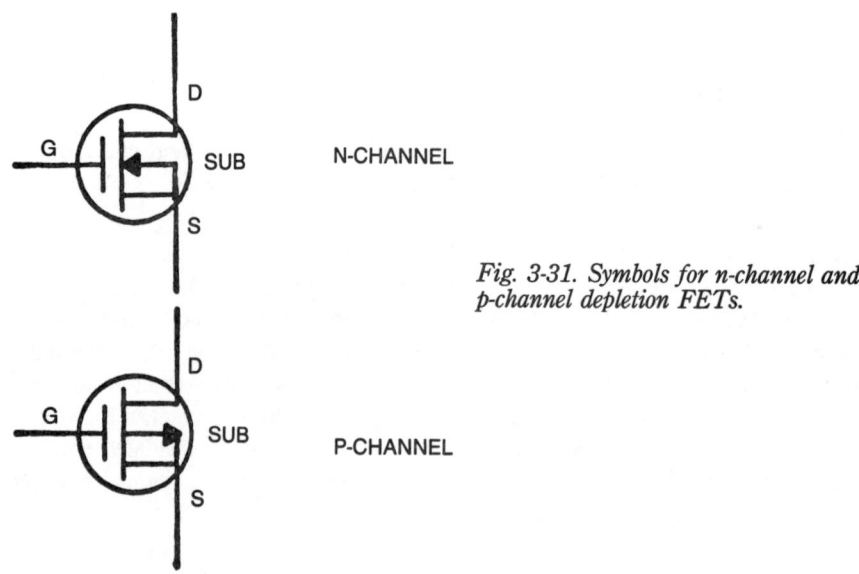

Fig. 3-31. Symbols for n-channel and p-channel depletion FETs.

Enhancement MOSFETs

The enhancement MOSFET is a normally-off device (exactly the opposite of the depletion type) in which it is necessary to apply a gate potential before channel conduction takes place. Again, both p-channel and n-channel devices are possible. The p-channel enhancement MOSFET requires a negative gate potential to begin conduction, while the n-channel requires a positive gate potential.

Figure 3-32 shows the construction of the enhancement MOSFET. In this case, an n-channel device is shown. In all MOSFETs, there is an inherent tendency for electrons to cluster close to the interface between the metal-oxide layer and the channel material.

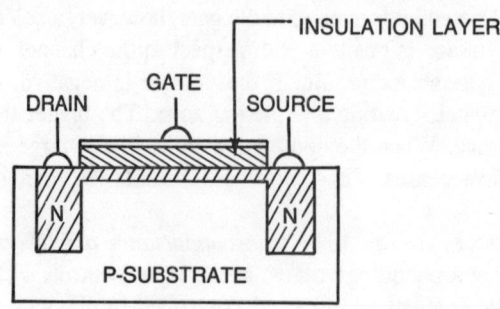

Fig. 3-32. Enhancement MOSFET with structure unbiased.

This will form a thin n-type region in the p-type substrate, located immediately beneath the insulating layer. The drain and source regions are n-type semiconductor material diffused into the p-type channel material. There is ordinarily no conduction between them, except for the tiny n-type region close to the gate structure. When the gate voltage is zero, the electrons of the n-channel are prevented from migrating out of the drain and source regions. But when a positive potential is applied to the gate (Fig. 3-33), the electrons from these regions are attracted to the insulator-substrate interface, causing a conduction zone to appear between the drain and source. Current can flow under this circumstance. The conduction of the enhancement MOSFET is controlled by varying the gate voltage. This increases or decreases the attraction of the drain and source electrons toward the gate region.

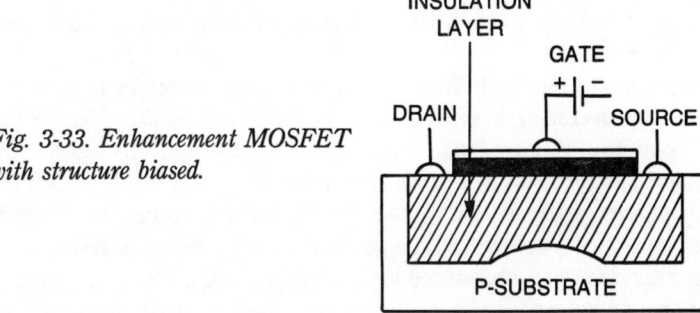

Fig. 3-33. Enhancement MOSFET with structure biased.

The circuit symbols for enhancement type MOSFETs are shown in Fig. 3-34. These symbols are the same as the depletion symbols, except that the channel bar is broken into sections to indicate the enhancement action taking place inside the transistor.

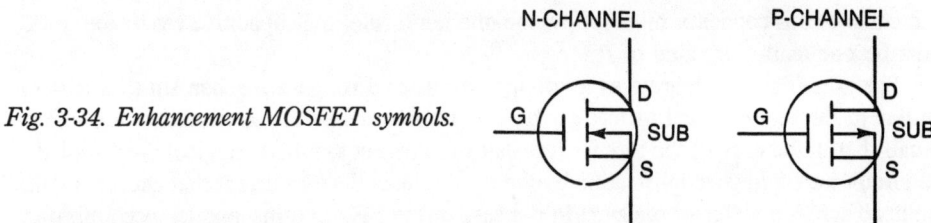

Fig. 3-34. Enhancement MOSFET symbols.

In both types of MOSFETs, the activity can be used as a switch. In both cases, one condition causes a low channel resistance, while another gate condition causes a high channel resistance. The on-resistance depends, in part, upon the signal voltage applied, but it is typically between 10 and 2000 ohms. Most have a value in the 100-ohm range. The variation of on-resistance is something of a problem and can lead to distortion of the output signal. But manufacturers have been able to limit this problem by correct FET switch topology. Designers can further limit the effects of the problem by using a load resistance that is high compared with the on-resistance of the channel. Load resistances between 10 and 100 kilohms are recommended.

The off-resistance of the channel is usually many megohms. The specification is usually given in terms of a *leakage current* rather than a resistance. In most switches, the leakage current under the off condition, called the *effective off-resistance*, will be from 0.01 picoamperes to 1000 picoamperes. Assuming a +10 V signal, and a "typical" 10 pA leakage current, then the resistance would be:

$$R_{off} = \frac{10 \text{ V}}{1 \times 10^{-8} \text{A}}$$
$$= 10^9 \text{ ohms} = 1000 \text{ megohms}$$

The on-resistance of the channel, on the other hand, drops to a much lower value, such as 100 to 10,000 ohms. This fact, incidentally, makes the FET almost ideal for use as an analog or digital electronic switch. A control voltage can be applied to the gate to obtain an expected switching action from the difference between on-resistance and off-resistance of the channel.

The primary use of the field-effect transistor is as an amplifier or as the amplifying active element in an oscillator or active filter. The input impedance of the FET is typically very high. For some JFETs on the low end of the price scale, this figure might be less than 1 megohm, but for most it is well over a megohm. For some MOSFET devices, the input impedance can exceed 1 terraohm (10^{12} ohms). In some circuit configurations, the high input impedance, coupled with a generally low output impedance, makes the device useful for service as an impedance converter. An amplifier for a signal source that has a high internal impedance can be made by using a field-effect transistor at the front end.

In Fig. 3-35, an elementary FET amplifier circuit is demonstrated. The circuit of Fig. 3-35 shows a common-source amplifier. The input impedance of this circuit is essentially the value of input resistance R1; the JFET input resistance is usually much higher than the value of this resistor. The input signal voltage is applied across the gate and source terminals of the JFET. The source terminal must be bypassed for ac by capacitor C3. The value of this capacitor must follow the one-tenth rule: the capacitive reactance of C3 must be one-tenth the value of R3.

The amplifier can operate as a voltage amplifier through a mechanism that is very similar to the scheme used to make a bipolar transistor operate as a voltage amplifier. Recall that in the case of the bipolar transistor (a current amplifier device) we "fooled" the circuit into voltage amplification by the use of the collector current that caused a voltage drop across a collector resistor. In the case of the FET (a transconductance amplifier device), we "fool" the circuit into operating as a voltage amplifier by connecting a similar resistance (R2) in series with the drain. The channel of the FET operates as a resistance that varies with the applied gate voltage. When this voltage increases in the positive direction, the channel resistance drops, so the voltage at the drain is reduced (the percentage of the terminal voltage dropped across R2 increases). Similarly, when the drain resistance increases, the voltage at the drain increases. If the bias on the transistor is set at a point where the drain voltage is half of V+, the output voltage will swing positive and negative about this potential. The bias is set by the voltage drop across resistor R3 (analogous to cathode bias in vacuum tubes). This voltage is equal to the product of the source-drain current and the resistance of R3. Normally, the gate should be negative with respect to the source (this is an n-channel device), so making the source more positive than the gate serves the same purpose.

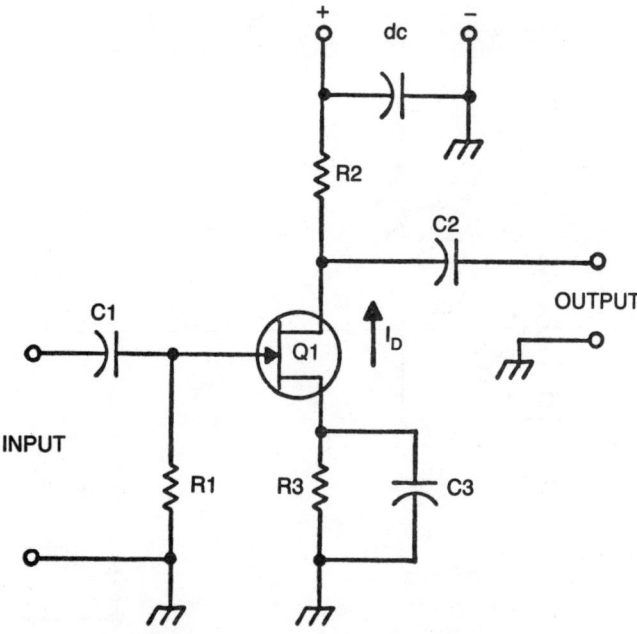

Fig. 3-35. JFET amplifier (common source).

DESIGNING WITH FETs

There are several different types of field-effect transistor. The two basic classes are the junction field-effect transistor (JFET) and the metal oxide-semiconductor field-effect transistor (MOSFET), also called the insulated gate field-effect transistor (IGFET). (These different types were explained previously in this chapter.)

TWO OPERATING MODES

The two different operating modes for field-effect transistors are depletion mode and enhancement mode. Many FETs can operate in either mode, depending on the operating conditions. In the depletion mode, the channel current is maximum when the gate-source voltage is zero. In the enhancement mode, the channel current is minimum when the gate-source voltage is zero.

There are two regions of operation for a field-effect transistor. In the ohmic region, which exists for low drain-source voltages (V_{DS}), the channel current is proportional to the drain-source voltage. In the pinchoff region, which is also called the constant-current region, the current will not increase for further increases in drain-source voltage. It is this latter characteristic that allows the use of the junction field-effect transistor as a constant current source. In this application, the source and the gate are tied together, and the device operates as a diode with a constant current characteristic.

A field-effect transistor can be biased in any of several ways, including self-bias, external bias, and certain combinations of self and external bias. Figure 3-36 shows two methods of FET biasing. Figure 3-36A shows an example of external bias. A reverse bias

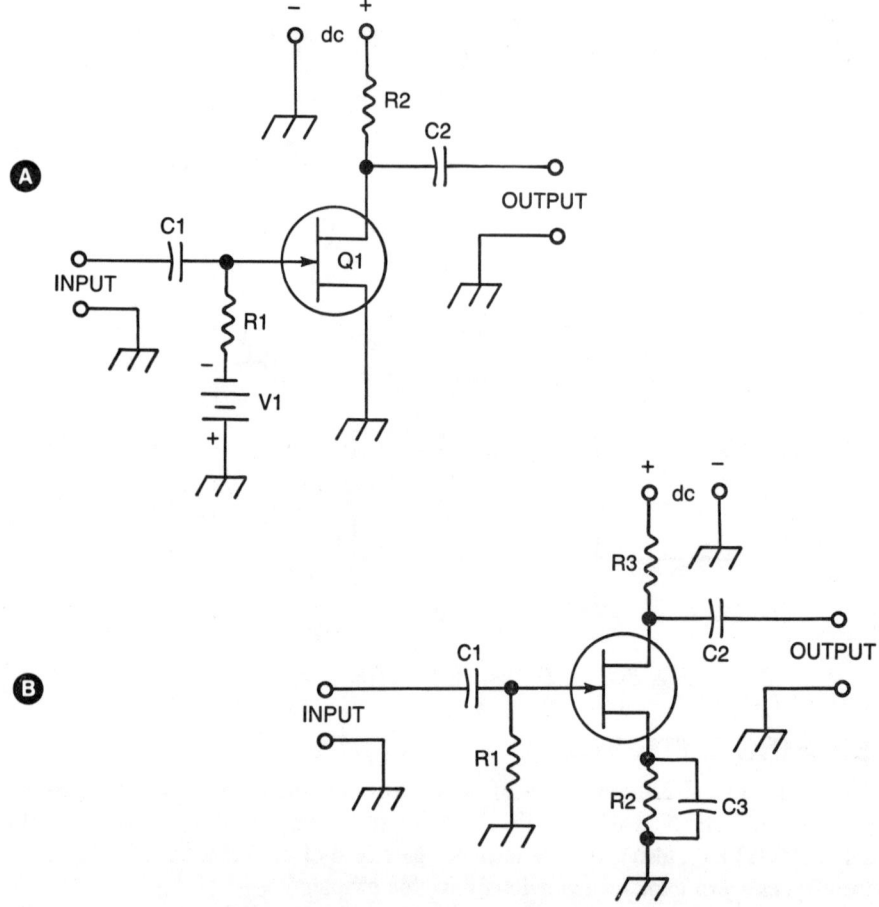

Fig. 3-36. (A) Battery bias for the FET; and (B) self-bias.

is applied to the gate of the field-effect transistor through resistor R1. The resistor is used to limit current and provide a load for the input signal. It also allows the discharge of electrons that build up on the gate structure. The level of bias is set by varying V1.

An example of self-bias is shown in Fig. 3-36B. A JFET circuit is shown in Fig. 3-37. In this case, the voltage drop across source resistor R2 is the bias voltage. The bias must make the gate more negative than the source. A negative voltage was applied to the gate in the previous case, but in this case the source is made more positive than the gate (conceptually the same thing). The value of the resistor must take into consideration the source-drain current of the FET and the desired level of bias. Apply Ohm's Law to find the gate-source voltage:

$$V_{GS} = -I_D \times R_S$$

where V_{GS} is the gate-source voltage, I_D is the drain-source current, and R_S is the source resistor.

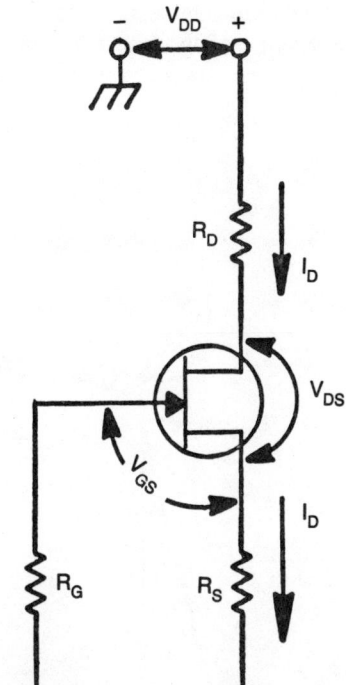

Fig. 3-37. JFET circuit.

A power supply potential must be supplied to the FET (V_{DD}), but only a portion of the power supply voltage appears across the drain-source (V_{DS}). The voltage drops around the circuit are:

$$V_{DD} = I_D (R_D + R_S) + V_{DS}$$

where V_{DD} is the power-supply voltage, I_D is the drain-source current, R_D is the drain resistor, R_S is the source resistor, and V_{DS} is the drain-source voltage.

You could select a value for the collector current (see the data sheet for the specific device) and a drain-source voltage. In most cases, the value of V_{DS} will be one-half of V_{DD}. For example, if 28 V was the supply voltage (V_{DD}) and 2 mA was the desired current amount to flow in the drain circuit, you would select values for the resistors consistent with this goal. As a rule of thumb, the source resistor should be approximately one-tenth to one-fifth of the drain resistor.

GRAPHICAL METHOD

One of the methods for determining the proper values for the source resistor is to use the graph of the I_D-versus-V_{GS} characteristic of the particular device under consideration. These graphs are printed in the manufacturer's data book and the specification sheets. An example is shown in Fig. 3-38. A line is drawn from the pinchoff voltage on the horizontal axis and the maximum drain current on the vertical axis. A proper drain current must be chosen. Then draw a line from the vertical axis at that point to the load line. The point at which this current line intersects the load line is directly over the proper gate-source voltage (V_{GS}). The slope of the line from the origin (0.0) to the *Q point* represents the resistance of the correct source resistor. In this case, the slope is:

$$S = \frac{(0.028 \text{ A})}{(1.6 \text{ V})}$$

$$= 0.0175 \text{ siemens}$$

$$R = 1/S = 57 \text{ ohms}$$

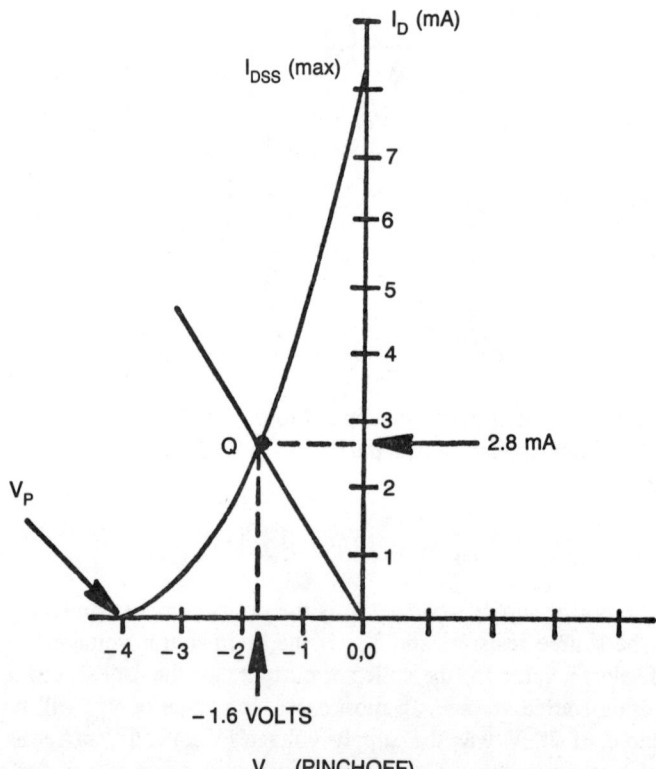

Fig. 3-38. Graphical solution for bias resistor.

BIPOLAR TRANSISTOR POWER AMPLIFIERS

The solid-state power amplifier solved the problem of size and bulk in amplifier designs. Vacuum-tube amplifiers required massive vacuum tubes, larger power transformers and, because of the high-voltage requirements, could not easily be operated in mobile or portable situations where only battery power was available. Vacuum tube amplifiers required ac inverters, often in the form of vibrator power supplies (does that tell the author's age?) that produced a large amount of spurious output signal called *hash*. Transistors not only made the actual amplifier smaller but could be operated directly from the 12-to-15-Vdc power supplies available in automobiles. It is in the power amplifier that the differences between the vacuum-tube design and the more modern solid-state design are really pronounced.

There is also a substantial improvement in the efficiency of the amplifier. The anode efficiency of the tube amplifier roughly approximates the collector efficiency of the transistor, but the transistor does not require filament (heater) power. The heater supply in the standard 25-to-50-watt audio power amplifier required another 25 watts just to heat up the vacuum tubes. Before considering some of the basic circuits that are now common, let's look at some transistor amplifier theory.

Figure 3-39 shows the current-versus-collector potential diagram for a bipolar power transistor. This graph shows the collector current caused by a given collector potential at various values of base current (I_{B1}, I_{B2}, and I_{B3}). The load line is drawn from a point where the current is maximum (only $V_{CE(SAT)}$ appears across the transistor collector-emitter terminals) to a point where the collector-emitter voltage is maximum ($I_c = 0$).

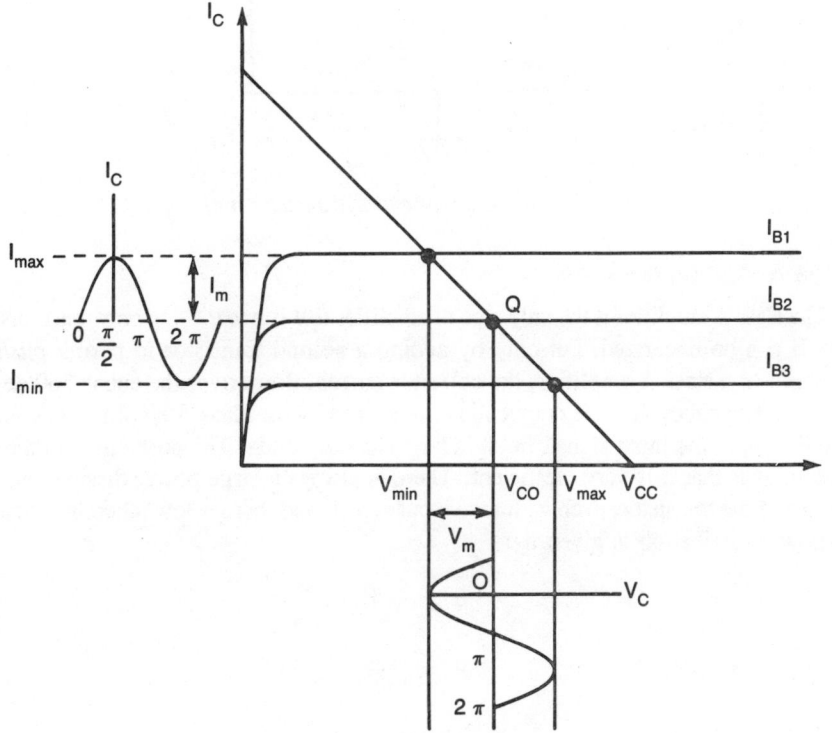

Fig. 3-39. *Power transistor operating curves.*

An operating point will be established where, under quiescent conditions (no signal), the transistor is operated at point Q (approximately midway). The various values used in this graph are depicted in the schematic of Fig. 3-40. When an excitation signal, V_i, is applied to the base of the transistor, the base bias will vary the base current. This base current variation causes the operating point of the transistor to vary back and forth along the load line. The amplifier is theoretically linear, meaning there is no distortion as long as the operating point is not shifted into one of the nonlinear regions of the family of curves.

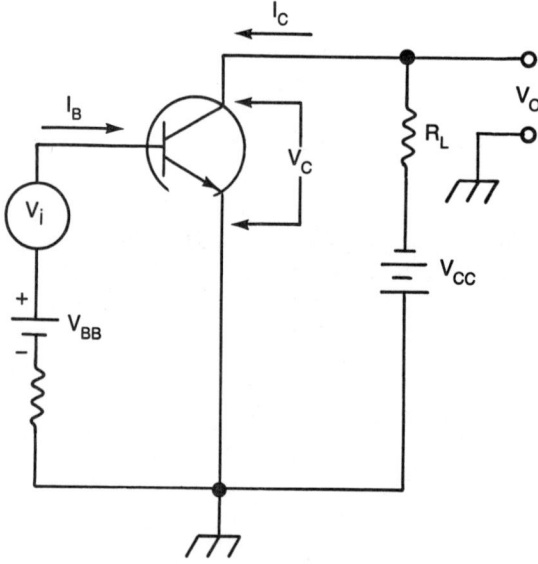

Fig. 3-40. Simple single-ended transistor amplifier.

CLASS A AMPLIFIERS

The class A amplifier is the only type of amplifier that is linear using only one transistor. Class B can be linearized, but only by adding a second transistor to permit *push-pull* operation. In a class A amplifier, the collector current flows over the entire 360 degrees of the input signal cycle. The output signal produced by the class A amplifier is a faithful reproduction of the input signal, but it is larger in amplitude. The problem with the class A amplifier is that it is very inefficient. There is always a large power dissipation in the collector of the transistor, even with zero input signal, and there is low inherent efficiency. The power dissipation is given by:

$$P = V_c I_c$$

or

$$P = I_c^2 R_L$$

where P is the power in watts (W), I_c is the rms collector current in amperes (A), and V_c is the rms collector-emitter potential in volts (V).

How do you express the power in terms of the maximum and minimum swings of the collector current and C-E (collector-emitter) potential? Assume the waveform is sinusoidal (makes the arithmetic easier) and label the peak current and voltage swings I_m and V_m, respectively. The rms current is the peak current divided by the square root of 2. Similarly, the rms collector-emitter voltage is the peak voltage divided by the square root of 2. By taking the difference between I_m at the positive peak and I_m at the negative peak, expressions for power in terms of the voltage and current swings can be derived:

$$I_c = I_m/(2)^{1/2} = (I_{max} - I_{min})/(2)(2)^{1/2}$$

and,

$$V_c = V_m/(2)^{1/2} = (V_{max} - V_{min})/(2)(2)^{1/2}$$

Assuming our symmetrical sine wave signal, we can write the power equation in the form:

$$P = V_m I_m/2$$

$$P = I_m^2 R_L/2 = V_m^2/(2)(R_L)$$

which reduces to: $$P = \frac{(V_{max} - V_{min})(I_{max} - I_{min})}{8}$$

The last equation, then, is the expression of power in a class A amplifier when the voltage and current extremes are known (as they can be found from a load line graph). The values calculated from the equation, however, are valid only for a *sine wave* input signal.

Efficiency of Class A Amplifiers

The efficiency of any amplifier is the ratio of the signal power delivered to the load to the dc power consumed from the power supply. This does not include such wasted power as the filaments of a vacuum tube, but only the power used in the anode of the tube, or the collector of the transistor. The conversion efficiency factor η is defined as:

$$\eta = (P_L/P_{dc}) \times 100\%$$

where η is the conversion efficiency in percent, P_L is the power delivered to the load, and P_{dc} is the dc power consumed from the power supply.

The equation shown above is a generalized case, useful for any amplifier. But, we are dealing with a specific amplifier—the bipolar transistor power amplifier. In that case, you can substitute the values for the current and voltage that make up the power expressions demanded:

$$\eta = (\tfrac{1}{2} V_m I_m)(100\%)/(V_{cc} I_c)$$

$$= (50)(V_m I_m)/(V_{cc} I_c) \text{ percent}$$

This expression tells something of the conversion efficiency, and from it you can make some inferences for the different types of operation. In the small-signal case, for example, the output power is very small, but the dc power consumed remains $V_{cc} I_c$, so the result is a very low conversion efficiency. The expression also tells something about the large-signal behavior of the class A amplifier. If the Q point is set exactly midway in the curve, then the collector current will swing from zero to the saturation current. In this case, I_m is equal to I_c. Similarly, the voltage between the collector and emitter will swing from a low of $V_{CE(SAT)}$, which is essentially zero for our present purposes, to V_{CC} when the collector is totally cut off. In that case, V_m is equal to one-half of V_{CC}. Substituting these values into the previous equation yields the maximum theoretical efficiency of the class A amplifier under discussion:

$$\max = \frac{(50)(V_m I_m)}{(V_{CC} I_{cq})} \text{ percent}$$

$$= \frac{(50)(\tfrac{1}{2} V_{CC})(I_C)}{(V_{CC} I_c)} \text{ percent}$$

$$= (50)(\tfrac{1}{2}) \text{ percent} = 25 \text{ percent}$$

But just what does this last equation mean in practical terms? Consider the now-standard 10 watts required of an ordinary automobile radio/stereo-tape unit. Many of these, especially those made as OEM (original equipment manufacturer) radios for the car makers, use class A audio power amplifiers. For every 10 watts delivered to the load, 30 watts of dc power will be wasted. The total power requirement will be 40 watts. Note that the unused 30 watts doesn't just evaporate; it is "used" to heat the collector of the transistor. As a result, class A power amplifiers require rather massive heatsinks for the transistors to dissipate the massive heat generated. In fact, car radio manufacturers sometimes try to avoid placing the transistors on the front panel of the radio, lest they heat up the dashboard and elicit customer complaints.

Class A Amplifier Circuits

The class A amplifier is capable of delivering ac signal power with just one transistor. Figure 3-40 shows a popular class A power amplifier used in the audio stage of many car radios. The output transistor is a pnp germanium device. Although these are now considered "old hat" in the light of new silicon devices, the pnp germanium transistor was the first power transistor commercially available in production quantities. In auto radios, there are only a couple of devices of any great popularity. The now obsolete Motorola 2N176 and its successors, the Delco DS-501 and the Delco DS-503, were among the more popular. (Delco replaced their versions, incidentally, with the DS-520 and DS-525 devices.)

The output portion of this power amplifier consists of an impedance-matching choke coil (L1) that is connected to the loudspeaker (LS). Some models, incidentally, connected the loudspeaker and the choke coil directly in parallel instead of using the tap arrangement shown. The collector of the pnp power transistor is kept close to ground for dc currents. The voltage drop across the coil, when the transistor is correctly biased and signal level is zero, will be around 1 to 1.5 volts. This particular circuit configuration was used in auto radios because the standard car electrical system has the negative terminal of the battery connected to chassis ground.

The signal is coupled to the base of the transistor through an interstage transformer (T1). This transformer is used to isolate the dc levels of the audio driver circuits from the base of the power transistor. An impedance transformer matches the relatively high impedance of the driver to the low impedance of the base circuit. This circuit is no longer used extensively, replaced by the direct-coupled circuits to be described in a moment.

The bias for the power transistor is supplied by a resistor network (R2, R3, and R4). Note that the 0.2- to 0.3-Vdc bias is a function of temperature, so resistor R3 is shunted by a high-value thermistor (R4). The value of the thermistor changes with changes in temperature, so it helps accommodate thermal tracking. Resistor R1 is used for two purposes. Because it is a very small value resistor (0.22 to 0.68 ohms), it produces a small amount of degenerative feedback (helps reduce distortion), and it acts as a fuse resistor. In fact, some manufacturers designate this resistor as a *fusistor*. It is designed to blow when the current level increases above a certain level, such as when the output current increases due to a shorted power transistor. Any 5 W power resistor in the small value range *can* be a fusistor, especially if it is labeled for use in auto radios. Capacitor C1 is a large value electrolytic and is used to keep the cold end of the transformer (T1) secondary at a low impedance to ground for ac, while retaining a high dc-to-ground resistance.

A direct-coupled class A power amplifier, also used as an auto radio output stage, is shown in Fig. 3-41. This circuit uses three transistors in a direct-coupled cascade arrangement. The output stage is a pnp germanium transistor that is very similar to the circuit in Fig. 3-42. The same tapped choke circuit is used to match the collector impedance to the loudspeaker impedance. Some of these circuits use an emitter fusistor as in Fig. 3-42; some do not.

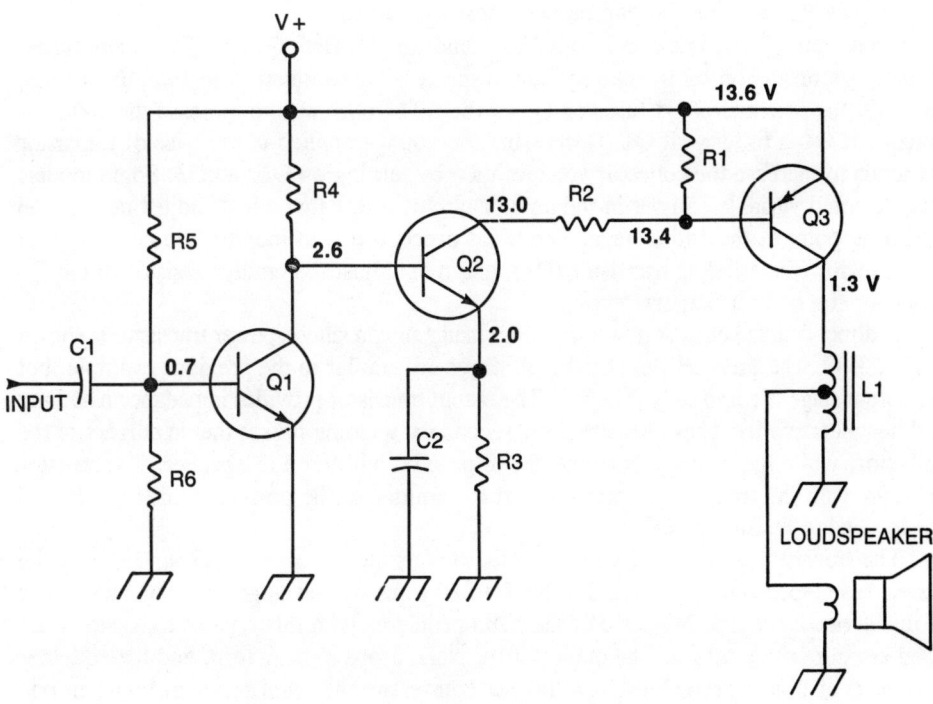

Fig. 3-41. Direct-coupled class A audio power amplifier.

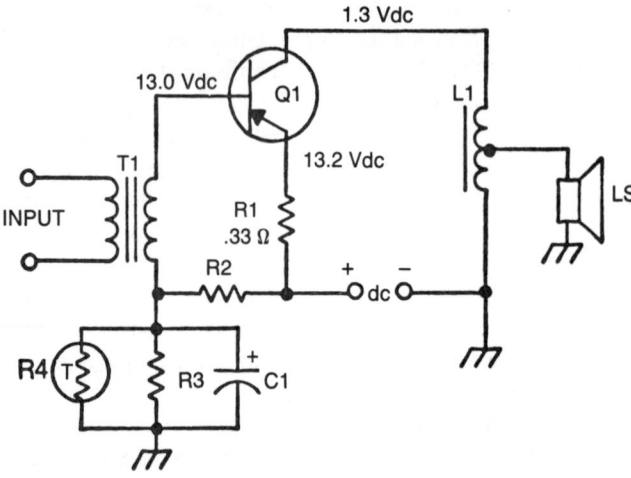

Fig. 3-42. Transformer-coupled power amplifier (class A).

Bias for transistor Q3 is set by the operation of transistor Q2. The collector current of transistor Q2 flows through resistors R1 and R2. It is the voltage drop across resistor R1 that sets the bias point of Q3. When Q2 is turned on further, the collector current will increase, which increases the voltage drop across resistor R1. Increasing the voltage drop across R1 turns on the pnp output transistor further.

Transistor Q2 is, in turn, controlled by transistor Q1. Both Q1 and Q2 are npn types, so they are turned on by increasing their respective base-emitter potentials. Increasing the collector potential of Q1 turns on Q2 further. The only way to increase the collector voltage of Q1 is to turn off Q1. Decreasing the voltage applied to the base of transistor Q1 tends to increase the collector voltage, thereby turning on Q2 and Q3. Some models place a small variable resistor in the emitter circuit of Q1 that allows adjustment of the operating point of the entire stage. The usual procedure is to monitor the dc voltage at the collector of the output transistor (Q3), and then adjust the emitter resistor in the Q1 stage for the correct output voltage.

A direct-coupled class A power amplifier using an npn silicon power transistor is shown in Fig. 3-43. The preamplifier and driver stages are similar to the previous example, but the output stage is completely different. The output transistor provides impedance matching and degenerative feedback. Note that the secondary winding passes the dc current to the collector, while the primary is connected in parallel with the loudspeaker. The resistor in series with the transformer primary and the emitter of the power transistor is one of the low-value fusistor types.

The transistor generally used in circuits such as the one shown in Fig. 3-43 are the plastic case types. These devices use the TO-220 transistor package. A typical transistor in this series is the 2N5249. One of the main problems with this type of transistor is an open emitter-base junction. The output of the stage drops to near zero, and the collector current is also near zero. This is in distinct contrast to the main problem found in pnp germanium power transistors: collector-emitter shorts. Many auto radios used this circuit. Auto radio technicians found early in the game that this transistor was unreliable in class

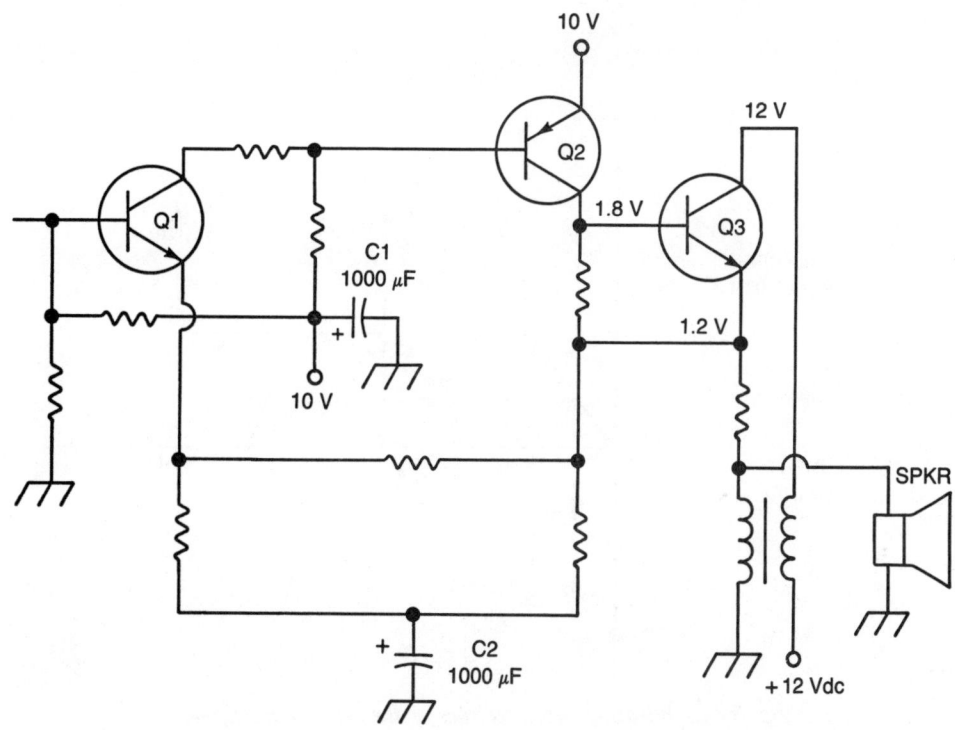

Fig. 3-43. Direct-coupled transistor class A amplifier using npn devices.

A power amplifier service at the auto radio power levels, so they would replace it with a 2N3055 or some other TO-3 npn silicon device. It was fortunate that many auto radio manufacturers used auto radio chassis that still had either TO-3 or TO-66 holes drilled or molded, so the conversion was easy.

CLASS B AMPLIFIERS

The class B amplifier is defined as one in which the collector current flows over 180 degrees of the input cycle. The transfer characteristic of a typical class B amplifier is shown in Fig. 3-44. The transistor receives no dc bias, so its operating point is set only by the input signal. This means that the collector current flows on only one-half of the input signal excursion. If the transistor is npn, then the collector current flows only during the positive half of the input cycle. If the transistor is a pnp type, the collector current flows on only the negative half-cycle of the input signal. The result is an output signal that closely resembles a half-wave-rectified, amplified version of the input signal. But this presents a problem. An amplifier that produces only a half-wave output is *not* a linear amplifier. It cannot, therefore, be used as a linear audio power amplifier unless a second power transistor is used to process half of the signal in one transistor and the other half in the second power transistor.

Figure 3-45 shows a complementary symmetry push-pull power amplifier. The name "push-pull" is derived from the fact that one transistor "pushes" while the other "pulls."

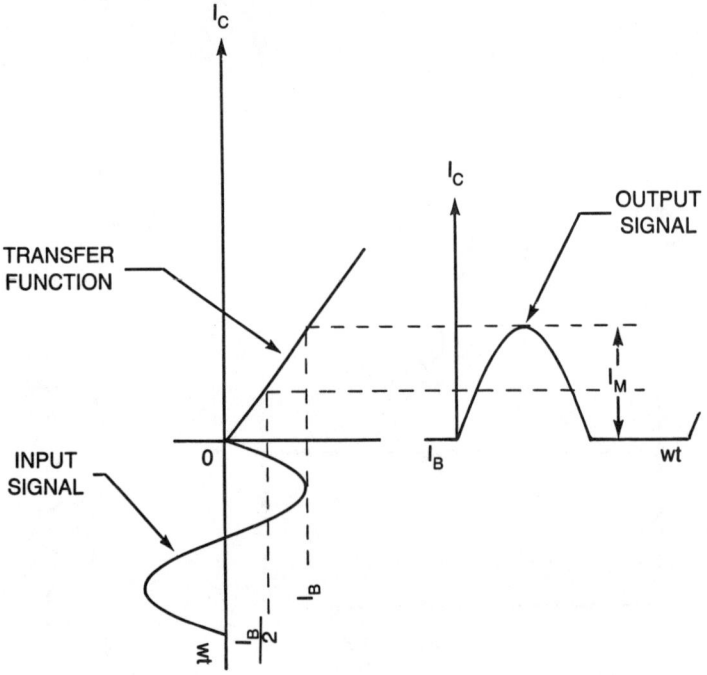

Fig. 3-44. Operating characteristics of the class B amplifier.

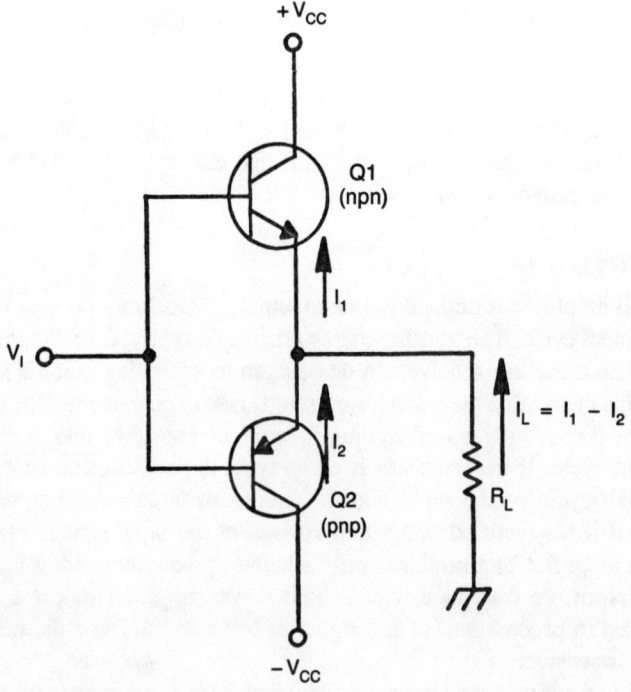

Fig. 3-45. Class B push-pull amplifier using complementary symmetry.

Note that two opposite polarity transistors are used: Q1 is an npn device and Q2 is a pnp device. The properties of the two transistors are exactly opposite. When the input signal is on the positive excursion of the cycle, npn transistor Q1 is turned on and pnp transistor Q2 is turned off. Similarly, on the negative half of the cyclic excursion, npn transistor Q1 is turned off and pnp transistor Q2 is turned on. This allows the outputs of the two devices to be summed, forming the full-wave output signal required for linear operation.

How much power can be obtained from a class B power amplifier? What type of efficiency is to be expected? First analyze each half of the push-pull pair in its turn. If the signal is sinusoidal, the dc current will be $I_{dc} = I_m/\pi$. This value is found using mathematics and is based upon the half sinusoid. Given the value of dc current, the power level can be written as:

$$P = \frac{2 I_m V_{cc}}{\pi}$$

There is a factor of 2 because two transistors are feeding power to load resistor R_L. This equation expresses the power. The efficiency is approximately 78.5 percent for the ideal class B power amplifier.

One advantage of the class B amplifier is that the current drain goes to zero when the signal level is also zero. This makes the class B push-pull power amplifier almost ideal for low-power portable projects and devices. Almost all transistor radios, for example, use class B audio power amplifiers. The maximum power dissipation in the collector is expressed by:

$$P_{max(c)} = \frac{2 V_{cc}^2}{\pi^2 R_L}$$

The maximum output power (when $V_m = V_{cc}$) is expressed by:

$$P_o = V_{cc}^2/2 R_L$$

Combining,

$$P_{max(c)} = 4P_o/(\pi)^2$$
$$= 0.4 P_o$$

CLASS AB OPERATION AND CROSSOVER DISTORTION

Transistors are not totally linear devices when small signals are applied. There is a small necessary bias potential that must be overcome before the device becomes linear. This potential, called the *junction potential*, is 0.2 to 0.3 volts in germanium devices and 0.6 to 0.7 volts in silicon devices. Figure 3-46 shows the actual transfer property for a class B push-pull power amplifier. The nonlinear region of the transfer curve produces the distortion shown in the output signal. This is called *crossover distortion* because it occurs when the signal is at a low level, or at the zero crossover points.

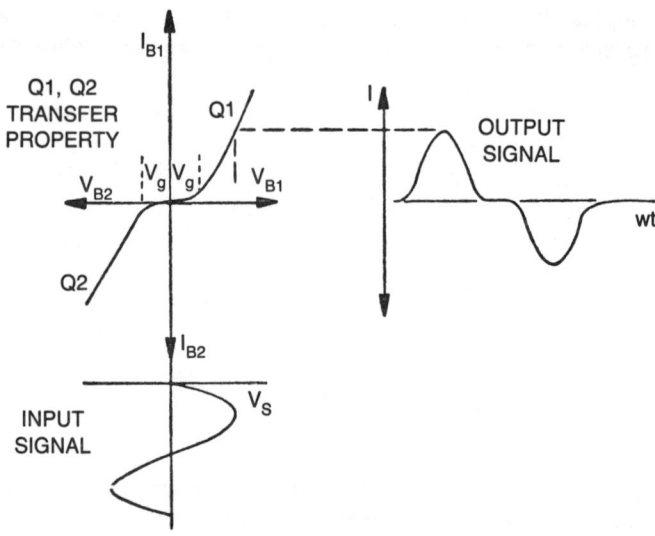

Fig. 3-46. Crossover distortion.

The crossover distortion problem can be overcome by using class AB operation. This type of amplifier is defined as one in which the collector current passes for more than 180 degrees of the input cycle but less than the 360 degrees required of class A. Class AB, therefore, operates the transistor in a region that is between class A and class B.

The practical implementation of Class AB operation is shown in Fig. 3-47. The transistors are given a small amount of forward bias by diodes D1 and D2. This will shove the transfer property farther up the curve into the linear region.

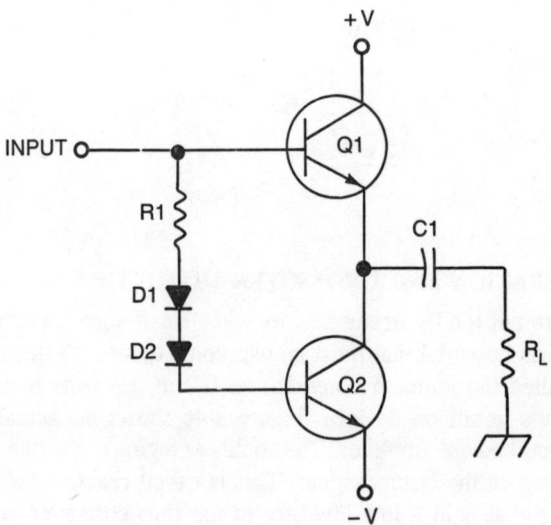

Fig. 3-47. Solution for crossover distortion.

The output load—usually a loudspeaker—is isolated from the transistors with a dc-blocking capacitor. The potential at the junction of the two transistors is approximately ½((V+) − (V−)). This capacitor has a value of 500 to 1000 µF in auto radios and up to 10,000 µF in high-power, high-fidelity amplifiers.

Class AB Push-Pull Power Amplifier Circuits

Several different "standard" class AB or class B push-pull amplifier circuits have been developed over the years. Some of these circuits are merely transistorized versions of ideas that were common in vacuum-tube days, while others are completely different from anything that was used in tube days.

Figure 3-48 shows a basic push-pull power amplifier that is a holdover from older tube designs. This particular circuit was popular in transistor portable radios and a few communications applications, often as a modulator in an AM transmitter. The heart of the circuit is a pair of transformers, T1 and T2. Transformer T1 is the input transformer, also called the *interstage transformer* because it connects the power amplifier input to the output of the preamplifier or driver stages. The secondary of the input transformer is center tapped. If we take the center tap as the ground (or common) point, then the signals at the two ends of the transformer will be out of phase with each other by 180 degrees.

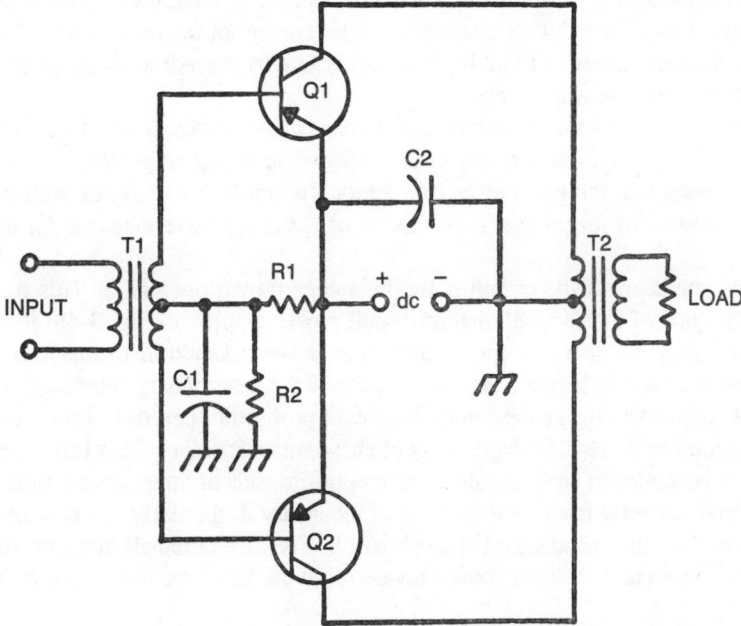

Fig. 3-48. Classical push-pull power amplifier circuit (class AB).

Capacitor C1 keeps the center tap of transformer T1 at ac ground potential while allowing the dc potential required for bias to be above ground. The signals fed to the two transistor bases from the input transformer will be 180 degrees out of phase, so the criteria for push-pull operation is satisfied.

The collectors of the two transistors are connected to the output transformer, T2. This transformer has a center-tapped primary, and the dc collector voltage is applied to the transistors through this center tap. Note that the collector voltage is negative with respect to ground, because the transistors used in the circuit are pnp types. This circuit is for negative-ground mobile operation. When there is no signal applied to the input transformer, the two transistors are equally biased, so they have equal collector currents. These currents will exactly cancel each other in the primary of T2, so the net output is zero. When there is a signal applied to T1, however, one transistor is cut off (during, say, the positive excursion of the input signal), and the other transistor is turned on. This unbalances the collector currents flowing in the primary of T2. The current balance is no longer equal, so there is a net output that is proportional to the difference in collector currents in Q1/Q2. Similarly, on the negative excursion of the input signal, the opposite situation occurs. The transistor that had been cut off is then turned on, and the transistor that had been on is turned off. In that case, there is still an unbalanced collector current, but it takes on the opposite polarity. The output magnitude, however, is the same.

Dc bias to the transistors is supplied through voltage-divider network R1/R2. The bias voltage is applied to the transistors through the secondary of transformer T1.

Capacitor C1 keeps the center tap of the transformer at ac ground potential while allowing the dc to remain above ground. Similarly, capacitor C2 keeps the emitters of the power transistors at ac ground potential while keeping the dc above ground. Capacitor C2 is sometimes called a *decoupling* capacitor because it bypasses to ground any power supply variations created by the changing emitter current of the transistors. These power supply variations are seen as a valid ac feedback signal by preceding stages of the amplifier that share the same power supply.

An example of a split-secondary *totem pole* push-pull power amplifier is shown in Fig. 3-49. This circuit is not like any of the standard push-pull amplifiers used in vacuum-tube days. Note that the two power transistors are connected in *series* with each other for dc currents. This means that the collector of Q2 is the current source for the emitter of Q1.

Again, the heart of the circuit is the interstage transformer (T1). This transformer differs from that of the conventional push-pull power amplifier (Fig. 3-48) in that it has two secondary windings (sometimes called a *split secondary* transformer). Dots on the component are used to indicate the phase sense of these secondary windings. The phase at the dotted ends is always the same. To obtain push-pull operation, however, the two transistors must be driven 180 degrees out of phase with each other. This is done by feeding the respective transistor base terminals from opposite ends of their secondaries. The base of Q1 is driven directly from the dotted end of secondary *A*. Similarly, the base of transistor Q2 is driven from the undotted end of secondary *B*. The base terminals of the two transistors are biased from separate resistor voltage dividers, but the technique is the same as previously seen.

There are two low-voltage resistors (0.33 ohms) used in this circuit. One 0.33-ohm resistor is placed in each transistor emitter circuit and serve two functions: they are fusistors (they act as fuses if the transistor shorts), and they supply a small amount of degenerative, or negative, bias to the transistors.

Output is taken at the junction between the collector of Q2 and the emitter of Q1. The power-supply dc voltage is split approximately equally between the two stages, so there is a dc potential at this point equal to $(V+)/2$. Such a potential would damage the loudspeaker system connected to the output, so a dc isolation capacitor (C3) is used in

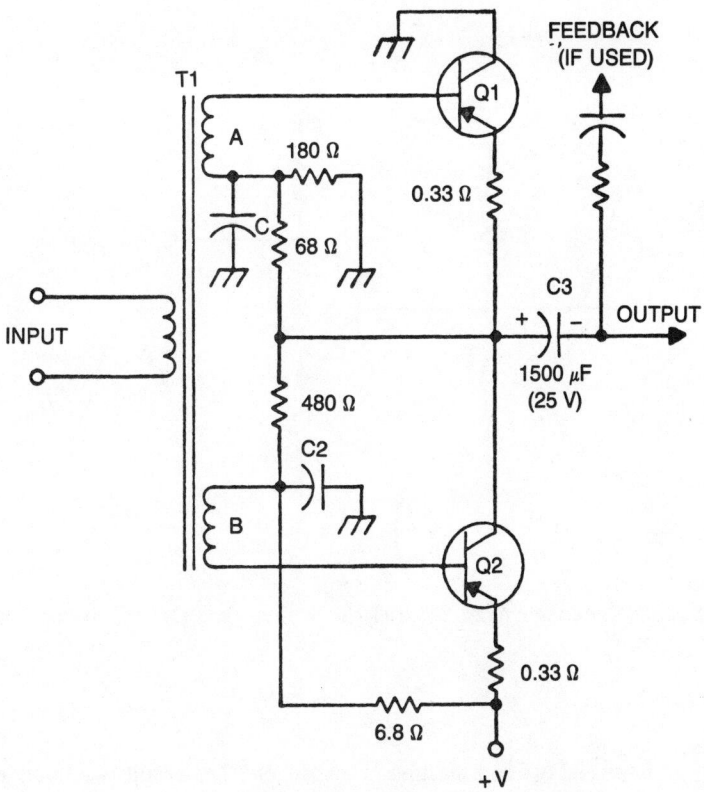

Fig. 3-49. Split-secondary totem-pole push-pull power amplifier (class AB).

series with the load. This capacitor limits the low-frequency response of the power amplifier, so its value must be low with respect to the load resistance at the lowest frequency of operation. This requirement can make for some very large capacitor values. In the case shown, 1500 µF is used. This value is sufficient for most communications and "low-fi" audio applications, but it is not sufficient for most hi-fi applications. Values from 5000 to 10,000 µF are common, but there are high-fidelity amplifiers on the market with values to 25,000 µF.

An example of a transistor phase-inverter circuit is shown in Fig. 3-50. This circuit is used to replace the interstage transformer and provides two output signals that are 180 degrees out of phase with each other. One signal is taken from the collector of the transistor, which means that it has a 180-degree phase shift with respect to the input signal. The other signal is taken from the emitter of the transistor (the stage operates as an emitter follower), so the signal is in phase with the input signal. This results in two signals that are 180 degrees out of phase with each other. It is necessary to adjust the values of the collector and emitter load resistors to make the amplitudes of the two signals equal.

In our initial discussion of class B push-pull power amplifiers, we used two transistors of opposite polarity: an npn and pnp transistor. These two transistor types have opposite actions when driven by the input signal. A positive-going excursion of the input signal turns the npn unit on and the pnp unit off. Similarly, on the negative half of the input

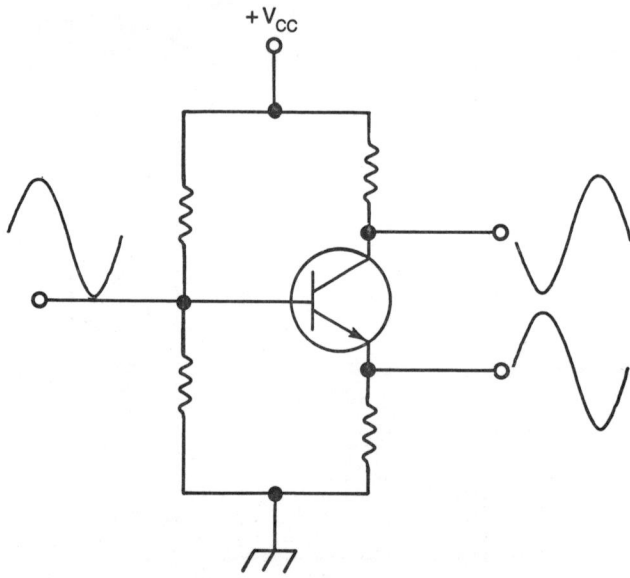

Fig. 3-50. Transistor phase inverter can replace the interstage transformer.

signal, the npn is turned off and the pnp is turned on. The result is a complete, almost linear reconstruction of the signal when the two emitter currents are summed together.

There is a problem in the design of this type of *complementary symmetry* amplifier. The two transistors are required to be *matched*— they must have the same properties except for polarity. This is not too much of a problem at low powers, but some early solid-state amplifier designers found only a few complementary pairs available from semiconductor manufacturers at the higher power levels. This situation resulted in the development of the *quasi-complementary symmetry* amplifier shown in Fig. 3-51. In this push-pull power amplifier, the output stage is a totem pole arrangement of two identical npn power transistors. The configuration is similar to that of the split-secondary totem pole amplifier shown in Fig. 3-49. The driver transistors, which operate at a lower power level, are in complementary symmetry. This is the circuit used in most hi-fi amplifiers on the market today and is even available in the form of integrated circuits and hybrid circuit modules.

Miscellaneous Topics

The power sensitivity of an amplifier is increased markedly by using a *Darlington pair* (Fig. 3-52) as the output transistor. In some cases—not power amplifiers—the two transistors are identical. In most power amplifiers, however, the output transistor (Q2) is a power unit, and the input transistor (Q1) is a driver transistor. The current gain (*beta*) of the transistor pair is the product of the individual *betas*:

$$H_{fe} = H_{fe(Q1)} \times H_{fe(Q2)}$$

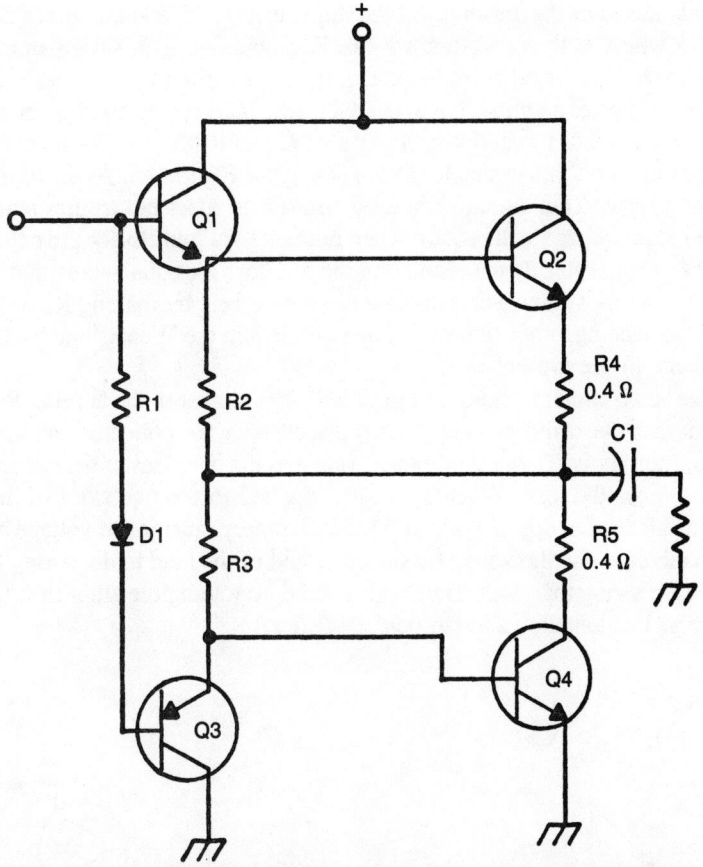

Fig. 3-51. Quasicomplementary symmetry push-pull power amplifier.

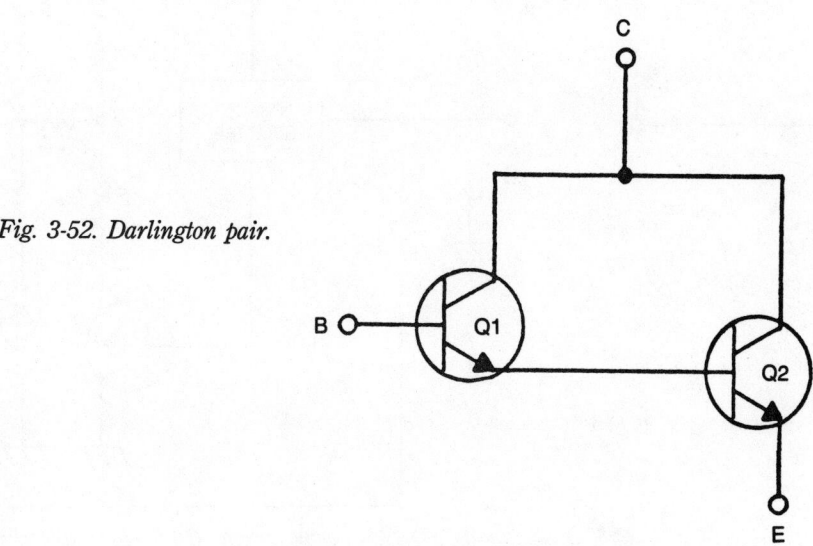

Fig. 3-52. Darlington pair.

This greatly increases the sensitivity of the stage. In some hi-fi amplifiers, a Darlington pair is used in which both transistors are inside a common TO-3 transistor package. Motorola has made these, and many manufacturers of hi-fi equipment use them.

A new form of power amplifier has appeared on the market called *bridge audio*. Delco Electronics, the electronics manufacturing arm of General Motors Corp., offers an IC power amplifier that uses this principle (Delco DA-101). Figure 3-53A shows a common dc Wheatstone bridge. This circuit should be familiar to most electronics students and workers. The output voltage will be zero when ratios R1/R2 and R3/R4 are equal to each other. Note that it is not strictly necessary that the resistors are equal—only that the *ratios* are equal. In many cases, however, simplicity is served best by making R1 = R2 = R3 = R4 = R. Unbalancing either or both voltage dividers in the Wheatstone bridge causes an output voltage to the present.

The bridge audio circuit is shown in Fig. 3-53B. It is similar to the familiar Wheatstone circuit, except that the resistors have been replaced with the collector-emitter paths of the four power transistors. The load is connected across the junctions of the two transistors, so E_o appears across the load. When a signal is applied to the two sides of the bridge, they will be unbalanced in opposite sense from each other, causing the voltage at one end of the load to increase and the voltage at the other end of the load to decrease. The result is a larger voltage swing than is possible with a standard totem pole amplifier. (Each half of the circuit is, incidentally, a totem pole amplifier.)

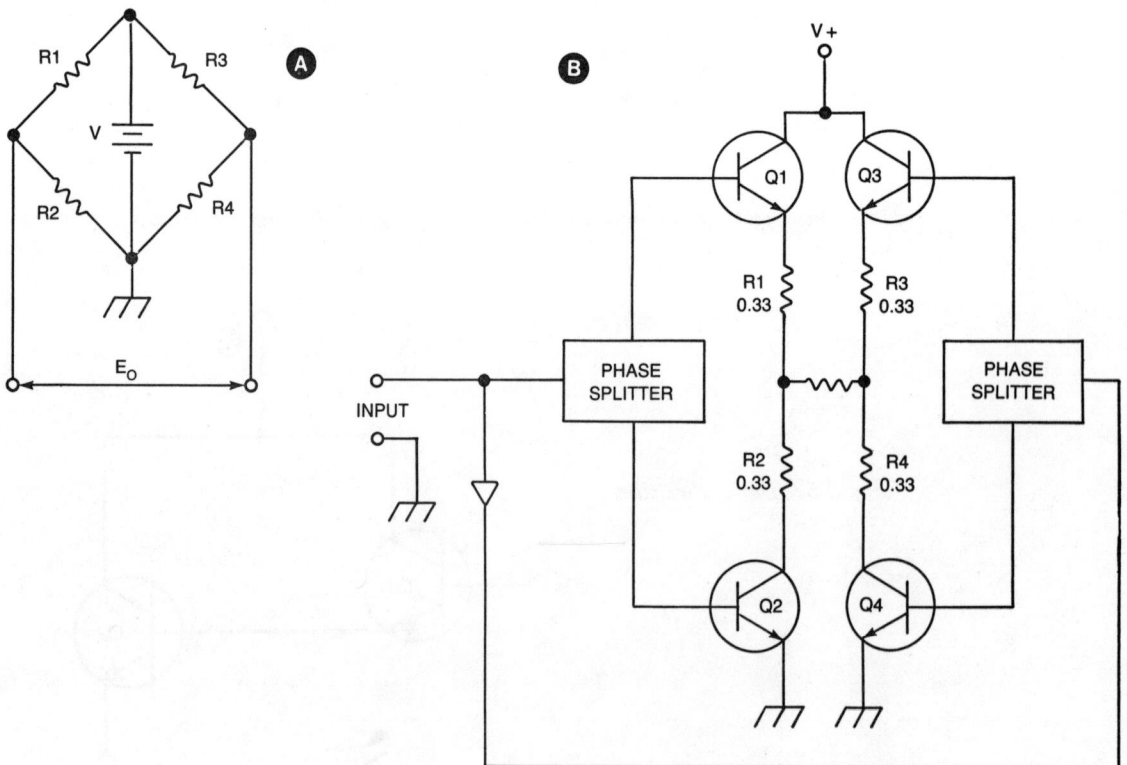

Fig. 3-53. Bridge audio (A) Wheatstone bridge; (B) bridge audio circuit.

TRANSISTOR RF POWER AMPLIFIERS

The vacuum tube held sway in the world of electronics for some 50 or more years. Beginning in the early- to mid-1950s, however, the transistor (having been invented a few years earlier) began to replace the tube in electronic circuits. It was in 1962 that the first automobile radio for a major auto maker (General Motors) was fully transistorized, followed in 1963 by the other car makers.

The transistor slowly eroded the position of the vacuum tube until now, when there are no longer any domestic tube makers producing receiving tubes. The only area where the vacuum tube is still king is in power amplifiers at radio frequencies. But transistors are catching up. Until just recently, the tube was the only way to generate radio-frequency power in the 1000-watts-and-up range. There now exists a commercial all-transistor RF power amplifier on the market that produces 5000 watts. It is already possible to buy an AM broadcast transmitter with 250 W driver and 5 kW final stages. Amateurs can buy all-transistor radios with transistorized power amplifiers producing to 1500 watts at 30 MHz. There are also 250 W-class, VHF transmitters available (to 200 MHz) that are excellent for land mobile operation because they need no turn-up.

SUITABLE AMPLIFIER CLASSES

The class A amplifier is sometimes used as a power amplifier in audio circuits. Indeed, automobile radios and some communications receivers use nothing but class A, single-ended power amplifiers. This is not usually done in the radio frequency range, however. These ranges use class B, class AB, and class C amplifiers. Class C amplifiers are used in cases where the amplifier need not be linear, and they take advantage of the higher efficiency of this circuit. Transmitters using the class C amplifier include FM transmitters, CW transmitters (radiotelegraphy transmitters) and those AM transmitters in which the modulated stage is the final amplifier. If the modulation is impressed on a stage prior to the final amplifier (except in the case of FM transmitters, the following amplifiers (including the final) must be linear amplifiers. In those cases, the class B or class AB push-pull arrangement is used. Some transistorized RF power amplifiers operate in class B simply because it is very easy to build. A class B transistor amplifier can be built with no bias resistor at all.

SIMPLE RF AMPLIFIERS

The simplest form of transistor power amplifier for radio frequencies is shown in Fig. 3-54. This circuit is often found in the buffer and driver stages of citizens band transmitters. The input circuit consists of a parallel resonant tank circuit that is tuned to the frequency of operation. The coil in the tank circuit is tapped to match the impedance of the driving source. For circuits conducting a power transfer, it is a fundamental property that maximum power transfer occurs only when the load and source have the same impedance. Signals are passed to the base of the power transistor (Q1) through coil L2, which serves as a transformer secondary to L1.

The output circuit of Fig. 3-54 is also a parallel resonant tank circuit. In this case, the tap on the inductor is connected to the collector of the transistor, which is now the source of power. This particular circuit is called a series-fed amplifier because the dc current

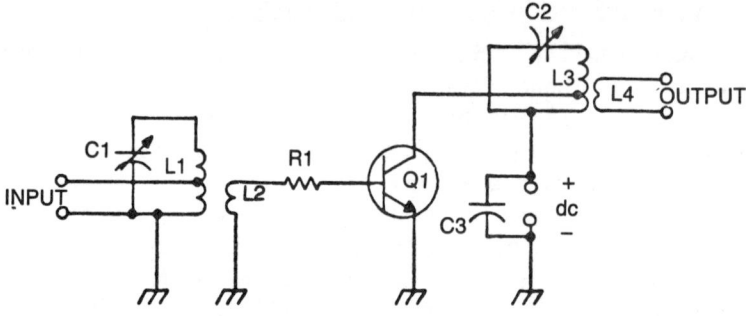

Fig. 3-54. Simple RF power amplifier.

passes through the inductor of the tank circuit before it reaches the collector of the transistor. A parallel-fed, or shunt-fed, amplifier would not have the collector current pass through the tank circuit coil. Output signal to the load is taken from a low-impedance transformer secondary. A power-supply bypass capacitor (C3) is connected across the power-supply terminals. This capacitor ensures that the power-supply impedance is low going to the resonant frequency of the tank circuit. The capacitor also provides decoupling to other stages that share the same power supply.

If you want the amplifier to operate class B, delete resistor R1. In that case, the transistor is zero-biased, so it will operate only on the positive half-cycles of the input signal. When the series resistor (R1) is in the circuit, the stage will operate class C. This is because a small current through the resistor sets up a dc bias voltage drop across the resistor. It is this voltage drop, which reverse biases the transistor, that makes the stage operate class C.

RF POWER TRANSISTORS

Some of the first commercially available RF transistors were intended for use in CB transmitters. These devices would produce several watts of RF energy at 27 MHz and were terribly expensive; $15 was not an uncommon price. To make matters worse, those transistors could not accept a high VSWR (voltage standing-wave ratio) in the transmitter/antenna circuit. The transistors would pop if an anomaly developed in the antenna circuit. Many CB operators lose their power transistors operating the transmitter with the antenna disconnected (a $35 mistake). Most modern transistors sustain a higher VSWR than the earlier models, generate far greater power levels, and operate at much higher frequencies. Figure 3-55 shows a typical RF transistor package. The threaded stud

Fig. 3-55. Typical RF power transistor.

is used to mount the transistor to the heatsink and is usually connected to either the collector or the emitter terminal of the transistor. The wings, which make this device look like a space satellite, are the collector, base, and emitter terminals of the transistor. The wide, flat shape of these wings is intended to keep the lead inductance low. It is the lead inductance that prevents many transistors from operating at high frequencies.

TYPICAL 30 MHZ CIRCUIT

Some solid-state RF power amplifiers are tuned, while others are broad banded. Figure 3-56 shows a circuit that is tuned to operate at approximately 30 MHz. The 50-ohm input impedance (standard for RF power amplifiers) is reduced to approximately 5 ohms at the base of the power transistor by an LC network consisting of C1, C2, and L1. The collector impedance-matching network consists of C5, C6, C7, and L4. The output impedance is 50 ohms.

The collector current passes to the collector of the power transistor through RF choke L3 and another RF choke L5. Together with capacitor C8, this circuit forms a low-pass filter to decouple the collector signal to ground before it reaches the power supply. RF choke L5 is actually a series of ferrite beads instead of an actual coil. Ferrite "donuts" are slipped over the wire that carries the collector current to the transistor. Such beads are known to possess inductance and are used extensively in HF and VHF transmitters.

Figure 3-57 shows a method for obtaining higher power levels. There are two basic ways to increase the power level in an amplifier: parallel-connect two or more devices,

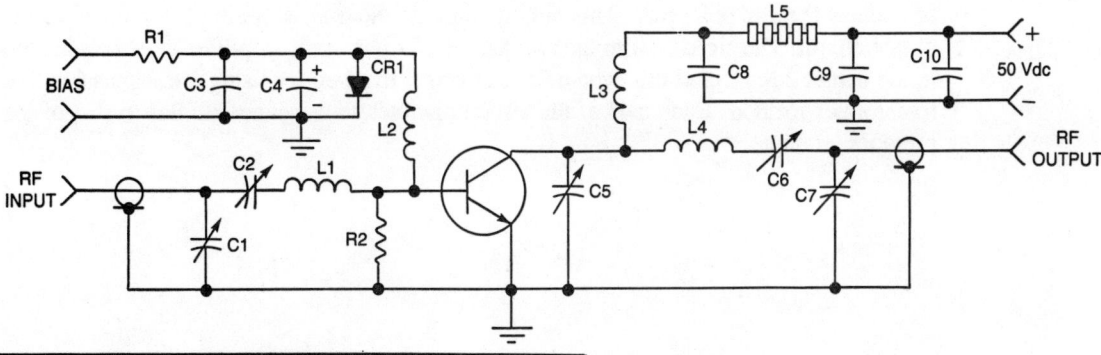

Component	Description
C1, C2, C7	780 pF (ARCO 469)
C3, C8, C9	0.1 µF, 100 V (ERIE)
C4	500 µF, 6 V
C5	180 pF (ARCO 463)
C6	480 pF (ARCO 466)
C10	30 µF, 100 V
R1	10 Ω, 10 W
R2	10 Ω, 1 W
CR1	1N4997
L1	3 turns #16 Wire, 5/16" I.D., 5/16" long
L2	10 µH molded choke
L3	12 turns, #16 enameled wire closewound, 1/4" I.D.
L4	5 turns, 1/8" copper tubing, 9/16" I.D., 3/4" long
L5	10 ferrite beads (FERROXCUBE x56-590-65/38)

Fig. 3-56. A 30 MHz transistor power amplifier.

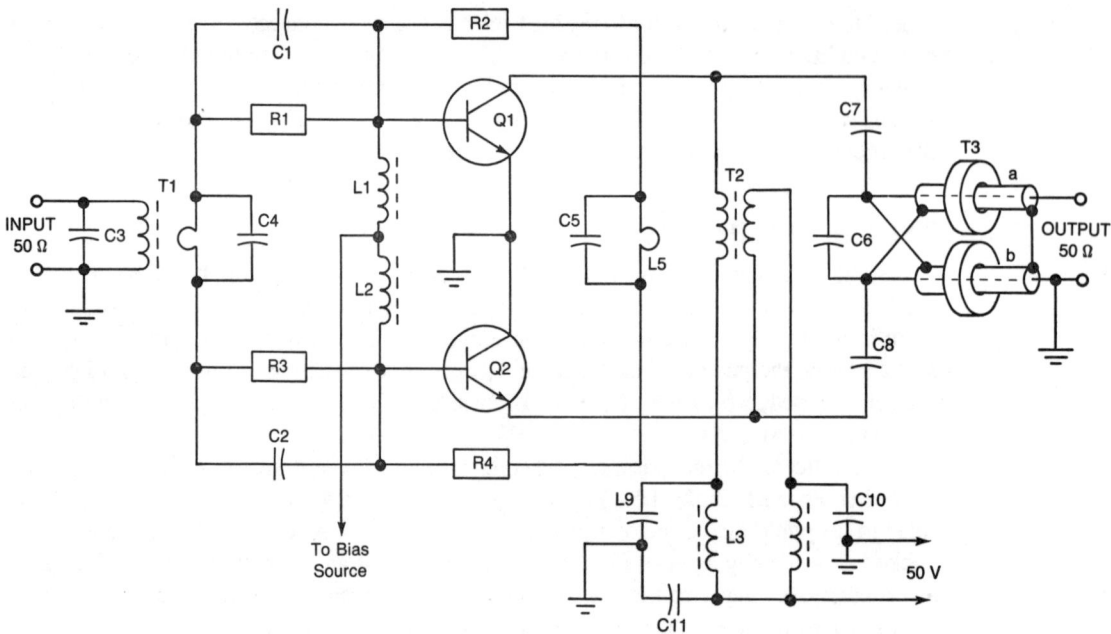

Fig. 3-57. A 300-watt HF power amplifier.

or connect them in push-pull. This circuit connects the transistors in push-pull. Note the output circuit. The signals from the two halves of a push-pull amplifier must be summed in the output circuit, and this type of circuit offers that service using a transmission line transformer method. Each half of the circuit operates in a manner similar to that of the previous example.

4

Basic Radiotelephone: Part II

This chapter covers miscellaneous concepts such as theory on instruments, oscillators, applications of amplifiers, and an introduction to transmitting and AM modulation.

INDICATING INSTRUMENTS

Most analog meters used for electronic measurements employ *D'Arsonval movement* as the basis for operation. With necessary modifications and by understanding its construction and operation, the entire field of understanding measuring instruments can be greatly simplified.

By adding resistances in series with the ordinary milliammeter or microammeter, you can make voltage readings. Or, by using a shunt resistance across the meter, its current measuring capability can be increased as desired. Although the meter might only be used to measure direct current or voltage, the addition of a small diode rectifier permits ac to be measured as well. The sensitivity of a meter is dependent on the current drawn by the meter movement for full-scale deflection; the less current required, the greater the sensitivity. Voltmeters are rated according to *sensitivity* in ohms per volt, and this is equal to the reciprocal of the current required for full-scale deflection. In other words, a voltmeter rated at 20,000 ohms per volt would have a current of 1 over 20,000, or 50 microamperes, for full-scale deflection. This is considered to be a good meter, but not good enough for measurements where circuit loading would result in inaccurate readings, such as most amplifier input circuits and many control circuits. In such cases, the volt-ohm-milliammeter (VOM) is no longer useful, but electronic multimeters (VTVM, TVM, DVM, DMM, etc.) are capable of accurately reading in the most sensitive circuits. Some electronic meters merely use an amplifier that boosts the sensitivity of the D'Arsonval meter movement and provides extremely high input impedance (usually 10 megohms or more). All but the highest impedance circuits are not loaded with such an instrument. Currently, there is considerable progress in meter design, with the new FET-input multimeter having a capability of measuring resistors, transistors, and other components in the circuit. Using the field-effect transistor as an amplifier, the input impedance ranges to well over 15 megohms. The high-low voltage arrangement in the ohmmeter section prevents transistors and diodes from conducting during in-circuit resistor measurement on *low power* and allows them to conduct on *high power* for accurate testing of the semiconductors.

OSCILLATORS

Oscillators are generators of alternating current signals with the output frequency dependent on the characteristics of the circuitry. The fact that they are capable of generating signals at various frequencies dictates their use in radio and television receivers and transmitters. Needless to say, oscillators have many other uses in electronic equipment, but the receiver and transmitter applications are of primary interest at this time.

Although oscillators fall into many categories, most oscillators used in radio communications are labeled *feedback* oscillators and obey similar basic principles. If a charged capacitor is connected across an inductor, the capacitor causes a current to flow through the coil from negative to positive and form an electromagnetic field around the inductor; energy is stored in that field. As the capacitor becomes fully discharged, the electromagnetic field around the coil collapses and causes a back EMF to recharge the capacitor in the other direction. As the capacitor is recharged, the field around the coil is set up again with the energy stored therein, which collapses and recharges the capacitor in the original direction. This current reversal in the circuit generates an alternating wave and the ability of the LC circuit at resonance to cause oscillations is called the *flywheel effect*. This action could continue forever except for the loss of power in the resistance of the coil and capacitor in the form of heat. The gradual decrease in amplitude of the oscillations in a parallel resonant circuit is called *damping* and the resulting wave produced is a *damped sine wave*.

The oscillating frequency depends on the values of the inductance and capacitance. The frequency is lowered as either or both values are increased. Because it takes longer for the capacitor to charge (and also longer to discharge) as the value of capacitance increases, it means fewer oscillations are possible per second. Naturally, decreasing the values of capacitance or inductance in the LC parallel resonant circuit results in less time to charge and discharge with more oscillations per second, or a higher frequency. The formula for determining the resonant frequency of an LC "tank" circuit is,

$$F_r = \frac{1}{2\pi \sqrt{LC}}$$

F_r is the resonant frequency in Hz, 2π equals 6.28, L is the inductance in henrys, and C is the capacitance in farads.

Because the LC circuit needs much more energy to be useful, power is furnished through an amplifier element so that it will not stop oscillating when loaded. By using an amplifier with the LC "tank" circuit, oscillations are sustained by "feeding back" a portion of the output signal to the input. Hence, *the oscillator is actually a self-excited amplifier* and no external signal or trigger is needed to start it. As soon as dc power is applied, circuit noise will be amplified and fed back, causing a weak signal at the input. The input signal is amplified to a strong signal and more is fed back until oscillations snowball to full strength.

AUDIO AMPLIFIERS

Audio or sound amplifiers increase the voltage, current, or power of audio frequency signals in the 20- to 20,000-Hz range. the level of the input is always very low, and one or two stages of voltage amplification are usually required to build the weak input signal to a sufficient amplitude to drive a power amplifier. When vacuum tubes are used, the voltage of the signal is increased, and with transistors, the current is increased. The end result is the greater signal amplitude necessary for the power stage. Impedance coupling, RC coupling, direct coupling, or transformer coupling can be used. Transistor amplifiers can be used in cascade for additional gain, and high power gain is available with the common-emitter arrangement.

In an audio amplifier, the output waveform must be the same as the input signal. Any deviation is called *distortion*. One common form, known as frequency response distortion, results when signal gain varies with frequency. Amplitude distortion is a variation in gain with amplitude and can result from a defective transistor, improper bias, too much drive, defective coupling capacitor, or low output impedance. The measure of distortion is the *linearity* of the amplifier, which relates how true the output signal is to the input signal. Several specific measures of linearity are available, including *total harmonic distortion* (THD) and *intermodulation distortion*.

Voltage or current amplifiers serve only limited output power needs and are usually operated class A, which is low in efficiency but high in linearity. Because these amplifiers operate with low power, the efficiency of operation is not important, but the quality of the output is extremely important. The output waveform should be an exact replica of the input waveform except for the amplitude. The power amplifier can only reproduce the applied signal; it cannot improve the quality. Power amplifiers are operated class B as a rule to provide the large output required. Another form of power amplifier is class AB, which is a hybrid of A and B.

The impedance of a power amplifier output must be matched to the impedance of the load in order to realize maximum power, and a transformer can be used to provide such a match. An output transformer should have the correct turns ratio to provide a correct impedance match. The correct turns ratio is equal to the square root of the ratio of the impedances. If the power amplifier has an output impedance of 8,000 ohms and the loudspeaker an impedance of 16 ohms, the ratio of the impedances is 8,000 to 16, which equals 400. So, the proper transformer turns ratio is the square root of 400, or 20. The primary of the output transformer must have 20 turns for every single turn in the secondary, for example, 1,000 to 50. The winding with the greater number of turns is connected to the higher impedance, and the lower number going to the lower impedance.

MICROPHONES

A microphone converts sound waves into electrical signals, and these tiny impulses are amplified thousands of times by audio amplifiers. There are several types of microphones, but all make use of the pressure provided by the sound wave against a diaphragm, ribbon, or plate. This pressure causes a variation in a mechanical-to-electrical transducer. One transducer method uses the piezoelectric effect, which produces an electrical voltage when a mechanical strain or pressure is applied. The ordinary crystal microphone operates this way by using two thin crystals cemented in an arrangement called a *piezoelectric bimorph* cell. Sound waves striking the diaphragm cause a twisting strain on the cell and that produces a tiny electrical output voltage. A third method uses a moving coil that cuts magnetic lines of force in a permanent magnet that produces a small voltage across that coil. In a dynamic microphone, sound waves strike a diaphragm attached to the coil, causing it to move in and out, thus generating a tiny voltage that corresponds to the sound-pressure changes. A more thorough coverage on microphones is given in the question and answer section, along with an evaluation of the different types.

RADIO FREQUENCY AMPLIFIERS

Radio frequency (RF) amplifiers are used to boost radio signal power. Normally, they operate in class C, because it is more practical due to the greater efficiency and power output. However, if a modulated signal is being amplified, a class B "push-pull" amplifier circuit must be used to avoid distorting the modulation. RF voltage amplifiers often operate class A, because small amounts of power are needed and even with the low efficiency, the losses amount to very little.

It is important to remember that plate current flows all the time in a class A amplifier, about half the time in class B, and only for short pulses in class C. The short pulses of the class C RF amplifier output are rounded into a clean sine wave by the flywheel effect of the plate tank circuit. Bias requirements for class C are never critical above a certain threshold level and are easy to supply. Grid-leak bias is often used, but a resistor and bypass capacitor should be included in the cathode to limit plate current in cases where grid drive is lost. Otherwise, a loss of grid drive would permit the plate current to rise to a degree where the vacuum tube could be damaged permanently.

TRANSMITTERS AND AM MODULATION

A transmitter often consists of a number of simple stages in which each performs a specific function. The *oscillator* generates the RF and applies it to the buffer stage, which isolates the load from the delicate oscillator and prevents that load from affecting frequency stability. The next stage is a *multiplier* to change the output frequency of the oscillator to the desired (higher) level, and sometimes more than one stage is necessary, because single stages rarely multiply more than four times (quadrupler). If we need to multiply the oscillator frequency by six, a doubler followed by a tripler would do the job well.

Because the oscillator determines the frequency of the transmitter, it must operate on the correct frequency at all times and cannot drift more than a slight amount from that frequency. *Crystal control* of the oscillator is best for this necessity of frequency stability. However, because even crystals change frequency slightly with temperature, many transmitters house crystals in temperature-controlled ovens.

In recent years, temperature-compensated crystal oscillators (TCXO) have replaced oven-controlled oscillators in many transmitters. The TCXO is capable of oven-like behavior but does not require a heater.

Power supply regulation is also a necessity for good oscillator frequency stability; therefore, a separate oscillator supply is desirable.

The primary purpose of the buffer amplifier is isolating the oscillator from the stages following, which could very well change the frequency (by load variations). Because the buffer operates class A or B, the input draws only negligible power from the oscillator.

Frequency multipliers are normally doublers or triplers, although the quadrupler is used at times. The multiplier stage or stages must offer some means of distorting the input signal. Distortion increases the harmonic content, allowing the harmonic energy to be used for the output. Transistors, pn diodes and varactor diodes are used in frequency multipliers.

The power amplifier provides the big boost in the signal and supplies the carrier that is coupled through the transmission line to the antenna. As a result of the large amount of power spurious oscillations frequently show up in this stage. In addition to wasting power, these oscillations can get to the antenna and cause interference, so it is important that they be eliminated. The Faraday shield forms a comb-like screen between primary and secondary of the output to reduce the transfer of harmonics.

Transmitter power output is determined by either the indirect method or the direct method with reasonable accuracy, but in some cases, such as AM broadcast transmitters, the direct measurement of output power must be used. Both methods are taken up in detail a little later.

While holding the carrier frequency constant, the amplitude is varied by the modulating signal. Amplitude modulation is accomplished by introducing the audio signal from the modulator to an element of the final power amplifier. In RF power amplifiers, vacuum tubes are still used extensively because of their superior power handling ability. The plate is the most popular, but the screen grid, control grid, suppressor grid, or cathode can be used. As the carrier varies in amplitude with the modulating signal two sidebands are generated with one above and one below the carrier. The bandwidths of the sidebands are limited by the FCC according to the service. For an AM broadcasting station, the limit is 5 kHz.

SINGLE SIDEBAND, SUPPRESSED CARRIER (SSBC)

Single sideband makes use of the fact that two of the three signal components normally radiated in AM transmission can be eliminated with additional advantages resulting. Because at least 50 percent of the radiated power is in the carrier (which carries no intelligence), it is suppressed in single-sideband (SSB) transmission. The two sidebands, the upper and lower, are the same except for the frequency. So, one is eliminated and the result is single-sideband, suppressed-carrier (SSBSC) transmission.

With SSBSC, there is a very large plus when compared to the conventional AM in efficiency. In an ordinary AM transmitter with a power of 1,000 watts, there is only about 250 watts in each sideband. So by using SSBSC transmission, it is possible to raise the power to four times that value, or 1,000 watts, for the single sideband radiated—without exceeding plate dissipation ratings or even the capacity of the power supply. The required bandwidth is halved, and this reduction means a better signal-to-noise ratio at the receiver. The BFO (beat frequency oscillator) of an ordinary superheterodyne receiver serves a useful purpose in detecting single sideband as well making reception of A-1 (code) transmissions easier and more pleasant to hear, as the pitch of the note is adjustable.

QUESTIONS AND ANSWERS

Q1: Make a sketch showing the construction of the D'Arsonval type meter and label the various parts. Draw a circuit diagram of an FET voltmeter and a wattmeter.

A: D'Arsonval Movement. The D'Arsonval movement is shown in Fig. 4-1A. In the diagram, only one turn of wire is shown, though in practice the coil consists of many turns of very fine wire, each turn adding more effective length to the coil. The coil is usually wound on an aluminum frame or bobbin to which the pointer is attached. Oppositely

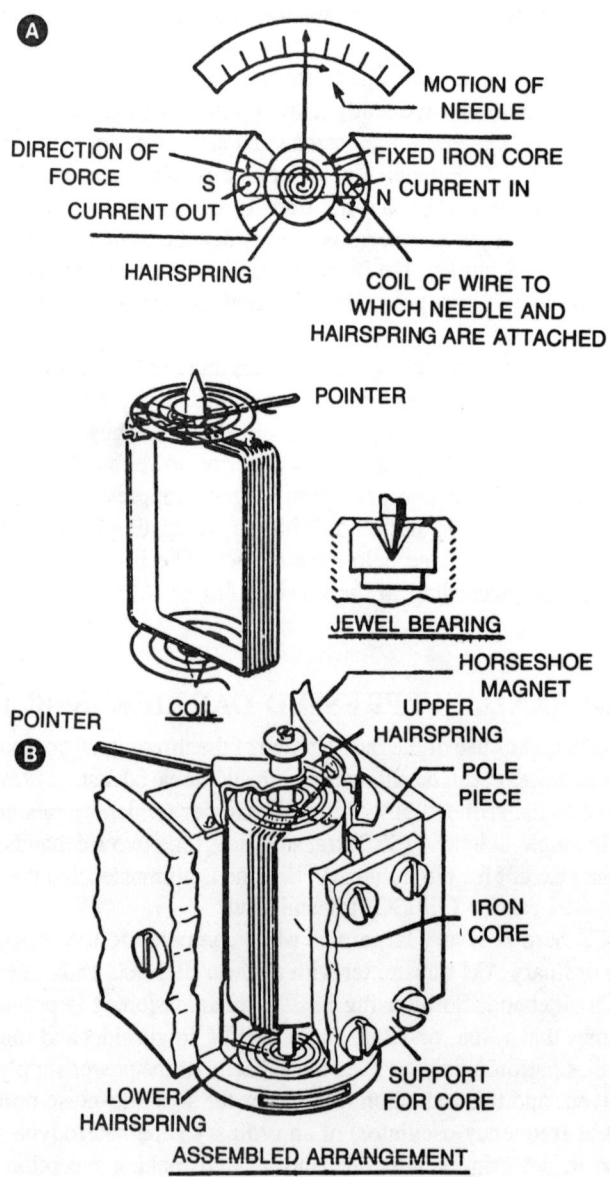

Fig. 4-1. (A) D'Arsonval movement; (B) detailed view of basic movement.

wound hairsprings are also attached to the bobbin, one at either end. The circuit to the coil is completed through the hairsprings. In addition to serving as conductors, the hairsprings serve as the restoring force that returns the pointer to the zero position when no current flows.

In the movement, the deflecting force is proportional to the current flowing in the coil. The deflecting force tends to rotate the coil against the restraining force of the hairsprings. The angle of deflection, then, is proportional to the deflecting force. When the deflecting force and the retraining force are equal, the coil and pointer cease to move further.

The deflecting force is proportional to the current in the coil; the angle of rotation is proportional to the deflecting force; thus, the angle of rotation is proportional to the current through the coil. When current ceases to flow in the coil, there is no longer a force to oppose the restraining force of the hairspring, and the pointer returns to its resting position. Figure 4-1B is a detailed view of the D'Arsonval movement.

Electronic Voltmeters. The dc electronic voltmeter circuit is shown in Fig. 4-2. The dc voltage to be measured is applied between the dc input terminal and ground. The dc input voltage is therefore applied through R1 to the gate of Q1. The source of Q1 is grounded. The meter is connected across a normally balanced bridge so that the application of the dc voltage unbalances the bridge and causes the meter to deflect. The bridge consists of R3, R4, and R5 plus Q2 and Q3. The calibration is in dc volts.

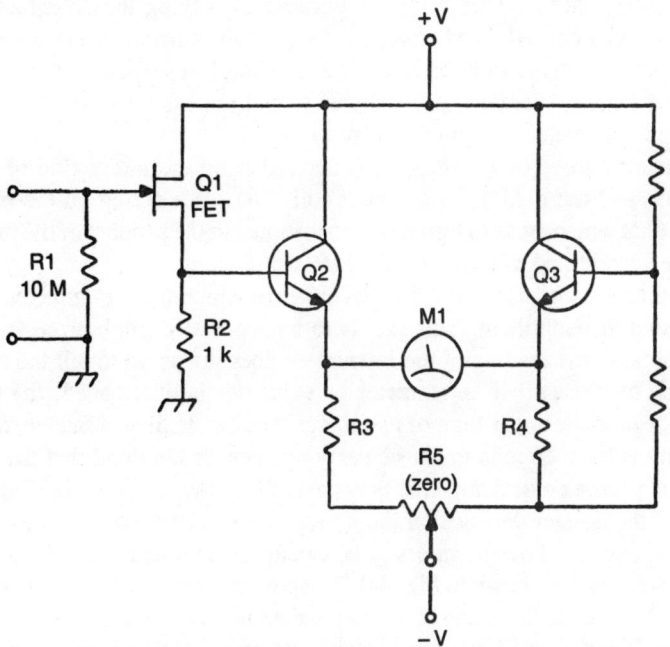

Fig. 4-2. Transistorized voltmeter circuit.

Wattmeter. Ac electric power is measured by means of a *wattmeter*. This instrument uses the *electrodynamometer* type of movement. It consists of a pair of fixed current coils and a movable potential coil, as shown in Fig. 4-3. The fixed coils are made up of a few

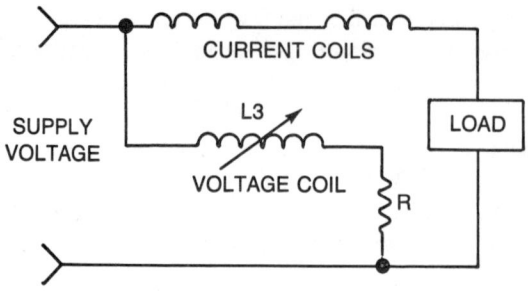

Fig. 4-3. Schematic of a typical analog wattmeter.

turns of comparatively large-conductor wire. The potential coil consists of many turns of fine wire; it is mounted on a shaft carried in jeweled bearings so it can turn inside the stationary coils. The movable coil is linked to a pointer that moves over a graduated scale. Flat coil springs hold the pointer to a zero position.

The stationary current coil of the wattmeter is connected in series with the circuit (load), and the movable potential coil is connected across the ac power line.

When line current flows through the current coil of a wattmeter, a magnetic field is set up around the coil. The strength of this field is proportional to the line current and is in phase with it. The potential coil of the wattmeter generally has a high-value resistor connected in series with it. This is for the purpose of making the potential-coil circuit of the meter as purely resistive as possible. As a result, current in the potential circuit is almost perfectly in phase with the line voltage. Therefore, when voltage is impressed on the potential circuit, current is proportional to and in phase with the line voltage. By definition, then, the meter measures true power.

The actuating force of a wattmeter is derived from the interaction of the field of its current coils and the field of its potential coil. The force acting on the movable coil at any instant (that which tends to turn it) is proportional to the product of the instantaneous values of line current and voltage (recall that $P = E \times I$).

The wattmeter consists of two circuits, either of which will be damaged if too much current is passed through them. This fact is to be especially emphasized in the case of wattmeters, because the reading of the instrument does not serve to tell the user that the coils are being overheated. If an ammeter or voltmeter is overloaded, the pointer will be indicating beyond the upper limit of its full-scale range. In the wattmeter, though, both the current and voltage circuits might be carrying such an overload that their insulation is burning, and yet the pointer might be only part of the way up the scale. This is because the position of the pointer depends on the *power factor* of the load circuit as well as on the voltage and current. Power can easily be calculated in a dc circuit by connecting an ammeter and voltmeter as shown in Fig. 4-4. The power is the product of the two readings.

Apparent power is the metered I times metered E and is true power when E and I are in phase. When E and I are out of phase, however, true power is less than EI by the power factor. This power factor is the cosine of the phase angle between E and I. In other words, $P = (\cos)\Theta\, EI$.

Q2: Show by a diagram how a voltmeter and an ammeter should be connected to measure power in a dc circuit.

A: See Fig. 4-4. The power of the circuit is determined by multiplying the voltage by the current as indicated by the meters. In a dc circuit, the power in watts is equal to the voltage in volts times the current in amperes (P equals E × I).

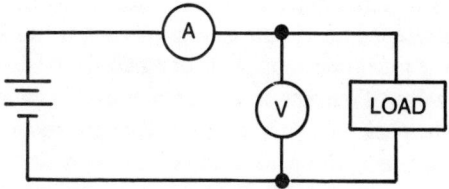

Fig. 4-4. Measurements used to determine power in a dc circuit.

Q3: If a 0-1 mA dc milliammeter is to be converted into a voltmeter with a full-scale calibration of 100 V, what value of resistance should be connected in series with the milliammeter?

A: This is easily determined from Ohm's Law: For 1 mA of current to flow (to deflect the meter full-scale) when 100 Vdc is applied, resistance is equal to 100 V divided by 0.001 ampere; thus, the value of the resistor would be 100,000 ohms.

Q4: A one-milliampere meter with a resistance of 25 ohms is used to measure an unknown current by shunting the meter with a 4-ohm resistor. It then read 0.4 milliampere. What was the unknown current value?

A: The unknown current was 2.9 milliamperes as determined by the basic formula $R_m I_m = R_s I_s$, where R_m is the meter resistance, I_m the current through the meter, R_s the resistance of the shunt and I_s the current through the shunt. The current and resistance through the meter ($R_m I_m$ equals E_m) and the voltage drop E_m is equal to the voltage drop across the shunt, because they are in parallel. Hence, I_s is E_s divided by R_s.

Q5: An RF VTVM is available to locate the resonance of a tunable primary tank circuit of an RF transformer. If the VTVM is measuring the voltage across the tuned secondary, how would resonance of the primary be indicated?

A: Resonance would be indicated by a peak reading (maximum voltage) on the VTVM. In a parallel LC circuit, impedance is greatest at resonance, and with a voltage drop proportional to impedance, it would be maximum at resonance also. Because the signal voltage across the primary is maximum at resonance, maximum voltage would be induced into the secondary, and tuning the primary for a peak reading of the VTVM in the secondary would show resonance in the primary tank circuit.

Q6: Define the following terms and describe a practical situation in which they might be used.
  (a) rms voltage
  (b) peak current
  (c) average current
  (d) power
  (e) energy

A: **RMS Voltage.** The abbreviation *rms* means *root mean square* and is applied to alternating voltages as a means of comparison with an equivalent dc voltage value. In ac,

143

then, rms is the equivalent dc voltage value required to deliver the same *effective* heating power to a load. The *effective*, or rms value of a sine wave is 70.7 percent (0.707) of the peak ac voltage value. Ac voltmeters are typically calibrated in rms volts.

Peak Current. One of the most frequently measured characteristics of a sine wave is its amplitude. Unlike dc measurements, the amount of alternating current or voltage present in a circuit can be measured in various ways. In one method of measurement, the maximum amplitude of either the positive or the negative alternation is measured. The value of current or voltage obtained is called the "peak" voltage or current. To measure the peak value of current or voltage, an oscilloscope or special meter (peak reading) must be used. The peak value of a sine wave is 100 when the average value is 63.7 and the rms value is 70.7. Peak current measurements are often made on antennas, transmission lines, audio circuits, and RF amplifiers.

Average Current. The average value of a complete cycle of a sine wave is zero, because the positive alternation is identical to the negative alternation. In certain types of circuits, however, it is necessary to compute the average value of one alternation. This can be accomplished by adding together a series of instantaneous values of the wave between zero degrees and 180 degrees and then dividing the sum by the number of instantaneous values used. Such a computation would show one alternation of a sine wave to have an average value equal to 63.6 percent of the peak value. Average values of current and voltage are useful in calculating the unfiltered output voltages and currents of rectifiers. The dc output of an unfiltered rectifier is equal to the average value of the applied voltage alternations.

Power. Power is a means for measuring the rate at which work is accomplished, and it may be calculated in purely resistive circuits by multiplying the load's current by its source voltage. In power measurement, 1 watt equals 1 volt times 1 ampere. If the source voltage is unknown, power can be calculated by multiplying the square of the current by the resistance of the load. Power is a common measurement in electronics, for it offers the principal means for measuring circuit efficiencies, circuit performance capability, and requirements of a source to adequately drive a load. The watt is the unit of electrical power, and is equal to work done at the rate of 1 joule of work per second. Typical practical applications for power measurements include RF output, measurement of audio output, and an almost unlimited number of other functions, both ac and dc.

Energy. Here the term *energy* represents the ability to perform work electrically. In other words, a certain amount of energy must be available in order for a source to be able to deliver a certain amount of power to a load. Energy, then, is the capacity to perform work, the basic unit of which is the joule. Other practical uses for the term occur in studies related to physics and in computing the value of laser discharges.

Q7: Draw circuit diagrams of each of the following types of oscillators (include any commonly associated components). Explain the principles of operation of each.
(a) Armstrong
(b) tuned-out, tuned-in (TOTI)
(c) Hartley (series- and shunt-fed)
(d) Colpitts
(e) multivibrator
(f) Pierce (crystal-controlled)
(g) Miller

A: Armstrong. The simplest of all of the oscillators is the Armstrong. The Armstrong circuit is shown in Fig. 4-5. L2-C1 forms the tank circuit, which determines the resonant frequency, and L1 is the feedback coil (often referred to as a *tickler*). (The bias circuit is not shown.)

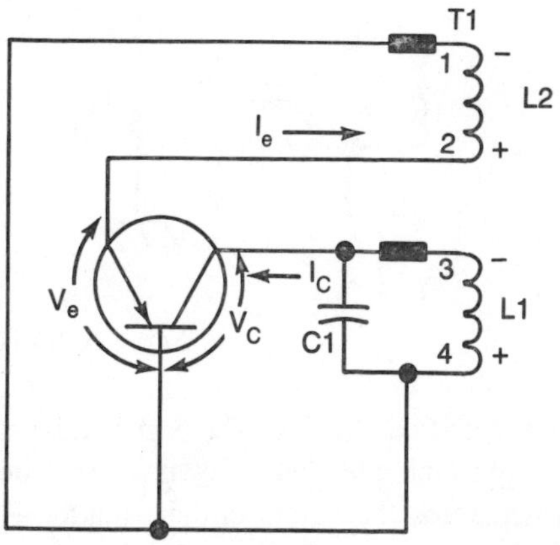

Fig. 4-5. Armstrong oscillator circuit.

Oscillations begin in the circuit when the bias conditions of the transistor are normal and power is applied. The amplitude of current flow in the circuit will increase, causing an expanding magnetic field around the tank circuit. This induces a voltage in the tank coil and charges the tank capacitor. The charge of the tank capacitor causes output current to increase by increasing the potential on the collector. This *regeneration* continues until the nonlinear characteristics of the amplifying device cause a difference in the rate of change of the output current.

When the induced voltage of the tank coil falls below the charge of the capacitor, the tank capacitor begins to discharge. The discharge of this capacitor causes the input potential to decrease, thereby decreasing output current. When the tank capacitor is completely discharged, the field of the tank coil collapses and charges the capacitor with the opposite polarity. Partway through this portion of the cycle of operation, the input potential will become sufficiently negative to cut off the transistor's conduction. When the field of the tank coil is completely collapsed, the capacitor will begin to discharge. As the tank voltage comes nearer the bias point, the input potential approaches the point where the amplifier comes out of cutoff (this oscillator is biased for class C operation). As the amplifier begins to conduct, regenerative feedback occurs and replaces the lost energy. Oscillations can now continue until dc power is removed from the circuit.

Tuned-Out, Tuned-In. In the TOTI oscillator, the input circuit (L1-C1, Fig. 4-6) is tuned to the resonant frequency desired. When the first surge of current starts this circuit oscillating, the oscillations appear at the grid and are amplified in the output circuit. This circuit consists of L2-C2. The feedback path in the TOTI oscillator occurs through the interelement capacitance of the amplifying device. Energy is coupled from the output circuit to the input circuit. If L2-C2 is tuned to the same frequency as L1-C1, the phase of the feedback is not proper to sustain oscillations; for this reason, the plate circuit is made inductive at the frequency of oscillation of the input circuit to make the feedback regenerative. This is done by tuning the output circuit to a slightly higher frequency.

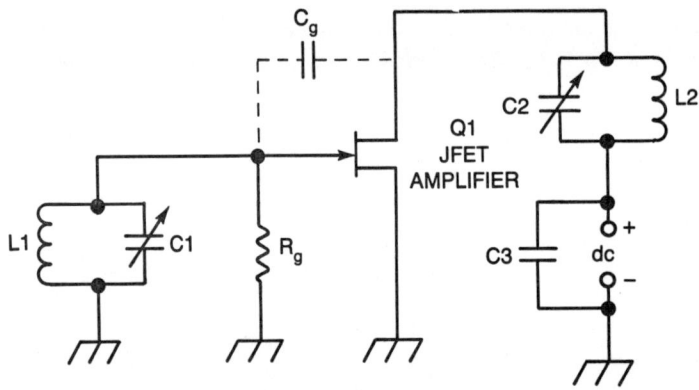

Fig. 4-6. TITO oscillator.

Series-Fed Hartley. The principal identifying characteristic of the Hartley oscillator is the split tank coil, half of which feeds the input of the transistor, and the other half of which feeds the output circuit. Figure 4-7A shows the basic series-fed circuit.

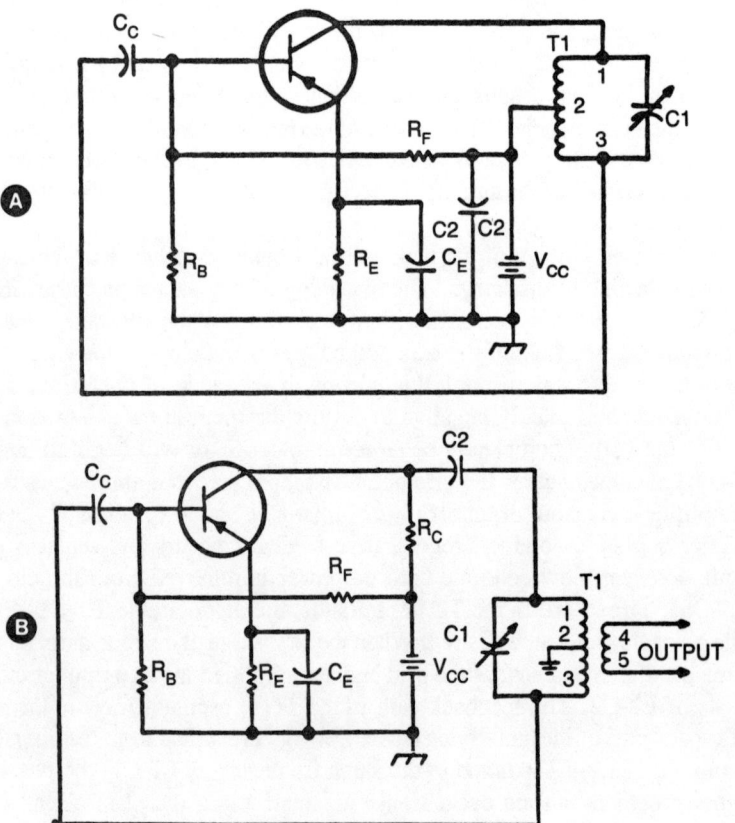

Fig. 4-7. (A) Transistor series-fed Hartley oscillator; (B) shunt-fed Hartley.

In the Hartley oscillator, one tank circuit is actually made to serve as both base and collector resonant circuits. The base is coupled to one end of the tank and the collector is connected to the other end. The emitter is attached to a point on the inductor. This divides the coil between the base and the collector circuits in the form of an inductive voltage divider, as shown. The voltage across L1 is between the base and emitter, thereby applying a signal to the base. The amplified voltage at the collector appears across L2. This provides the necessary feedback.

In the Hartley, the dc collector current must pass through inductor L2 before it can return to the emitter. The disadvantage in this arrangement is that the collector supply is placed at a high ac potential with respect to the emitter. Also, the supply has a large distributed capacitance to ground, and this capacitance is shunted across the tank coil (L2).

In the series-fed Hartley, resistors R_B and R_F provide the necessary bias for the base-emitter circuit. Collector bias is obtained through transformer winding 1-2. Capacitor C_E provides an ac bypass around the emitter swamping resistor (R_E). The feedback is obtained from the induced voltage in winding 2-3 coupled through capacitor C_C to the base of the transistor. Capacitor C2 places terminal 2 of the tank coil at ac ground potential.

Shunt-Fed Hartley. The disadvantage of the series-fed circuit (discussed above) can be overcome by keeping the dc collector supply and the oscillating current separate. This is accomplished in the shunt-fed Hartley (Fig. 4-7B). The collector current oscillations are coupled to the split-inductance tank by means of capacitor C2. The capacitor prevents the dc current from returning to the cathode through the tank. The collector current, therefore can return only through the choke in series with the V_{CC} source. This choke represents its reactance from appearing in the supply because its reactance is very large.

Resistors R_B, R_C and R_F provide the necessary bias conditions for the circuit. The frequency-determining network consists of the series combination of windings 1-2 and 2-3 in parallel with capacitor C1. Because this capacitor is variable, the circuit can be tuned through a wide range of frequencies. Capacitor C_E provides an ac bypass around the emitter swamping resistor (R_E).

The coil functions as an autotransformer to provide the regenerative feedback signal. The feedback signal is obtained from the induced voltage in winding 2-3 coupled through capacitor C_C to the base of the transistor. By shunt-feeding the collector through resistor R_C, direct current flow through the tank coil is avoided.

Colpitts. The Colpitts, like the Hartley, is a split-tank oscillator; the difference is that the Hartley incorporates a split inductance, whereas the Colpitts uses a split capacitance in the tank circuit. The capacitance of the tank circuit in the Colpitts (Fig. 4-8) is provided by capacitors C1 and C2, which form a capacitive voltage divider between the base and collector circuits. By adjusting C1 and C2, it is possible to control the frequency and amount of positive feedback.

Figure 4-8 illustrates the transistor Colpitts oscillator circuit. Regenerative feedback is obtained from the tank circuit and applied to the emitter of the transistor. Base bias is provided by resistors R_B and R_F. Resistor R_E develops the emitter input signal and also acts as the emitter swamping resistor. The tuned circuit consists of the capacitors C1 and C2 in parallel with inductor winding 1-2. Capacitors C1 and C2 form the voltage divider. The voltage developed across C2 is the feedback voltage.

Multivibrator. One of the simplest oscillators that can be used as a frequency divider is the synchronized multivibrator. There are many varieties of multivibrator circuits, but

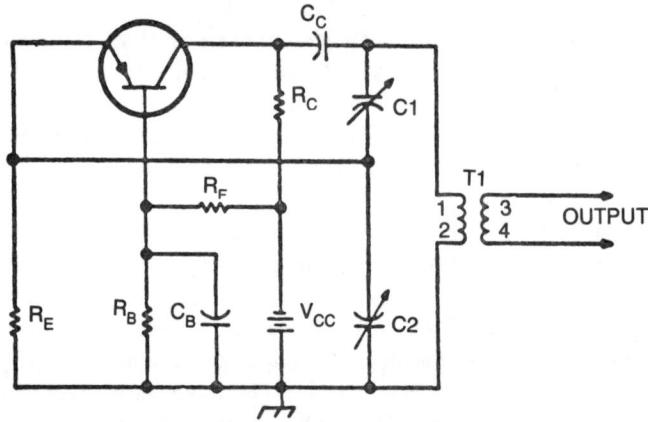

Fig. 4-8. Colpitts oscillator.

essentially they are all modifications of a two-stage, resistance-coupled amplifier circuit with the output fed back to the input.

A small amount of voltage applied to base circuit can be used to trigger oscillation. Any voltage that is an integral multiple of a natural frequency of the oscillator provides this triggering action. The frequency can be much higher than the actual frequency of operation of the oscillator. The output from one multivibrator controlled in this manner can be ten times less in frequency than the controlling voltage. The output of this multivibrator can be connected to another multivibrator that also divides by a like amount, providing division by 100. In this way, the high frequency of the crystal oscillator and the master oscillator in an FM system can be reduced to a frequency that falls in the audio range.

The basic transistor version of the multivibrator is shown in Fig. 4-9. This is a two-stage, RC-coupled, common-emitter amplifier with the output of the first stage coupled to the input of the second stage, and the output of the second stage coupled to the input of the first stage. Because the signal in the collector circuit of a common-emitter amplifier is reversed in phase with respect to the input of that stage, a portion of the output of each

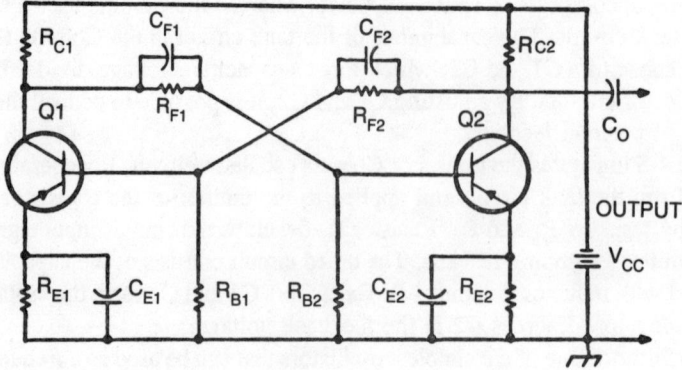

Fig. 4-9. Transistor multivibrator.

stage is fed to the other stage in phase with the signal on the base electrode. This regenerative feedback with amplification is required for oscillation. Bias and stabilization are established identically for both transistors.

Pierce. The Pierce oscillator, also frequently known as the ultraudion, is shown in its JFET form in Fig. 4-10. This is considered to be the simplest of the crystal oscillators, for all that is required are a few resistors and capacitors. As with the Colpitts oscillator, oscillation occurs because of the voltage feedback provided by the voltage divider formed by the interelement capacitance.

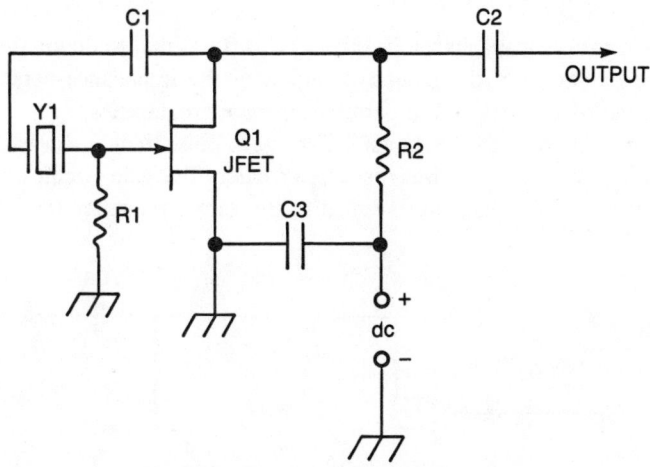

Fig. 4-10. Pierce crystal oscillator.

The bipolar transistor version of the Pierce oscillator is shown in Fig. 4-11. If the crystal were to be replaced by its equivalent LCR circuit, the functioning of the circuit would become analogous to that of the Colpitts oscillator. The circuit of Fig. 4-11 shows the common-base configuration with the feedback supplied from collector to emitter through

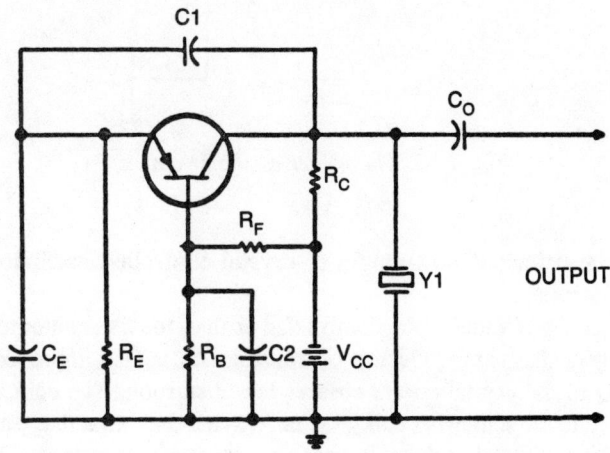

Fig. 4-11. Colpitts crystal oscillator, collector-emitter regeneration; Pierce crystal type.

149

capacitor C1. Resistors R_B, R_C, and R_F provide the proper bias and conditions for the circuit. The emitter resistor is the emitter swamping resistor (R_E). Capacitors C1 and C_E form a voltage divider connected across the output. Capacitor C2 is an ac bypass around base-biasing resistor R_E. Because no phase shift occurs in this configuration, the feedback signal must be connected so that the voltage across the emitter capacitor returns to the emitter with no phase shift occurring. The oscillating frequency of this circuit is determined not only by the crystal but by the parallel capacitance offered by capacitors C1 and C_E. These are normally made large to swamp both the input and output capacitances of the transistor and make the oscillations comparatively independent of changes in transistor parameters.

Because the parallel capacitance of C1 and C_E affects the oscillator frequency, the operation of the crystal is in the inductive region of the impedance-versus-frequency characteristic between the series- and parallel-resonant frequencies.

Miller. The Miller oscillator is shown in Fig. 4-12. This circuit is a crystal-controlled, version of the TOTI oscillator that was discussed earlier. The gate circuit is tuned by the piezoelectric crystal, while the drain is tuned to the same frequency by C1 and L1.

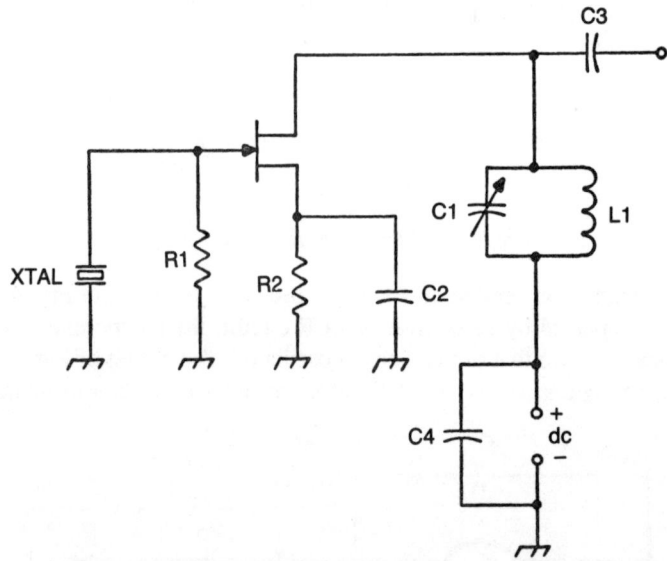

Fig. 4-12. *The Miller oscillator circuit.*

Q8: What are the principal advantages of crystal-controlled oscillators over tuned circuit oscillators?
A: Stability is improved by the crystal control that enables the transmitter to the "locked" closer to its assigned frequency. Quality of transmission is also improved, because the extremely high Q of the crystal circuit ensures less distortion. The compact size of the crystal, as opposed to the somewhat bulky LC tank, is another attractive feature. The only disadvantage is the additional problem in changing operating frequencies, and this would be a consideration only in some transmitters. As a rule, the disadvantage would not apply.

Q9: Why should excessive feedback be avoided in a crystal oscillator?
A: Excessive feedback could cause overheating that can crack the crystal. Frequency stability depends on the level of feedback; abnormal amplitudes could cause frequency deviation that could not immediately be corrected without changing crystals.

Q10: Why is a separate source of drain or collector power desirable for a crystal oscillator stage in a radio transmitter?
A: This ensures against frequency shift or dynamic instability during modulation that causes a change in load on the supply, and can, in turn, affect the oscillator frequency. Increasing the oscillator drain or collector voltage makes the oscillator frequency higher due to the decrease in input capacity. The frequency stability of the transmitter is considerably improved by providing a separate supply for the crystal oscillator transistor drain or collector.

Q11: What could result if a high degree of coupling exists between the output and input circuits of a crystal-controlled oscillator?
A: This could result in excessive feedback and the usual undesirable effects of dynamic instability.

Q12: Explain some methods of determining if oscillation is occurring in an oscillator circuit.
A: Occasionally, when testing RF amplifiers and transmitters, it is necessary to determine whether a radiated signal is being generated in the transmitter's oscillator circuit. In this case, you can rule out the oscillator by merely unplugging the crystal to see if the oscillation ceases. If it does not stop when the crystal is removed, parasitic oscillations and possibly inadequate neutralization of a stage or stages are indicated.

Other methods for checking oscillation of a stage include listening to the frequency on a receiver or other device capable of tuning to the signal frequency of the oscillator, monitoring the oscillator at close proximity with a diode connected across a sensitive meter, and holding a small fluorescent lamp near the tank coil (lamp will fluoresce if circuit is oscillating in high power stages).

Q13: What is meant by parasitic oscillations? How can they be detected and prevented?
A: Parasitic oscillations are signals generated within a transmitter other than by design. Most parasitics are not related to the operating frequency or its harmonics.

The most noticeable features of parasitic oscillation in an amplifier are erratic tuning and the radiation of spurious signals at other than the design frequency. When an RF power amplifier is operating properly, the tube plate current dips sharply as the tank circuit is tuned through resonance. This plate-current dip also corresponds to maximum power output and (usually) maximum grid current into the final amplifier. If a tetrode tube is operating normally, the plate current change might not be too great, but the screen current dip will still be significant. With parasitic oscillation, the plate current might not dip at all; the minimum might not correspond to maximum power output; several dips could appear in the tuning range; or grid current to the final amplifier will not coincide with the dip in the final amplifier plate current reading. Parasitics can be cured sometimes by inserting a parallel inductance and resistance in the grid or plate lead. This detunes one of the parasitic circuits sufficiently to prevent oscillation. (Usually, the process is no more complex than wrapping a few turns of solid wire around a carbon resistor so that both the resistance and the inductance are paralleled.) Another method is to insert a small resistance in series with circuit leads to introduce sufficient loss to stop oscillation. A third alternative is to incorporate a tuned parallel-resonant trap that actually inserts a very high impedance in the parasitic frequency path. In addition to the trap circuit, it is common to find small high-frequency capacitors connected from plate and control grid to cathode; these bypass the harmonic path.

Q14: What determines the fundamental frequency of a quartz crystal?
A: The fundamental is the lowest frequency of vibration for a specific mode of operation and is determined by the thickness, type of cut, substance, and type of mounting. Thicker crystals have a lower vibrating frequency than thinner ones. The modes of operation are *flexure*, *longitudinal*, *face shear*, *thickness shear* and *third-overtone*. Crystal substances are quartz, tourmaline, and rochelle salts, all of which exhibit piezo-electric powers. Capacity of the holder temperature both affect the crystal frequency, although the latter can be closely controlled by the use of a constant-temperature oven.

Q15: What is meant by the temperature coefficient of a crystal?
A: Temperature coefficient expresses the change of oscillation frequency due to changes in temperature and can be either positive or negative. A low temperature coefficient indicates that the crystal frequency varies only slightly with larger changes in temperature—a most desirable factor. A negative temperature coefficient signifies an increase in crystal frequency with a decrease in temperature and positive means the opposite. The pertinent information is normally printed on the crystal holder and refers to plus or minus parts per million per degree centigrade.

Q16: What are the characteristics and possible uses of an overtone crystal? A third-mode crystal?
A: The overtone crystal is ground to oscillate at an odd harmonic of its fundamental frequency, such as the third, fifth, seventh, etc., harmonic. This procedure permits control at much higher frequencies than would be possible otherwise, and at the same time reduces the number of frequency multiplier stages required at high frequencies, such as VHF and UHF. Needless to say, specially designed overtone crystals are far superior to the usual fundamental type crystal, although the latter will operate in such a circuit.

The third mode crystal is one that is ground for the third overtone of the fundamental and performs well at three times the fundamental.

Q17: Explain some of the factors involved in the stability of an oscillator (both crystal and LC-controlled).
A: Several important factors are involved:

(a) *Stray capacitance* changes the total capacitance of the tank circuit and causes the oscillator to "drift." Included are amplifier device interelement capacitance and reflected reactance, both of which should be held as low as possible to ensure a high C-L ratio in the tank circuit.

(b) *loading* of the oscillator lowers the Q of the tank and reduces stability. Isolation of the oscillator from its load as provided by a buffer amplifier between the oscillator and load is desirable.

(c) *Voltage* must be constant and preferably regulated to keep it that way.

(d) *Temperature* should be constant with high temperature coefficient crystals enclosed in a temperature-controlled oven. Temperature compensating components should also be utilized.

(e) *Q* should be high for good stability but not too high (which causes excessively sharp tuning); keep the resistance of the tank coil low by using heavy wire.

(f) *Shielding* with material having good electrical conductivity reduces stray fields, humidity, and air.

(g) *Bias resistors and capacitor values* must be suitable for stable operation. The Q value is considerably higher with a crystal type oscillator than one of the LC type, and stability of the former circuit is far superior to that of the latter as a result.

Q18. Is it necessary or desirable that the surfaces of a quartz crystal be clean? If so, what cleaning agents can be used that will not adversely affect the operation of the crystal?

A: The crystal surfaces must be clean. Dirt, lint, or grease will interfere with proper operation. Even grease or oil from the skin is harmful. A soft tissue along with soap and water is best for cleaning, followed by a thorough rinsing. Carbon tetrachloride is also an excellent cleaner and will ensure good contact with the holder. If the crystal is hermetically sealed in its holder, foreign material cannot easily reach the crystal and there should never be a problem with dirty surfaces.

Q19: What is the purpose of a buffer amplifier stage in a transmitter?

A: A buffer amplifier improves the frequency stability of the oscillator by isolating it from the load. It acts as a buffer by lessening the effect of the load on the critical oscillator output and presents a high-impedance load on the oscillator with little or no effect on circuit Q. Turning of the final amplifier, antenna circuit or swinging of the antenna could cause the oscillator frequency to shift without the buffer stage to separate it from such loading effects.

Q20: Draw simple schematic diagrams illustrating the following types of coupling between audio amplifier stages and between a stage and a load.
    (a) Resistance coupling.
    (b) Impedance coupling

A: (a) Resistance coupling between stages is shown in Fig. 4-13. R1, Rg and C form the coupling network. The reactance of C must be low at the lowest frequency amplified to avoid poor low-frequency response due to excessive loss across C.

(b) Impedance coupling is clarified by Fig. 4-14. The coupling network consists of L, C and Rg. Here again, proper frequency response depends on the ability of coupling capacitor C to pass the desired frequencies without loss.

Q21: What would probably be the effect on the output amplitude and waveform if the emitter resistor bypass capacitor in a common emitter audio stage were removed?

A: The bypass capacitor (Fig. 4-15) places the emitter at ground potential with respect to audio passing through the stage. The audio component must be removed from the emitter stage if the bias on the amplifier is to remain constant. Removing the bypass capacitor would cause the audio ac voltage to appear across the emitter resistor, thus causing a varying bias on the amplifier. Without a fixed bias point, the amplifier would no longer be capable

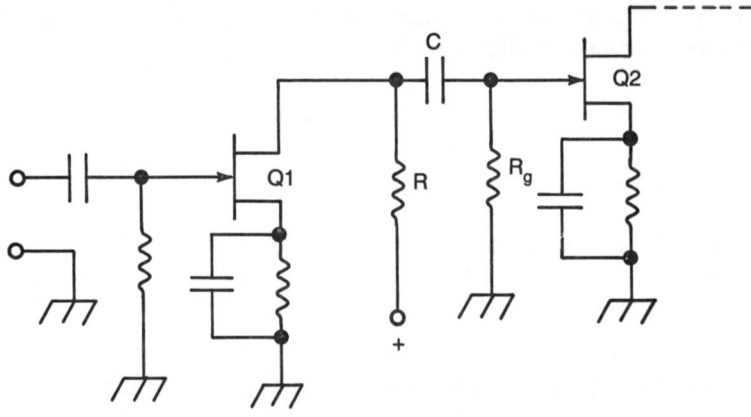

Fig. 4-13. Two-stage amplifier showing RC coupling.

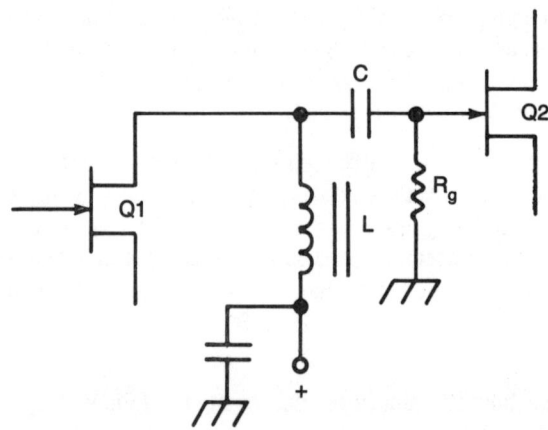

Fig. 4-14. Impedance coupling.

of operating in the center of its linear region. However, because current flow through an unbypassed resistor develops a voltage that varies at the same rate as the collector voltage, negative signal feedback is introduced. Negative feedback has a canceling effect, which negates any distortion introduced by the stage and reduces the overall amplitude drastically. Thus, while the operating point tends to drift, the amplitude of the signal will be low enough to keep it within the linear region, and distortion will be minimal.

Q22: Why do small-signal vacuum tubes produce random noise?
A: Random noise in vacuum tubes in generated by electron emission irregularities as the electrons are emitted from the cathode at random rather than in a smooth, continuous stream. This is known as random or shot noise and it results in thermal noise in the control grid circuit and partition noise from the variations in the division of screen and plate currents. Microphonic noise is caused by mechanical vibration of the cathode, grids, and plate inside the tube but is normally triggered by external sound waves.

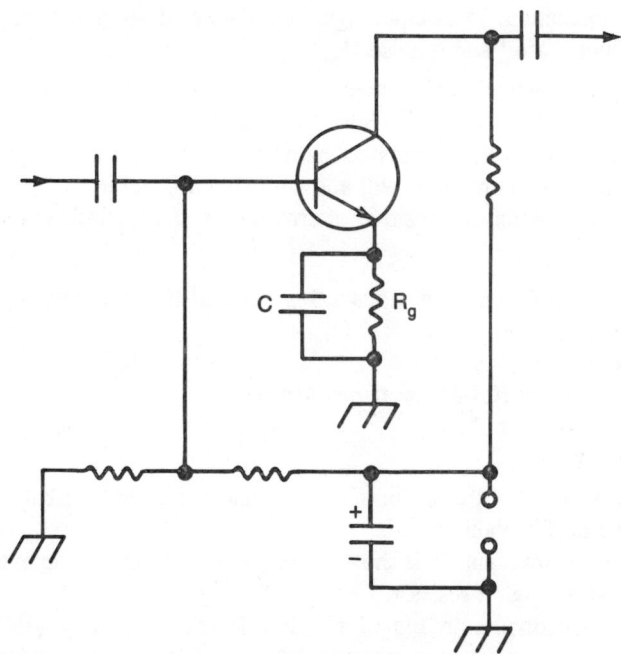

Fig. 4-15. Transistor amplifier stage.

Q23: Why are decoupling resistors and capacitors used in stages having a common power supply?
A: Decoupling networks prevent multistage amplifiers from oscillating due to unwanted signal voltage passing from one circuit to another through the common power source. Signal voltage variations from the output stage across the V+ or V− bus are fed back into the input stage, causing it to oscillate.

Q24: How would saturation of an output transformer create distortion?
A: Saturation causes the output transformer to operate in a nonlinear way, because an increase in the primary current no longer increases the flux, and the secondary signal is not an accurate pattern. Actually, the secondary response is flattened and severe audio distortion results. Excessive primary signals may drive the core flux to saturation and the dc drain (VFET) or collector (npn/pnp) current in the primary often adds to this primary signal, causing the core to saturate at a lower signal amplitude than it would otherwise. Push-pull operation permits much stronger signals to be handled before saturation, because dc currents cancel in the split primary winding. Summarizing, saturation of the output transformer reduces its inductance causing a lower load impedance and amplitude, especially at lower frequencies. Solid-state amplifiers are available that offer push-pull operation without transformers, so avoid this problem.

Q25: Why is noise often produced when an audio signal is distorted?
A: Noise is produced when an audio signal is distorted due to the presence of harmonics which, if sufficient in strength, cause the original signal to be noisy. When nonlinear audio amplification appears, the resulting amplitude distortion produces harmonics of the original

wave and intermodulation. The output signal consists of desired and undesired signals; the latter, of course, are "audio noise."

Q26: What factors determine the correct bias voltage for the base of a transistor?
A: In an audio amplifier, the bias is chosen so that the transistor operates as closely as possible to the center of its linear transfer characteristic during the no-signal state. When properly biased, a class A amplifier will allow maximum-level input signals to approach saturation of the transistor on positive alternations and approach cutoff on negative alternations.

Q27: Draw a schematic diagram illustrating each of the following types of biasing and explain its operation.
  (a) Battery
  (b) Power supply (tube-type transmitters)
  (c) Voltage divider
  (d) Source bias

A: (a) Figure 4-16 shows battery bias with a capacitor across the battery to furnish a low-impedance path for audio signals.

(b) Bias from a power supply is shown in Fig. 4-17. The center tap of the transformer returns to ground through a resistor.

(c) The voltage divider in Fig. 4-18 provides the necessary grid bias and this arrangement is suitable when two or more different voltages are required.

(d) A basic method of cathode bias is illustrated in Fig. 4-19 with the cathode returned to ground through a resistor. As drain current flows, a voltage drop across the resistor places the source positive with respect to the control grid. A capacitor across the resistors provides a low-impedance path for the signal voltage and prevents variations in the voltage drop.

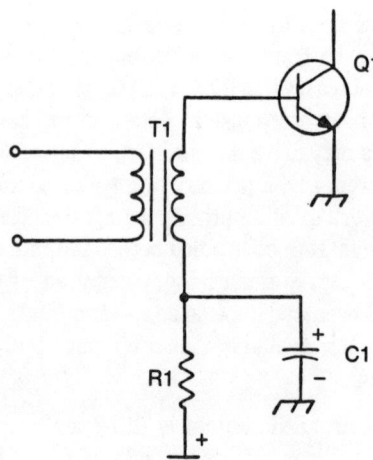

Fig. 4-16. Battery bias circuit.

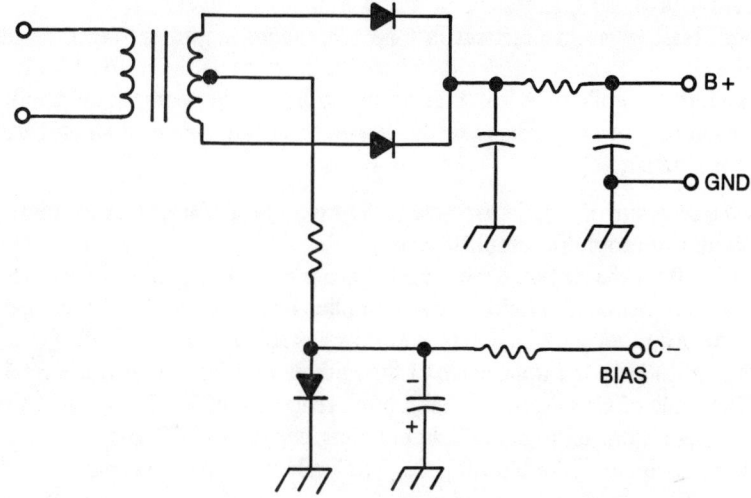

Fig. 4-17. Bias from a dc power supply.

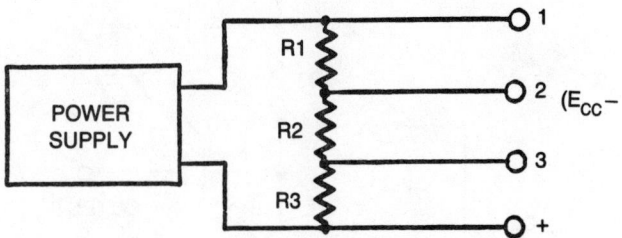

Fig. 4-18. Voltage-divider bias circuit.

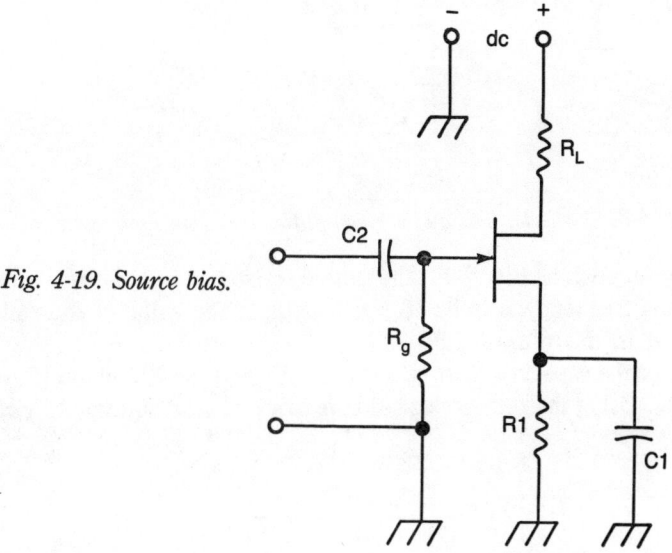

Fig. 4-19. Source bias.

Q28: Is grid-leak biasing practical in audio amplifier stages?
A: Grid-leak biasing is not practical in tube type audio amplifier stages because they normally operate without drawing grid current. A constant value of grid bias is required for audio amplifiers while grid-leak bias varies with the signal amplitude applied to the grid. This would cause more bias to be developed on strong signals than on weaker ones and distort the output.

Q29: Draw a diagram of a collector-base bias pnp audio preamplifier as might be used for audio microphone amplification.
A: See Fig. 4-20 for the circuit. Transistor Q1 is a pnp small-signal unit that is connected in the common-emitter configuration. Bias is supplied by the collector-base method through resistor R2, with emitter stabilization provided by resistor R1. The emitter bypass capacitor, C1, is used to keep the impedance of the emitter to ground low while retaining the dc level. The value of C1 is set such that it has a reactance of R1/10 or less at the lowest frequency of operation. Input and output coupling capacitors (C2 and C4, respectively) are used to pass signal while blocking dc. The collector load is provided by R3.

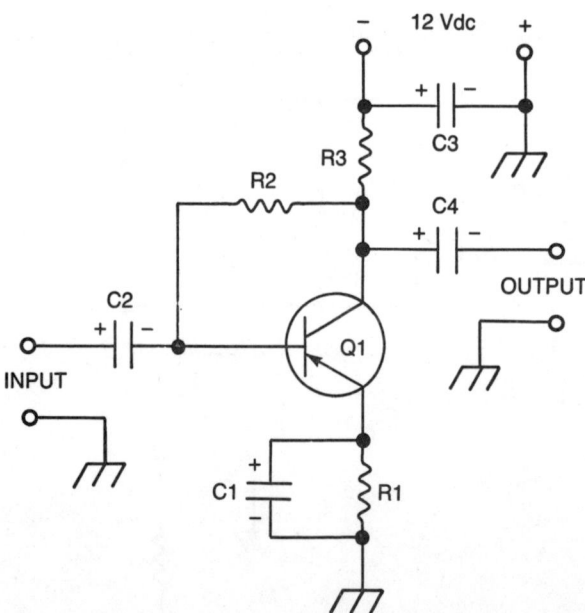

Fig. 4-20. Collector-bias audio preamplifier based on npn device.

Q30: In a circuit such as Fig. 4-20, the source resistance is 220 ohms and the lowest operating frequency is to be 30 Hz. Calculate the value of the emitter bypass capacitor in microfarads (μF).

A: The value of the capacitor must be such that X_{C1} will be 220 ohms/10, or 22 ohms, at 30 Hz. Using the standard equation rearranged (and adjusted to yield μF in the result):

$$C(\mu F) = \frac{1{,}000{,}000}{(2)\,(3.14)\,(F)\,(X_{C1})}$$

$$C(\mu F) = \frac{1{,}000{,}000}{(2)\,(3.14)\,((30\text{ Hz})\,(22\text{ ohms})}$$

$$C(\mu F) = \frac{1{,}000{,}000}{4145}$$

$$C(\mu F) = 240\ \mu F$$

Q31: Why does a class B audio-frequency modulator power stage require considerably more driving power than a class A amplifier?
A: Bias is adjusted in class A operation so that base current does not flow at any time and the input is a high impedance. Class B amplifiers are biased so that base current flows during the positive peak of each cycle of the input waveform, which represents appreciable power loss in the base circuit. Consequently, more driving power is required to overcome the current loss in the class B input.

Q32: Show by use of a circuit diagram two ways of using single-ended stages to drive a push-pull output stage.
A: Figure 4-21 shows an interstage transformer coupling a push-pull output stage to the driver with a center-tapped secondary winding. This provides signals to the push-pull bases that are 180 degrees out of phase as required. The inverter circuit in Fig. 4-22 provides signals from the single-ended stage that are 180 degrees out of phase to feed the push-pull stage.

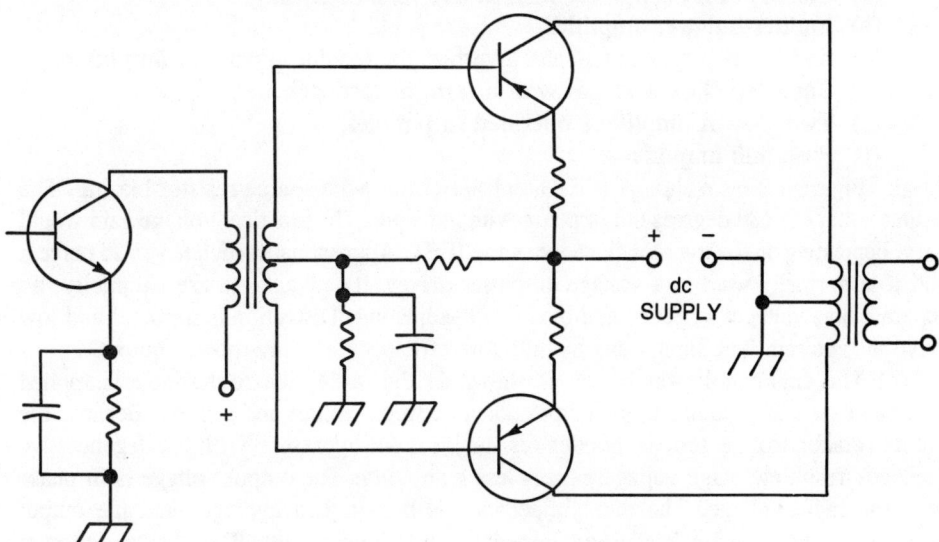

Fig. 4-21. Interstage transformer coupling.

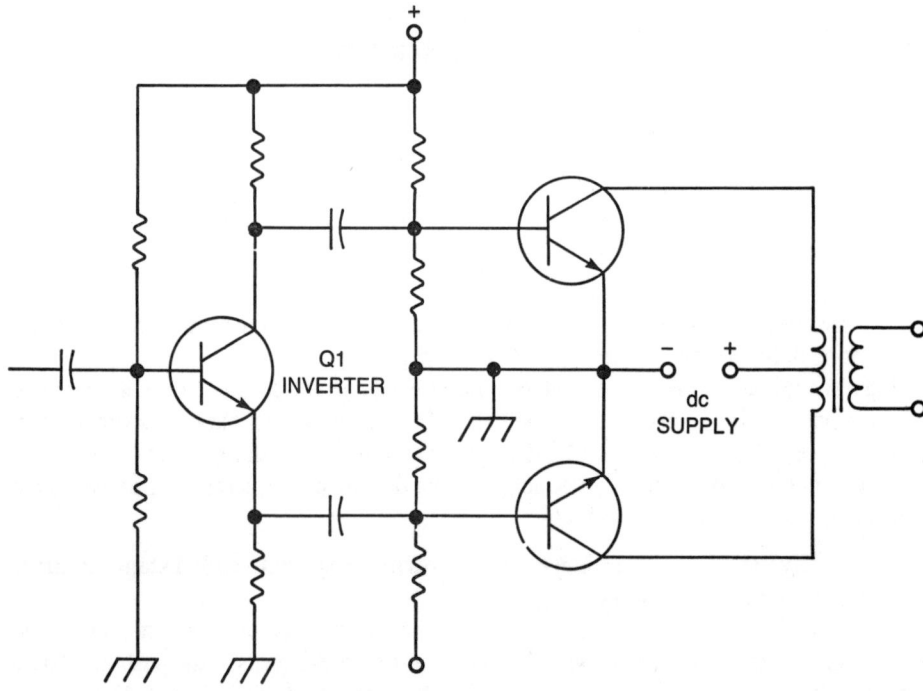

Fig. 4-22. Inverter circuit designed to feed push-pull output stage.

Q33: Draw circuit diagrams and explain operation (including input-output phase relationships, approximate practical voltage gains, approximate stage efficiency, uses, advantages, and limitations) of each of the following types of audio circuits.
(a) Class A JFET amplifier with source resistor biasing.
(b) Emitter-follower amplifier.
(c) At least two types of phase inverters for feeding push-pull amplifiers.
(d) Cascaded class A stages with a form of feedback.
(e) Two class A amplifiers operated in parallel.
(f) Push-pull amplifier.

A: (a) Figure 4-23 is a class A JFET amplifier circuit with source resistor biasing. The output voltage is 180 degrees out of phase with the input. The practical voltage gain would vary, depending on the transconductance of the JFET. Approximate efficiency is 25 percent and it is normally used as a voltage amplifier driver. Its advantages are simplicity, no separate bias voltage is needed and bias is self-adjusting. Distortion is minimal and low drive is required. The limitations include low efficiency and low power output.

(b) The emitter-follower circuit is shown in Fig. 4-24, where the input is applied between base and ground and the output taken between emitter and ground. Because the output is taken from across the emitter resistor, it is not bypassed. With the degenerative feedback resulting, stage voltage gain is less than unity. The output voltage is in phase with the input voltage. The input impedance is higher than average, and the output impedance low, making it a useful impedance-matching circuit. The emitter-follower operates as a class A amplifier; therefore, efficiency is usually about the same—25 percent.

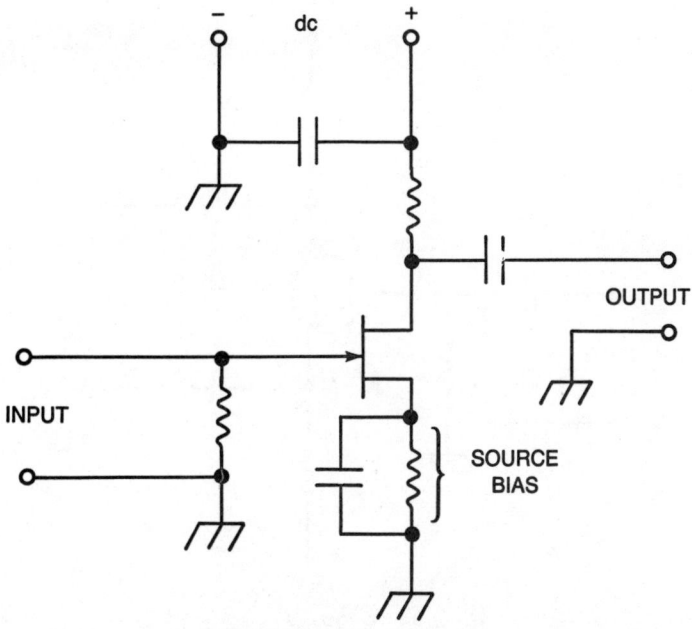

Fig. 4-23. JFET stage with source bias and source bypassing.

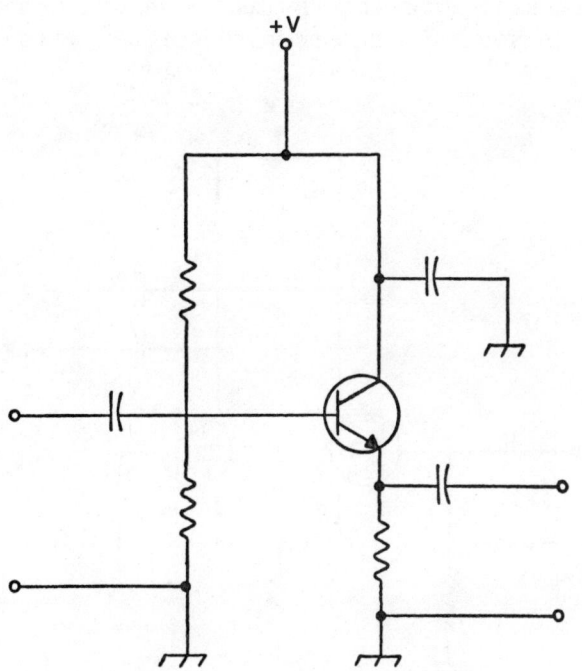

Fig. 4-24. Emitter-follower circuit.

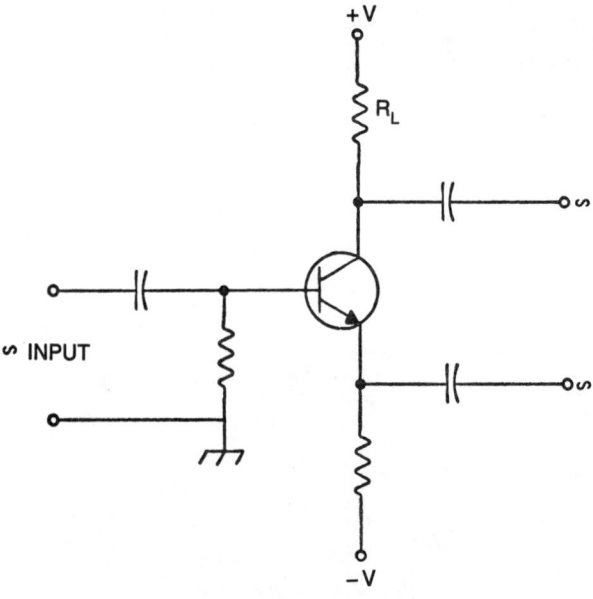

Fig. 4-25. Split-load inverter.

Applications include that of a matching or isolation stage between high- and low-impedance circuits and its only important limitation is that no power gain is offered.

(c) Figures 4-25 and 4-26 show two basic types of phase inverters. The first is a single-transistor split-load type. One output is taken from the collector and is 180 degrees out of phase, and the other is from the emitter and is in phase with the input. Voltage

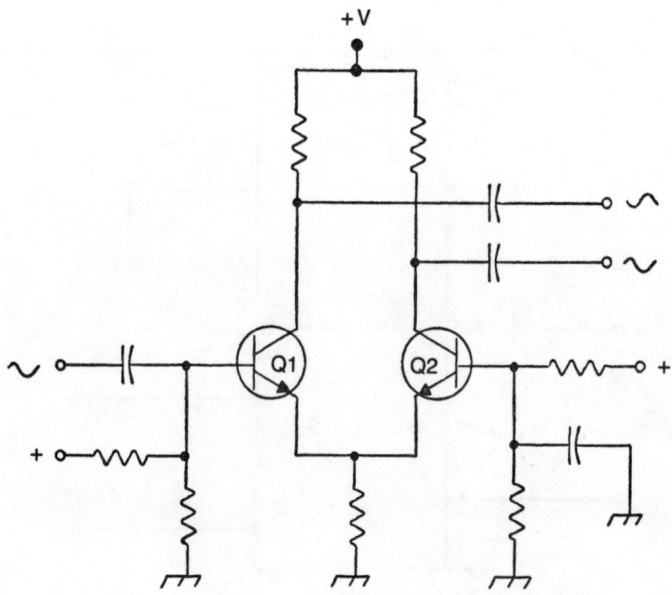

Fig. 4-26. Differential phase inverter.

gain is less than unity because of negative feedback, and the efficiency is 25 percent as usual with class A. Simplicity, excellent balance, and frequency response are advantages worthy of note, and a lack of voltage gain is the only significant disadvantage.

The second phase inverter (Fig. 4-26) is a differential type. The output of Q1 is 180 degrees out of phase with the input. Voltage gain is equal to about half the normal gain for each stage.

(d) Figure 4-27 is a cascaded two-stage amplifier with feedback. Because the signal has a phase reversal of 180 degrees in each stage, the output signal is in phase with the input signal after the double reversal. Voltage gain overall is equal to the product of the two stages, and because it operates class A, the efficiency is 25 percent. The popular application is a voltage amplifier with reduced distortion and good frequency response as a result of the feedback. The only limitation is that gain is somewhat reduced in comparison to a conventional circuit.

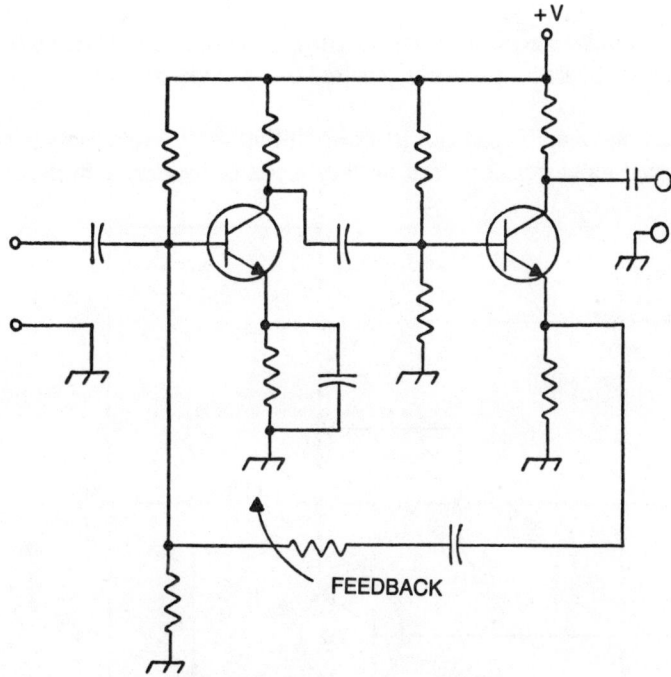

Fig. 4-27. Cascaded two-stage amplifier circuit.

(e) Figure 4-28 illustrated the use of two class A amplifiers in parallel. Operation is basically the same as with a single transistor. A varying input signal to the base results in corresponding collector current variation. There is 180-degree phase reversal from input to output, and power gain is obtained with the usual 25 percent efficiency. Power amplification is the usual application, and double the single-transistor output is obtained, but there is more distortion than would be found in the more popular push-pull circuit. The only possible advantage of the parallel hookup is the elimination of the phase splitter that is essential for push-pull. Disadvantages of consequence are the special output

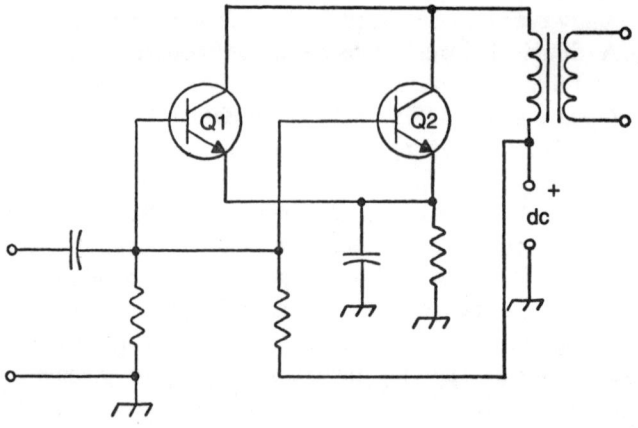

Fig. 4-28. Two class A amplifiers connected in parallel.

transformer requires double the dc collector current, there is no reduction in distortion, and a larger emitter bypass capacitor is required because the emitter resistor is lower in value.

(f) A push-pull amplifier circuit is shown in Fig. 4-29. Proper inputs are supplied by a center-tapped input transformer instead of a phase inverter. Power gain is double

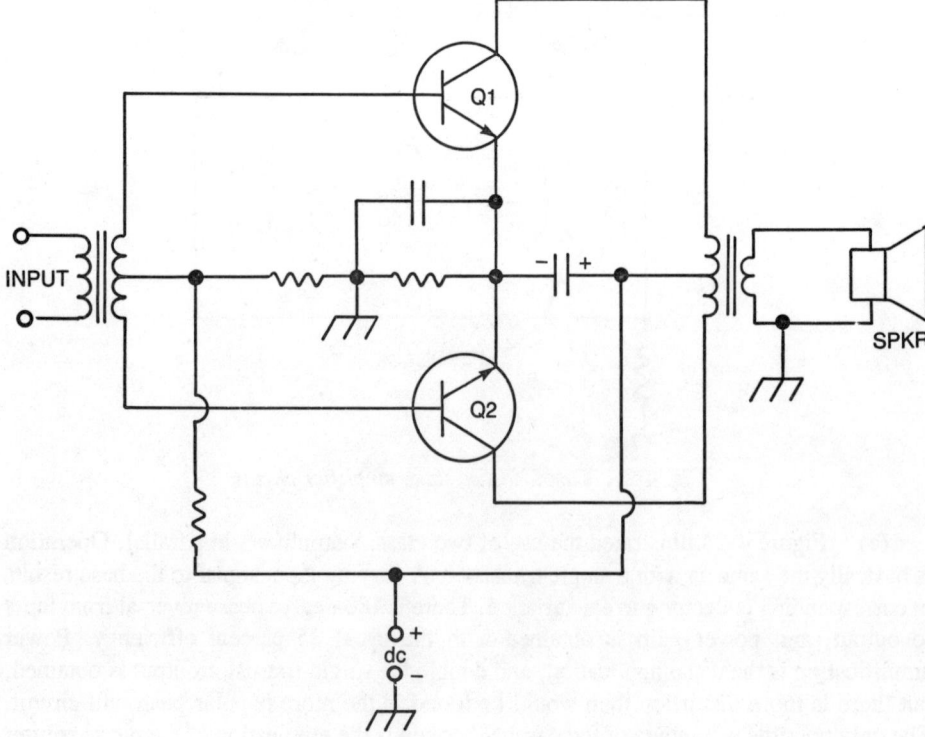

Fig. 4-29. Conventional push-pull amplifier.

that of a single device circuit, and efficiency is about 25 percent. Its most popular application is as a power amplifier in an audio output stage, and it displays many advantages in such an arrangement. Even harmonic distortion is eliminated, hum and regenerative feedback are reduced, there's no dc core saturation in the output transformer, and no emitter bypass capacitor is required. However, in the case of the latter, improvement is sometimes possible by using the bypass capacitor to compensate for a lack of balance in the transistors. Limitations are few. There might be a need for bias controls to ensure exact balance, plus matched transistors for improved operation, and the need for out-of-phase input signals.

Q34: Draw circuit diagrams and explain the operation of two commonly used tone-control circuits.

A: Two control circuits are presented in Fig. 4-30. The circuit shown in sketch A is a tone control for bass signals and that at B is the equivalent circuit for treble signals. Note that one circuit is the complement of the other in component placement. The first circuit (for bass signals) uses a high-pass filter network to pass high frequencies while shunting lower frequencies through an attenuator. Conversely, the circuit in B uses a low-pass filter so that low frequencies are passed to the amplifier stage directly while the high-frequency signal components are shunted to an attenuator.

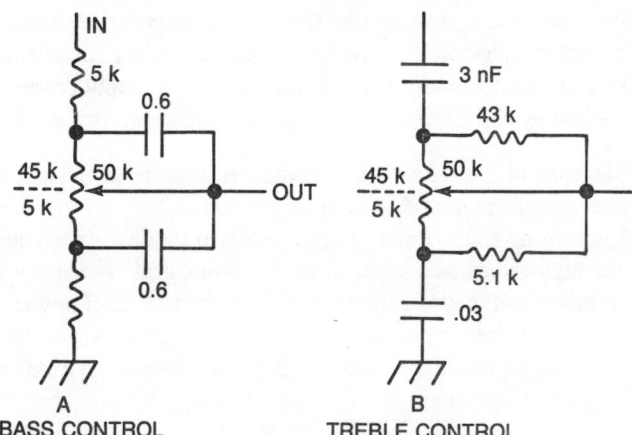

Fig. 4-30. Bass and treble controls.

Q35: Name some causes of self-oscillation in an operational amplifier hi-fi audio preamplifier with a gain of 1000.

A: The operational amplifier must be stabilized for use. In some cases, an "unconditionally stable" operational amplifier is used and only rarely oscillates spuriously. These amplifiers achieve unconditional stability by sharply reducing the gain bandwidth product of the device. Regular op-amps, however, must be frequency compensated in order to achieve stable (non-oscillating) behavior. Assuming that layout problems don't exist, look for +V and −V bypassing; use a 0.1 μF and 4.7 μF (or higher) capacitor on each lead. Place the 0.1 μF capacitor as close to the op-amp terminal it bypasses as possible. Next, use the lag or lead compensation techniques to tailor the frequency response of the device.

Q36: What factors should be taken into consideration when ordering a class A audio transformer; a class B audio output transformer feeding a speaker of known ohmic value?

A: Considerations for class A audio output transformers include: Normal operating power rating, direct current in the primary, frequency response, primary and secondary impedance, and shielding.

Factors for class B include the above, plus the fact that class B audio amplifiers require a center-tapped secondary for the input transformer and a center-tapped primary for the output transformer because push-pull operation is always used. In push-pull amplifiers, lead inductance in the output transformer primary must be held to a minimum to avoid distortion.

The importance of sufficient power-handling capability cannot be emphasized enough, because overheating will result in a breakdown of insulation and permanent damage. Even a slight overload in signal power can cause distortion on peaks. In single-ended class A stages, the dc flowing in the primary must never exceed the transformer's rated specifications, or core saturation will result. The frequency response of the transformer is marked by a drop-off in extremes at the low and high ends with a flat response between the two. Needless to say, the quality of the amplifier of system will only be as good as the ability of the transformer to respond to the frequencies involved, and the curve should be carefully noted when making a selection. The primary impedance should match or load the amplifying device coupled to it properly and the secondary impedance must match the speaker impedance exactly to afford maximum transfer of output power. The magnetic and electrostatic shielding of the transformer should be sufficient for the intended location.

Q37: Draw a diagram of a single-button carbon microphone circuit, including the microphone transformer and source of power.

A: In Fig. 4-31, a step-up transformer is used to match the low impedance of a carbon microphone to the high impedance input of an amplifier grid. A battery in series with the transformer primary and the microphone provides the source of power. As the sound vibrations produce a variation in pressure on the carbon granules, the resistance of the button is varied accordingly, resulting in changes in current flow in the transformer primary.

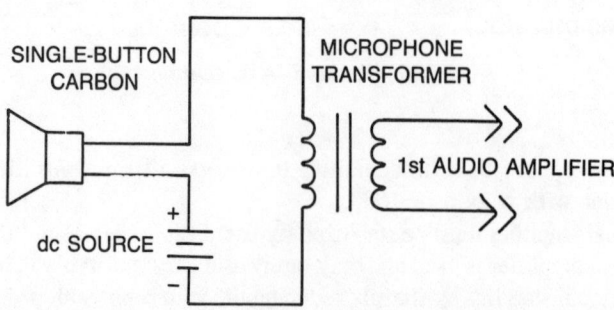

Fig. 4-31. Carbon microphone circuit.

Q38: If low-impedance headphones in the order of 8 ohms are to be connected in the output of a VFET amplifier, how should this connection be made to permit most satisfactory operation?

A: The drain circuit of a VFET is one characterized by a very high impedance value; 8 ohms represents a low value. If the drain of the VFET were coupled directly into the headphones, an impedance mismatch would occur that might serve to introduce distortion and compromise circuit efficiency. One approach is to use a transformer whose primary impedance matches the characteristics required by the VFET and whose secondary impedance is 8 ohms.

Q39: Describe the construction and explain the characteristics of a crystal-type microphone; of a carbon-button type microphone.

A: Crystal Type. When a quartz crystal is subjected to a mechanical pressure, the crystal produces a minute alternating voltage that can be measured with ordinary test equipment. Similarly, a crystal can produce mechanical energy when an electrical signal is applied to it. (This is called the *piezoelectric effect*.) Because the electrical energy emitted by the crystal is proportional to the mechanical energy required to drive it, the crystal is indeed an energy converter—or *transducer*.

The crystal responds to sonic vibrations by producing a voltage that reverses polarity in exact correspondence with the physical movement; low-frequency sounds produce voltages that alternate slowly, and high-pitched sounds produce output voltages that shift polarity with high frequency. Sound vibrations are coupled to the crystal by means of a metallic diaphragm. The electrical signals generated by the crystal can be used to drive a high-impedance amplifier directly.

The frequency response of the crystal mike is not uniform. This is due to the inertia of the crystal itself; however, for applications such as amateur radio or any other noncritical application, they are widely used. Because of its high impedance, the crystal microphone can be directly connected to the input of the grid circuit of a speech amplifier. The output of the crystal microphone is on the order of about -55 dB, although this figure tends to vary from manufacturer to manufacturer, depending on quality of microphone and other factors. The diaphragm type has a frequency response of 80 to about 6000 Hz. The type of crystal most widely used is Rochelle salt, because its sensitivity is somewhat higher than that of natural quartz.

Carbon Type. If a piece of pure carbon were split up and made granular, the resulting pile would possess a value of specific resistance and would pass a current if the pile were enclosed in an insulated container. If a variable pressure is applied to the pile, its density changes. If the source of the pressure variations is the human voice, the compressions and rarefactions of air applied to the carbon pile cause its resistance to vary at an audio rate. The variable-resistance characteristic of the carbon pile is the underlying principle on which the operation of the microphone is based.

In the single-button carbon microphone, the granules are placed in a cup-like button and are permitted to make contact with the suspended perpendicular element called the "diaphragm." If a stress is placed on the diaphragm, the pressure exerted on the carbon granules is increased and the resistance of the carbon pile decreases. Relaxation of the pressure restores the button's resistance to its original value.

The microphone button is placed in series with a battery, as shown in Fig. 4-31. Any current flowing in the microphone circuit will also flow through the primary of the transformer.

Q40: What precautions should be observed when using and storing crystal microphones?

A: Ordinary crystal microphones using Rochelle-salt crystals should be stored in a cool, dry place and not subjected to mechanical vibrations or physical abuse. Even though a measure of protection against moisture is provided by the wax seal, moisture-proof wrapping is standard procedure for storage. Protection from high temperatures is also mandatory during storage, because the salt crystal will not withstand temperatures in excess of 120 degrees F. The ceramic type can be subjected to much higher temperatures with no problem.

Q41: What is an RFC? Why is it used?

A: RFC is the abbreviation for *radio-frequency choke*. Such chokes are used to prevent the flow of (choke out) radio-frequency signals in a dc circuit. Depending on the value of the RFC, the device will impede the flow of RF at certain frequencies without interfering in any way with the passage of direct current through the series-connected winding.

Q42: What are the advantages of using a small-value resistor in series with the cathode of a class C RF power amplifier tube to provide bias?

A: The use of a cathode resistor in a class C RF power amplifier protects the tube against damage in the event excitation is lost or interrupted. If grid-leak bias only is used in the stage, a loss of grid drive due to interruption of the input signal will reduce grid bias to zero, causing excessive dc plate current to flow and destroy the tube. By having a portion of the bias obtained from the cathode resistor, bias will never be reduced to the danger zone because the dc plate current flowing through the cathode will result in a voltage drop across the resistor and produce some negative bias. Because plate current only flows during a small portion of the cycle in class C operation, the voltage drop across the cathode resistor is available only during that period and might supply only a portion of the grid bias as a safety measure.

Q43: What is the difference between RF voltage amplifiers and RF power amplifiers in regard to bias?

A: The RF voltage amplifier is normally operated class A but the RF power amplifier is in class B or class C. Power amplifier devices are often larger to dissipate and properly handle the additional power.

Q44: Describe a grounded-grid RF power amplifier and explain its operation.

A: In a grounded-grid RF amplifier with the signal fed to the cathode and the output taken at the plate, the grid at RF ground acts as a shield between the input and output circuits, thus reducing cathode-plate capacitance to an insignificant value and eliminating the need for neutralization. Although more input power is required with this circuit, most of it appears in the plate circuit as output power. The triode is less noisy than the tetrode or pentode.

Q45: Explain the principle involved in neutralizing an RF stage.

A: Neutralization in a radio-frequency amplifier involves the reduction to an absolute minimum, by cancellation, the signal transferred between input and output circuits through grid-to-plate capacitance, thus eliminating the tendency toward self-sustained oscillations. By feeding back a voltage from the output circuit to the input circuit about equal the opposite in phase to that fed through the grid-plate capacitance, the voltages cancel and the stage is neutralized. Grounded-grid power amplifiers rarely require neutralization.

Q46: Explain, step-by-step, at least one procedure for neutralizing an RF amplifier stage.

A: There are three common neutralization methods—plate, grid, and push-pull. The step-by-step procedure for plate neutralization, sometimes referred to as the Hazeltine method, follows:

1. Remove plate voltage from the tube so that any signal appearing in the plate circuit will be due to grid-plate capacitance.

2. Tune the preceding stages and the grid of the stage to be neutralized for maximum signal.

3. Connect or loop-couple an RF indicator to the plate tank circuit (oscilloscope, ac-VTVM, neon bulb, or even a flashlight bulb attached to a loop of a few turns of wire).

4. After tuning the plate tank to resonance or maximum indication, adjust the neutralizing capacitor until minimum RF is indicated. Retune the grid circuit for any additional RF indication, and recheck the plate tuning for any possible increases. Then, retouch the neutralizing capacitor setting for minimum or zero reading, after which the stage can be considered to be neutralized. In the event that the final indication is not satisfactory, the complete procedure should be repeated (preferably with a more sensitive indicator).

The undesirable effects of oscillation in the RF amplifier are many, such as possible tube failure from overheating, excessive plate current, component damage, generation of spurious frequencies, and modulation distortion.

Q47: What class of amplifier is appropriate for an RF double stage?

A: A class C amplifier is best because its output is rich in harmonics.

Q48: Draw a circuit diagram of a push-pull frequency multiplier and explain its principal of operation.

A: Figure 4-32 illustrates a push-pull multiplier with the bases connected in push-pull and the collectors connected in parallel. The balanced base circuit applies out-of-phase voltages to the bases of the transistors. The base with the positive voltage conducts while the other with a negative voltage is cut off. When the excitation impulse is reversed, the cut-off transistor receives the positive pulse at the grid and conducts while the other transistor receiving the negative is promptly cut off. As the collectors are connected in parallel, a single cycle of the input frequency produces two cycles in the collector circuit. Thus, the frequency is automatically doubled, and the output circuit can be tuned to any *even* harmonic. In the push-pull circuit, the fundamental and all *odd* harmonics are eliminated.

Q49: Push-pull frequency multipliers normally produce what order of harmonics—even or odd?

A: Push-pull frequency multipliers generally produce odd-order harmonics and operate class C for best results. Because collector current output is badly distorted, even harmonics are canceled because the output from each device is opposite in phase.

Q50: State some indications of and methods of testing for the presence of parasitic oscillations in a transmitter.

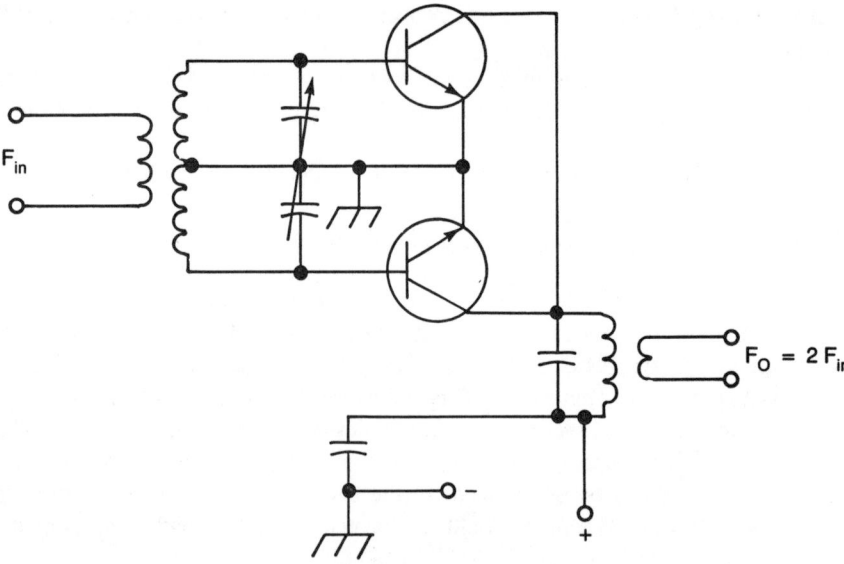

Fig. 4-32. Push-pull frequency multiplier circuit.

A: Not all combinations of input and output circuits can be used together successfully, because some of them permit the amplifier stage to oscillate at frequencies relatively unrelated to the frequency to which it is tuned. These parasitic oscillations are distinct from the sort of oscillation that occurs in an amplifier that is improperly neutralized or one in which the input circuit is not shielded sufficiently from the output. They are undesirable because they cause the transmission of spurious signals, thus impairing the efficiency of the amplifier.

The most noticeable features of parasitic oscillation in an amplifier are erratic tuning and the radiation of spurious (unwanted) frequencies. When an amplifier is operating properly, the dc plate current dips sharply as the tank circuit is tuned through resonance. This plate-current minimum also corresponds to maximum power output. If a tetrode is operating normally, the plate-current change may not be too great, but the screen current's dip will be significant. With parasitic oscillation, the plate current might not dip at all; the minimum might not correspond to maximum power output; or several dips could appear in the tuning range. Because the symptoms presented by a stage that is not properly neutralized are somewhat similar, it is difficult to tell the two effects apart unless neutralization is checked first.

Q51: Draw a schematic diagram and explain the operation of a **harmonic generator** stage.

A: A harmonic generator, or frequency multiplier stage, is shown in Fig. 4-33. The circuit uses two tuned resonant tank circuits, L1/C1 and L2/C2. The input tank circuit, L1/C1, is tuned to the input frequency, F1, while the output tank circuit is tuned to the output frequency which is an integer multiple of F1. In other words, F2 = NF1 where N is an integer (i.e. 2, 3, 4, 5 etc.). All waveforms except the pure sine wave are harmonic rich. Distorting the sine wave produced across L1/C1 generates a harmonic-rich, non-sinusoidal waveform. The function of diode D1 is to generate the distortion that creates harmonics

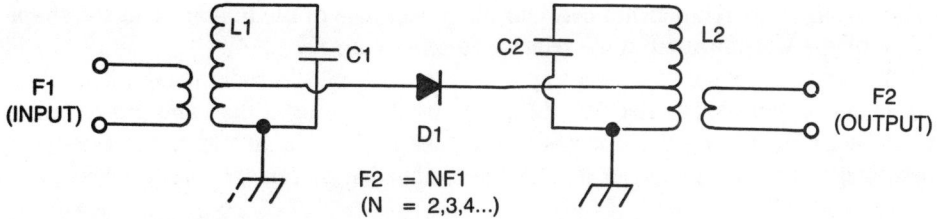

Fig. 4-33. Frequency multiplier or harmonic generator circuit.

that can be used to "ring" the output tank circuit. There is loss in this circuit, so some designers replace diode D1 with a single unbiased RF transistor in the common-emitter configuration.

Q52: What is the meaning of the term carrier frequency?
A: The frequency of the unmodulated RF carrier wave as assigned for transmission.

Q53: If a carrier is amplitude modulated, what causes the sideband frequencies?
A: Whenever modulation of the carrier takes place, as in AM, there are new frequencies present at the transmitter output in addition to the carrier frequency. These are known as *sidebands* and are formed by the distorting of the sinusoidal carrier wave to produce a nonsinusoidal wave that is equal to the sum of three sinusoidal waves—carrier frequency, carrier plus audio frequencies and carrier minus audio frequencies. Because many audio frequencies are present in the complex speech waveform, a group of sideband frequencies are produced with an upper sideband above the carrier frequency and a lower sideband below the carrier.

Q54: What determines the bandwidth of emission for an AM transmitter?
A: The actual width of the channel of frequencies of a radiated AM signal is equal to twice the highest audio modulating frequency. For an AM station with audio modulation at frequencies from 50 to 20,000 Hz, the upper sideband extends 20 kHz above the carrier and the lower sideband 20 kHz below it. Assume the carrier frequency as 600 kHz; the upper sideband would extend to 620 kHz and the lower to 580 kHz. Thus, the bandwidth in this exaggerated example would be 40 kHz. Once again, it should be noted that the term cycles per second (cps) comes up time after time and could pop up often in your FCC examination, so it is important to use it when asked to do so and at the same time remember that it is synonymous with hertz (Hz).

Q55: Why does exceeding 100 percent modulation in an AM transmitter cause excessive bandwidth?
A: It is first important to understand that some waveforms are more efficient harmonic generators than others. Square waves, in particular, are very effective in creating harmonic signals. An overmodulated AM signal produces a modulation envelope that is clipped off at the negative half-cycle. This clipped waveform resembles a square wave and is rich in harmonics of the modulating signal. Because the harmonic of a signal appears at a multiple of the fundamental signal—and because the bandwidth of a carrier is a function of the frequency content—it stands to reason that an overmodulated signal produces a larger proportion of sidebands over a broader range of frequencies than a 100-percent-modulated RF carrier.

Q56: What is the relationship between the percentage of modulation and the shape of the waveform envelope relative to carrier amplitude?

A: The amplitude of the transmitter output varies according to the audio voltage; it increases above the unmodulated carrier level during the positive swing of the audio and decreases below it during the negative excursion. The peak modulation amplitude expressed as a percentage of the actual carrier level is the modulation percentage, the formula:

$$A_m = \frac{M \times A_c}{100} = \frac{A_m}{A_c} \times 100$$

A_c is the level of the carrier, A_m is the maximum increase or decrease, and M is the modulation percentage.

During 100 percent modulation, the transmitter output drops to zero on negative peaks in the audio cycle and, because increasing modulation above 100 percent can't drive the transmitter output below that, the zero time is extended, and the envelope flattened during negative peaks.

Q57: Draw a simple schematic diagram showing a method of coupling a modulator tube to an RF power amplifier tube to produce grid modulation of the amplified RF energy. Compare some advantages or disadvantages of this system of modulation with those of plate modulation.

A: Figure 4-34 is a schematic showing a modulator tube connected to an RF amplifier to produce control grid modulation. The modulator output is coupled in series with the

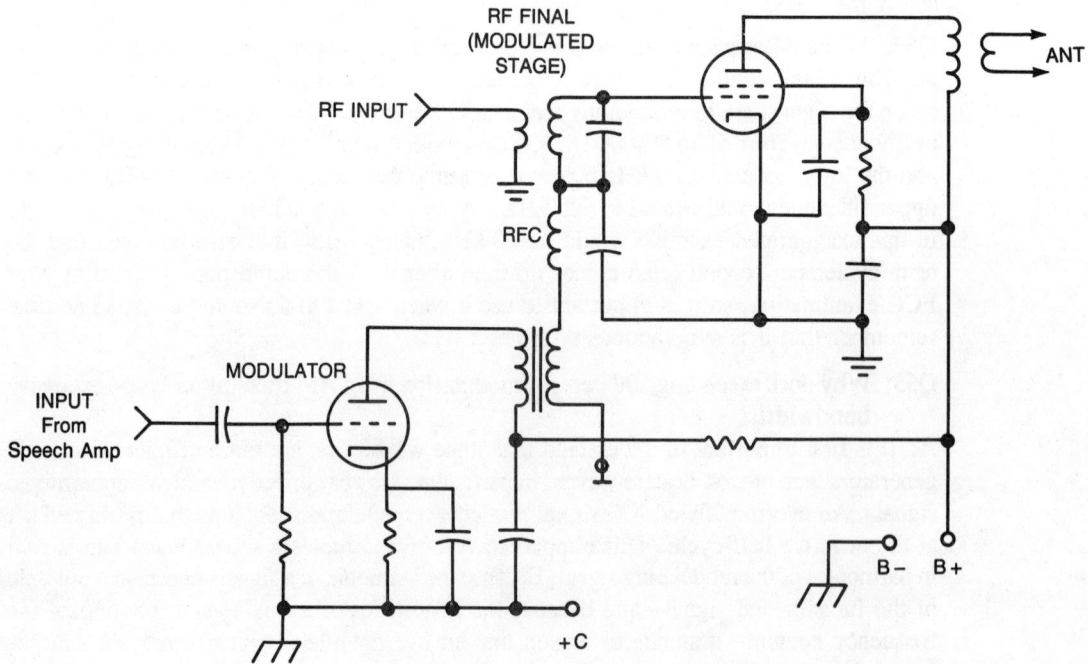

Fig. 4-34. Schematic of the connection used for grid modulation.

amplifier grid bias by the modulation transformer. Much less power in the modulator is needed for this form of modulation, thus there is a considerable saving realized with a smaller tube and modulation transformer. The disadvantage of note is lower efficiency because the RF output from the power amplifier is much lower with the same amount of power drawn from the supply as would be required for plate modulation. Unless grid modulation is very carefully handled, appreciable distortion will result.

Q58: What is the relationship between the average power output of the modulator and the plate circuit input of the modulated amplifier under 100 percent sinusoidal plate modulation? How does this differ when normal voice modulation is employed?

A: In order to get 100 percent modulation with a sine wave, the modulator must be capable of supplying 50 percent as much power as is normally required by the modulated power amplifier plate. If the power amplifier draws 500 watts, the modulator would need 250 watts of audio power to modulate it 100 percent with a sine wave. A speech wave has a lower average power, and because peaks might not exceed 100 percent, less average power is required from the modulator. The modulator supplies power for the sidebands and the dc supply the power for the carrier.

Q59: What is the relationship between the amount of power in the sidebands and the intelligibility of the signal at the receiver?

A: The greater the amount of power in the sidebands, the more intelligible the signal is at the receiver. Unfortunately, it is impossible to get more power into the sidebands than an amount equal to 50 percent of the unmodulated carrier . . . that is, at 100 percent modulation. Sideband power decreases as modulation percentage decreases. Total RF power decreases and the signal distorts when modulated percentage is increased beyond the 100-percent point.

Q60: What might cause FM in an AM radio-telephone transmitter?

A: Dynamic instability resulting from a common power supply, and if the modulator is not operating class A, undesired frequency modulation can result, because the oscillator frequency will shift with changes in loading.

Q61: What is meant by *frequency shift* or *dynamic instability* with reference to a modulated RF emission?

A: Frequency shift in an AM transmission is a deviation from the desired carrier frequency as a result of changes in the dc plate current of an RF oscillator. In severe cases, particularly where the AM transmitter is mobile and the plate current is affected by variations that are attributable to insufficient isolation between stages or oscillator instability, frequency drift can be extreme and rapid enough to appear as FM.

Q62: What would cause a dip in the antenna current when AM is applied? What are the causes of carrier shift?

A: This condition, commonly experienced by radio amateurs, is referred to as "downward modulation." It can be caused by a number of problems, including:

    1. Inadequate regulation on a power supply used to provide voltages to all transmitter stages.

    2. Insufficient RF drive to the final amplifier.

3. Soft tube in the final amplifier stage.

4. Inability of the final RF amplifier to handle peak currents of twice the normal unmodulated value (that could be attributable to a wide variety of problems, including poor design).

5. Modulation transformer problems, including arcing because of excessive voltage or RF leakage.

Q63: Explain the principles involved in a single-sideband suppressed-carrier (SSBSC) emission. How does its bandwidth and required power compare with that of full carrier and sidebands?

A: The principle of single-sideband suppressed-carrier (SSBSC) emission is to transmit only one sideband while suppressing the carrier and one of the sidebands. Each sideband carries the same information, so only one is needed, and the carrier contains no intelligence, thus most of the power in AM transmission is wasted. Because one third of the power in conventional AM is in the sidebands with two thirds in the carrier, only one sixth of the power to the antenna is necessary for communication. In SSB transmission, all the power is fed into one sideband, which affords a considerable power advantage. The single-sideband carrier requires only one half of the bandwidth of the AM signal and offers a gain of many dB over AM. Noise pickup would naturally be reduced by one half at the same time.

Q64: Draw a block diagram of a SSBSC transmitter (filter type) with a 20-kHz oscillator and emission frequencies in the range of 6 MHz. Explain the function of each stage.

A: A filter type SSBSC transmitter block diagram is illustrated in Fig. 4-35. The 100 kHz oscillator generates a reference signal that goes to a balanced modulator. Combining the usual audio (30 to 10,000 Hz) with the 100 kHz oscillator signal balances the carrier, and only the two sidebands remain. These are applied to the sideband filter, which passes only the upper sideband while eliminating the lower.

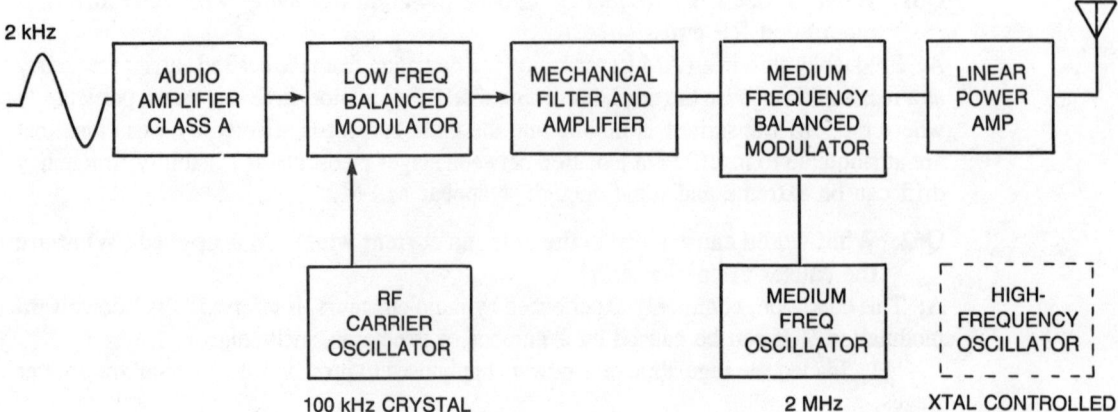

Fig. 4-35. SSB transmitter block diagram.

Q 65: Explain briefly how an SSBSC emission is detected.

A: The SSSBC signal is detected by mixing the low-frequency local oscillator output with the single-sideband output of the last IF amplifier in the second detector. An AM signal is thus produced and detected as usual. Although single or double conversion can be used, the single-sideband receiver differs primarily in the way detection takes place. The reinserted carrier at the receiving point must have the identical frequency of the transmitter carrier so that normal AM detection is possible.

Q66: Draw a block diagram of a single-conversion superheterodyne AM receiver. Assume an incident signal and explain briefly what occurs in each stage.

A: The block diagram of a typical superheterodyne receiver is shown in Fig. 4-36. Below corresponding sections of the receiver are shown the waveforms of a signal at the point. The RF signal from the antenna passes first through an RF amplifier where the amplitude of the signal is increased. A locally generated unmodulated RF signal of constant amplitude is then mixed with the carrier frequency in the mixer stage. The mixing (heterodyning of these two frequencies produces an intermediate frequency (IF) signal that contains al of the modulation characteristics of the original signal. The intermediate frequency is equa to the difference between the station frequency and the oscillator frequency associated with the mixer. The intermediate frequency is then amplified in one or more IF amplifiers and fed to a detector for recovery of the audio signal. The detected signal is amplified in the AF section and then fed to a headset or loudspeaker.

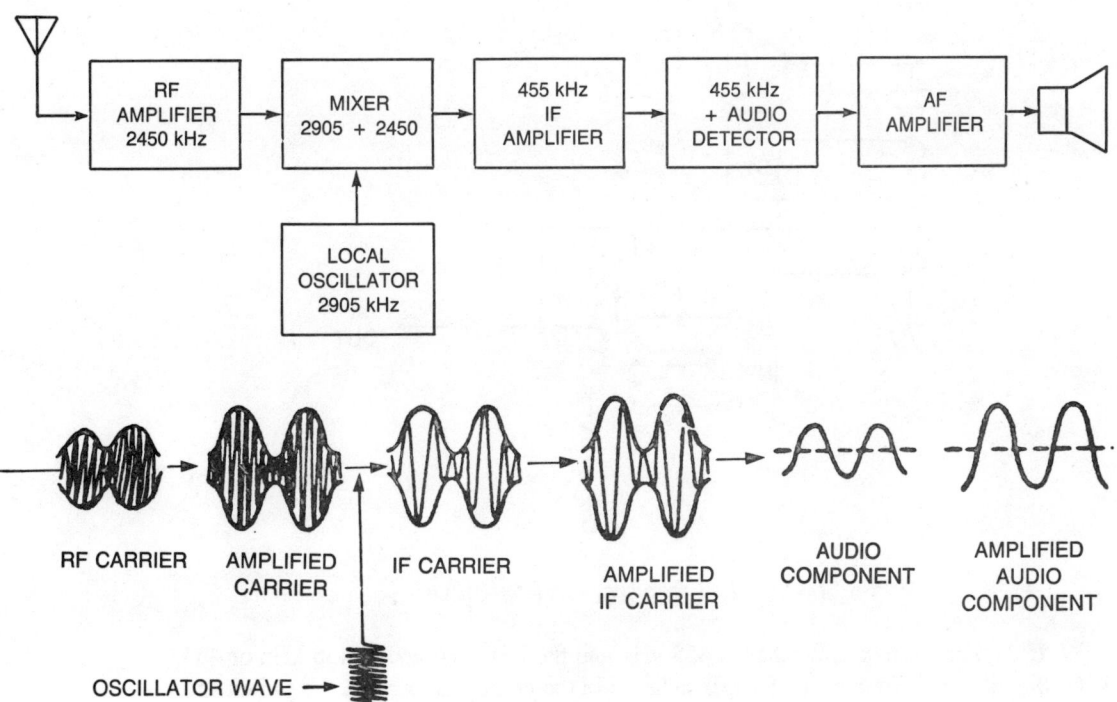

Fig. 4-36. Block diagram of a superheterodyne receiver.

Q67: Explain the relation between the signal frequency, oscillator frequency, and the image frequency in a superheterodyne receiver.

A: If the oscillator is tuned above the incoming frequency, it equals the sum of the incoming frequency and the intermediate frequency. The *image frequency* is the incoming frequency plus twice the intermediate frequency. When the local oscillator is below the incoming frequency, as in VHF, the oscillator frequency is equivalent to the incoming minus the intermediate frequency. The image frequency now appears at the incoming frequency minus two times the intermediate frequency.

Q68: Draw a beat frequency oscillator (BFO) circuit diagram and explain its use in detection.

A: The beat frequency oscillator (BFO) is necessary when CW signals are to be received, because these signals are not modulated with an audio component. In superheterodyne receivers, the incoming CW signal is converted to the intermediate frequency at the first detector as a single frequency signal with no sideband components. The intermediate-frequency signal is heterodyned (with a separate tunable oscillator known as the beat frequency oscillator) at the second detector to produce an AF output. In the circuit shown in Fig. 4-37, the Hartley oscillator (BFO) is coupled to the detector diode by capacitor C3.

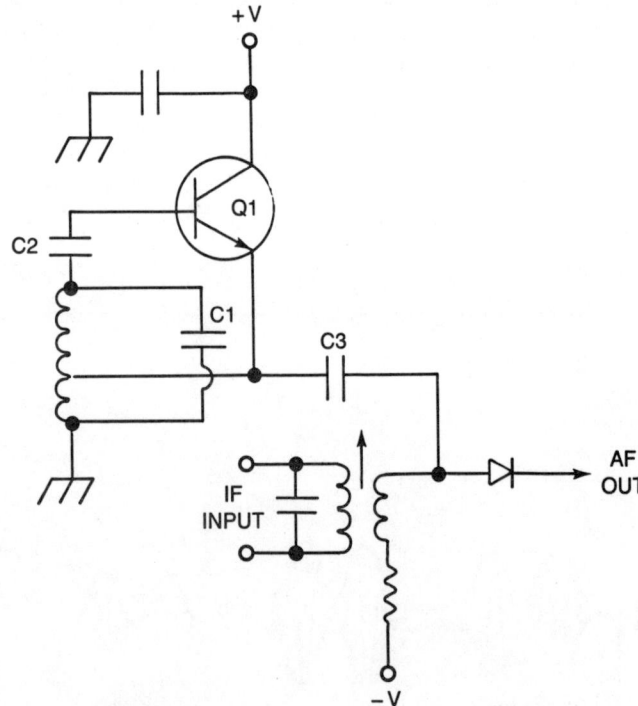

Fig. 4-37. IF beat frequency oscillator (BFO).

If the intermediate frequency is 455 kHz and the BFO is tuned to 456 kHz or 454 kHz, the difference frequency of 1 kHz is heard in the output. Generally, the switch and capacitor tuning control are located on the front panel of the receiver.

The BFO should be shielded to prevent its own output from being radiated and combined with desired signals ahead of the second detector. If AVC voltage is to be used, it should be obtained from a separate diode isolated from the second detector. One way is to couple the output of an IF amplifier stage ahead of the second detector to the AVC diode. Otherwise, the output of the BFO would be rectified by the second detector and would develop an AVC voltage, even on no signal.

Q69: Explain, step-by-step, how to align an AM receiver using the following instruments. In addition, discuss what is occurring during each step.
 (a) Signal generator and speaker.
 (b) Signal generator and oscilloscope.
 (c) Signal generator and VTVM.

A: The alignment procedure is the same regardless of the type of indicating instrument used. The signal generator must be AM modulated and should be reasonably well calibrated, especially if the receiver to be aligned is badly out of adjustment. Although the speaker can be used to indicate maximum response of the tuned circuits in a preliminary way, small signal changes are just about impossible to detect by ear. This is where the oscilloscope or VTVM prove quite useful.

The oscilloscope should be connected across the voice coil of the speaker. A VTVM connected between AVC (also called AGC) and ground is satisfactory for the dc type or across the speaker voice coil if an ac type meter is used. Maximum indication on any device used signifies correct alignment at that point. The usual IF frequency is 455 kHz, so the signal generator should be set to this frequency and connected to the grid of the last IF stage through a .01 μF capacitor. Using the minimum output necessary, peak the primary and secondary windings of the IF transformer for maximum indication on the meter or scope.

Now, move the generator to the input of the next to last stage of the IF and peak that transformer for maximum, just as was done on the last stage. Continue by moving the generator to the input of the IF stage being aligned until the first stage has been completed. It is necessary to keep reducing the generator output as alignment progresses, always maintaining the lowest level possible to ensure proper performance of the amplifiers.

The IF transformer coupling the output of the mixer to the input of the first IF amplifier is the last to be aligned and the signal generator should be connected to the mixer input in this case. It should be remembered that the generator output will have to be increased at this point because the mixer is not tuned to the IF frequency.

After completing the alignment of the IF section, proceed with RF alignment by connecting the signal generator to the antenna terminal through a small series capacitor and ground the other side to the chassis. Begin with a setting at 1400 kHz (both the generator and receiver), and adjust the oscillator trimmer for maximum indication on the scope or VTVM. Continue by adjusting first the mixer trimmer and then the RF trimmer for peak response. Reset the tuning dial to 600 kHz and the generator to the same frequency. Adjust the oscillator and low-frequency trimmer for maximum reading. Repeat the complete RF alignment procedure for possible improvement of your original settings. As each circuit is adjusted for resonance at the proper settings, stage gain should reach maximum and the receiver will be completely aligned.

Q70: What are the advantages and disadvantages of a bandpass switch on a receiver?
A: The quality of a received signal is proportional to the bandwidth of that signal, because higher frequencies require wider passbands. Narrowing the bandpass of a receiver allows a receiver's selectivity to be increased substantially, but it often causes intelligibility of a received signal to deteriorate, particularly when the frequencies contained in the received signal are relatively high. It must be remembered that all the intelligence of an AM signal is contained in the sidebands, and the sidebands are displaced from the carrier frequency by an amount of spectrum that is equal to the audio frequency being transmitted.

Q71: Explain sensitivity and selectivity of a receiver. Why are these important quantities? In what typical units are they usually expressed?
A: *Sensitivity* is the strength of a signal, usually measured in microvolts at the receiver input terminals, capable of producing a certain audio output. Its importance stems from the fact that it gives the receiver a merit rating according to its ability to perform adequately under weak signal conditions. *Selectivity* is the ability of the receiver to separate the selected frequency distinctly from adjacent undesirable frequencies and is usually expressed in Hz or kHz, which is the bandwidth, and measurement can be quoted in points where response is down a definite number of dB in relation to the IF center frequency (usually 6 dB and 60 dB).

5

Basic Radiotelephone: Part III

At this point in this book, most of the basic material has been covered and the study ahead should come much easier as reference is made to the material previously discussed. The first subject in this chapter is *frequency modulation* (FM).

FREQUENCY MODULATION SYSTEMS

Because static, lightning, and other electrical disturbances create amplitude-modulated interference to a radio signal, frequency-modulated waves can be received with comparative freedom from such annoyances. The carrier of the FM transmitter remains at its center or resting frequency without modulation. As modulation is introduced, however, the carrier varies in frequency by swinging higher and lower on either side of the normal center or resting point. Meanwhile, the amplitude of the wave does not vary, and the antenna current of the FM transmitter is unchanged with modulation.

The transmitter is constructed so that the frequency swing of the carrier is proportional to the amplitude of the modulating voltage. The frequency variation of the carrier is in step with the audio voltage, so the rate of carrier swing is actually equal to the modulating voltage. The *reactance modulator* varies the frequency or phase of the oscillator signal in the FM transmitter in step with the audio modulating voltage. Frequency multipliers enable a small deviation in the oscillator frequency to cause a much larger deviation in the output frequency. If you shout into the FM microphone, the deviation or swing of the carrier becomes greater, but as you raise the *pitch* of your voice, the *rate* of deviation becomes greater. FM receiver alignment is detailed in the Q & A section along with the correction of interference problems that can occur with mobile installations.

ANTENNAS

An antenna is an electrical conductor that radiates or picks up electromagnetic and electrostatic fields. It varies in size according to the operating frequency and may be mounted horizontally or vertically. The polarization of the radio wave is horizontal if the electrostatic field is traveling parallel to the surface of the earth, or vertical if it is traveling perpendicular to the earth's surface. The polarization of the wave depends on the *position* of the radiating antenna; i.e., vertical radiators produce vertical polarization, and horizontal radiators produce horizontal polarization. The receiving and transmitting antennas must be polarized or positioned similarly for best results. FM and television waves are usually horizontally polarized because the annoying reflections from tall buildings and man-made types of interference are not as likely to cause difficulties as with vertical polarization. The characteristics of several antennas are discussed later.

TRANSMISSION LINES

A *transmission line* is simply a means of transferring RF energy from the source to the load with maximum efficiency. Because transmission lines are made up of two conductors, inductance results. The lines also have capacitance, but neither the inductance nor capacitance results in a power loss. The real losses are due to RF resistance, leakage paths across the lines, and radiation. A line terminated in its characteristic impedance acts like a line of infinite length regardless of its actual length. Thus, it can be termed an *untuned line* because its action does not depend on its length. Untuned lines have numerous advantages and are frequently used to transfer RF because of the absence of standing waves. Current-handling capability is maximum and power losses are much smaller. The line performs equally well at all frequencies.

If a line is terminated in a value other than its characteristic impedance, the wave could be partially or completely reflected from the far end rather than being absorbed by the load.

FREQUENCY MEASUREMENTS

There are several methods of rough frequency measurement, including the simple dip meter, which is merely a transistor oscillator with a low-range current meter (mA or μA) to indicate collector current. The absorption frequency meter has a parallel resonant circuit that is inductively coupled to the circuit to be measured. It indicates a maximum reading on the milliammeter when tuned through resonance.

The heterodyne frequency meter offers excellent accuracy in determining frequency and uses a calibrated variable frequency oscillator (VFO) and a nonlinear mixer amplifier. Its output contains the input frequencies as well as the sum and difference. Because the difference frequency is in the audio range, earphones or a speaker can be used to monitor the output.

A vernier scale assists in reading a dial with much greater accuracy; it is merely a short scale, graduated so that 11 of its divisions are equal to 10 divisions of the scale being read. By sliding the zero of the vernier scale to the large scale point between the divisions being read, the decimal reading is indicated by the point where a small-scale division coincides with large-scale division.

An FM deviation meter consists of an FM receiver with a good peak-reading voltmeter to sample the FM transmitter output and combine it with the local oscillator in the mixer to produce the difference frequency. The deviation meter reads FM modulation in terms of deviation from carrier frequency.

FORMULAS

The following is a summary of the most important formulas for electronic calculations.

SINUSOIDAL SIGNALS ONLY

Effective value: 0.707 × peak
Average value: 0.637 × peak (for half-cycle)
Peak value: 1.414 × effective
Effective value: 1.11 × average
Peak value: 1.57 × average
Average value: 0.9 × effective

RESONANT FREQUENCY

$$F_r = \frac{1}{2\pi\sqrt{LC}} \quad \text{or} \quad \frac{.1592}{\sqrt{LC}}$$

$$L = \frac{1}{4\pi^2 F^2 C} \quad \text{or} \quad \frac{25,330}{F^2 C}$$

$$C = \frac{1}{4\pi^2 F^2 C} \quad \text{or} \quad \frac{25,330}{F^2 L}$$

REACTANCE

$$X_C = \frac{1}{2\pi FC} \quad \text{or} \quad C = \frac{1}{2\pi F X_c}$$

$$X_L = 2\pi FL \quad \text{or} \quad L = \frac{X_L}{2\pi F}$$

X   in ohms
C   in farads
L   in henrys
F   in hertz

QUESTIONS AND ANSWERS

Q1: Draw a schematic diagram of a frequency-modulated oscillator using a varactor modulator. Explain its principle of operation.
A: Figure 5-1 is a varactor FM oscillator which is one basic way of producing direct frequency modulation. Frequency is variable by adjusting the main tank circuit capacitor. With the modulator connected to the Hartley-type oscillator, the varactor functions as ca-

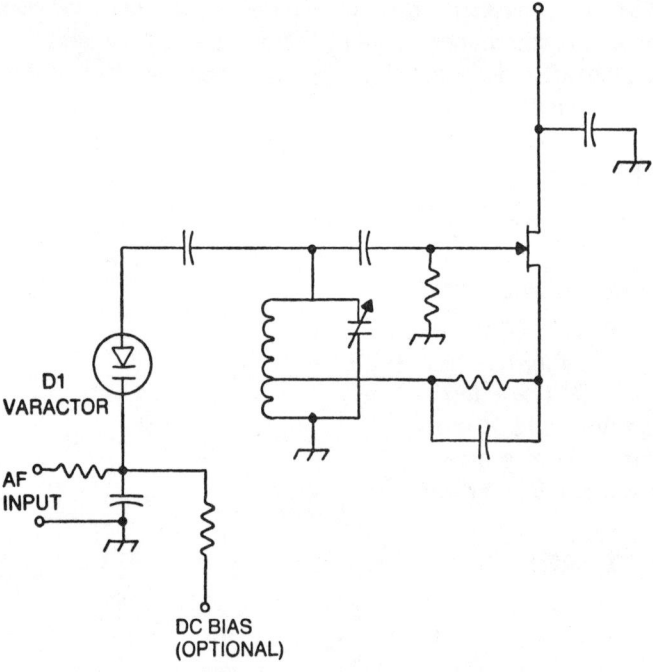

Fig. 5-1. FM oscillator circuit.

pacitance in parallel to the regular tank capacitor. Therefore, the oscillator frequency is dependent on the sum of the two capacitance values. Because the varactor capacitance varies with the audio amplitude, the frequency of the oscillator will vary according to the audio modulating input.

Q2: Discuss the following with reference to frequency modulation:
  (a) The production of sidebands.
  (b) The relationship between the number of sidebands and the modulating frequency.
  (c) The relationship between the number of sidebands and the amplitude of the modulating voltage.
  (d) The relationship between percent modulation and the number of sidebands.
  (e) The relationship between the modulation index or deviation ratio and the number of sidebands.
  (f) The relationship between the spacing of the sidebands and the modulating frequency.
  (g) The relationship between the number of sidebands and the bandwidth of emission.
  (h) The criteria for determining bandwidth of emission.
  (i) Reason for pre-emphasis.

A: (a) Sidebands are those groups of modulating frequency components arranged above and below the carrier. In AM, the upper sideband is formed by the modulating frequencies plus the carrier frequency; the lower sideband is formed by the modulating frequencies subtracted from the carrier frequency. The carrier frequency is constant in amplitude modulation and only the amplitude of the transmitted wave is varied according to the modulating voltage. In FM, the modulating signal causes the carrier frequency to vary in frequency according to the frequency of the modulation. As modulation results in slightly higher or slightly lower excursions of the carrier frequency, sidebands are produced by this distortion of the carrier wave. The actual number of sidebands produced in the case of FM varies according to the degree of carrier deviation or swing; the greater swing produces the greater number of significant sidebands. However, sidebands at greater distances from the center frequency are too weak to be considered.

(b) The number of sidebands is directly proportional to the modulation index and inversely proportional to the modulating frequency. Adjacent sidebands are separated by the frequency of the modulating signal. For example, consider an FM transmission on 108 MHz with a 5 kHz audio tone modulation. Sidebands would be produced above the center frequency (108 MHz) at 108.005, 108.010 and 108.015 MHz and below the center frequency at 107.995, 107.990, and 107.985 MHz. The lower audio frequencies form more sidebands near the center carrier frequency and as a result, these sidebands are more significant in amplitude. Because higher audio frequencies produce sidebands farther away from the center frequency, they carry much less energy and can usually be ignored.

(c) The number of sidebands bears a direct relationship to the amplitude of the modulating voltage. As the modulating voltage is applied to the reactance modulator, it causes the oscillator frequency to vary. The higher that voltage, the greater the deviation and the greater the number of sidebands produced.

(d) The relationship between the percentage of modulation and the number of sidebands is governed by the fact that the higher the percentage of modulation, the greater the frequency swing. This, in turn, results in a greater number of sidebands of significant amplitude. Therefore, the number of sidebands is directly proportional to the modulation percentage. Regular FM broadcast service at 100 percent modulation has a bandwidth of 150 kHz or 75 kHz above and below the carrier (center) frequency.

(e) The number of sidebands is directly proportional to the modulation index. The higher that index or deviation ratio, the greater the number of significant sidebands. The modulation index is equivalent to the carrier frequency deviation divided by the audio modulating frequency causing that deviation.

(f) The relationship between the spacing of the sidebands and the modulating frequency produces a spacing equal to the frequency of the audio modulation. If the sidebands are spaced 2 kHz apart, you could assume that the audio modulating frequency is 2000 Hz.

(g) The bandwidth of emission, which must contain 99 percent of the radiated power, depends on the number of significant sidebands multiplied by the modulating frequency. Although sidebands exist beyond this width, their power is small enough to be ignored.

(h) The criteria for determining the bandwidth of emission are the modulation index and frequency, which determines the frequency deviation. The bandwidth of emission is twice the sum of the frequency deviation and modulating frequency. Bandwidth of emission is always wider than the frequency swing as can be noted by FCC regulations for commercial transmitters operating below 450 MHz. Authorized bandwidth is 20 kHz, frequency deviation 5 kHz, or a frequency swing 10 kHz. Remember that the authorized bandwidth contains 99 percent of the power, and any sideband having at least ¼ of 1 percent of the total radiated power must be considered as part of that bandwidth.

(i) The reasons for pre-emphasis can be explained by the fact that very little energy is contained in higher frequencies of the audio range, and as a result they can be lost in the noise. Their importance should not be underestimated, because they improve speech quality and add to the identity of various musical instruments. Pre-emphasis is amplifying the high frequencies more than the low before modulation and enables them to override the noise. Overmodulation is not caused by this amplification, because the additional amplification of highs does not raise their signal level above that of the lower. The opposite of pre-emphasis used in an FM transmitter is de-emphasis in a receiver to bring the amplitude of the higher frequencies down to its normal level with regard to the lower frequencies.

Q3: How is good stability of a reactance modulator achieved?
A: An automatic frequency control (AFC) circuit is used to maintain the carrier frequency within tolerance. In the AFC system, the phase detector receives signals from the modulated oscillator and the crystal reference oscillator for comparison. In the event the modulated oscillator differs, a dc error voltage is produced that is fed to the modulator. The error voltage corrects the error by bringing the modulated oscillator to its proper frequency at which time the error voltage is reduced to zero. As long as the modulated oscillator remains on frequency, the error voltage is zero, but by drifting in either direction, an error voltage is developed that biases the modulator and causes the master oscillator to return to its center frequency. When the modulated oscillator is locked on center frequency, the phase detector output will not be zero because it varies at the audio rate with an average value of zero. In order to prevent this output from upsetting the modulated oscillator frequency through the modulator, an RC low-pass filter is connected between the phase detector and reactance modulator, which prevents the audio output of the phase detector from reaching the modulator.

Q4: Draw a diagram of a phase modulator, explain its operation, and label adjacent stages.
A: An example of a phase modulator is shown in Fig. 5-2. The modulator is placed at the output of the frequency generation circuits, be they phase-locked loop or simple crystal oscillator. In actuality, this circuit shunts the signal line between the frequency generator and the stages that follow (usually multipliers).

Two vector components of the signal are at the drain of transistor Q1. One is the direct signal through capacitor C3, and the other is the signal that is phase-shifted through Q1. The Q1 signal is phase-shifted 180 degrees, but its amplitude varies proportionally to the audio signal at the input. The two signals combine vectorially at the drain of the FET to produce the phase variations required for PM.

The principal reason that this is considered superior for most applications is that the carrier frequency can be easily generated in a well-controlled signal source such as a crystal oscillator or a PLL.

The varactor oscillator method is useful but tends to produce deviation figures that are different for different crystals. This means that its major use is in single-frequency transmitters. The reactance modulator, on the other hand, yields more consistent results when frequency changes are desired.

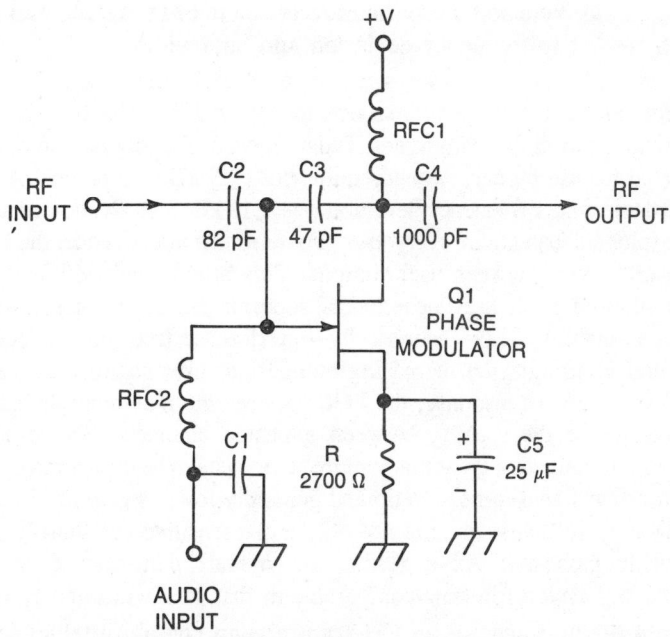

Fig. 5-2. Reactance modulator for indirect FM.

Q5: Explain, in a general way, why an FM deviation meter (modulation meter) will show an indication if coupled to the output of a transmitter which is phase-modulated by a constant-amplitude constant audio frequency. To what will this deviation be proportional?

A: The meter reading is proportional to either the frequency or phase change. Phase modulation always produces what looks to the meter as a frequency swing, and the meter reading is proportional to the amplitude and frequency of the modulation. Phase modulation of the same amplitude will read just twice as much for a 4 kHz signal as it will show for 2 kHz on the meter. The FM deviation meter, like the FM superhet receiver, uses a local oscillator, mixer, IF amplifier, and discriminator. If the transmitter is not modulated, the discriminator reads zero, providing the transmitter is not off frequency. However, center frequency drift will result in an output voltage from the discriminator, and this voltage is applied to the modulation meter through a rectifier and dc amplifier. This provides instant response to the slightest deviation voltage. Normally, the modulation meter is calibrated in percent and decibels; 100 percent or 0 dB usually indicates maximum authorized swing.

Q6: Explain briefly what occurs in a waveform when it is phase modulated.

A: Shifting the RF waveform in phase by audio modulation in a phase-shifting network causes the phase to be varied smoothly from the original. If the network phase shift causes it to lag the original wave, the resulting wave takes a longer time to reach its peak. As its wavelength is increased, frequency is decreased. If the phase shift leads the original, it peaks sooner and wavelength is decreased with frequency.

Q7: Discuss wideband and narrowband reception in FM voice communication systems with respect to frequency deviation and bandwidth.

A: Wideband FM communication systems are a "luxury" of the past. Because of increased demands for more and more spectrum space, the two-way radio bands have been undergoing a "squeezing" process for many years. Today, most of the popular communications bands are restricted for use by narrowband equipment (± 5 kHz deviation). A few years ago, the standard transmitter frequency deviation was ± 15 kHz, which allowed a total bandwidth of 30 kHz (plus sidebands) for every user. To minimize interference, the FCC maintained a spacing of 50 kHz between user stations. This broad bandwidth and wide deviation capability allowed such user benefits as superior signal-to-noise ratio, high-fidelity transmission capability, and—theoretically—interference-free communications capability.

Several years ago, the increasing numbers of user stations dictated the need for additional spectrum. In response, the FCC lowered the maximum deviation standard to ± 5 kHz and split the spacing between allocated channels. The bandwidth of user transmitters thus halved, a great many more stations have been accommodated. The early problems of high signal-to-noise ratio and generally lower signal level have been solved by the advent of receivers that are not only more sensitive but sharply selective within the allowable passband. As a result, the overall performance of a narrowband communications systems in now comparable to that of a wideband system.

Q8: Could the harmonic of an FM transmission contain intelligible modulation?

A: Because there is harmonic radiation of the carrier, there is harmonic radiation of the sidebands. Adjacent sidebands are separated by the harmonic multiple of the audio modulating frequency that makes the intelligence less realistic and subject to much distortion if the receiver is not able to pass the wider bandwidth. This could be two or three times as wide as the fundamental, depending on the harmonic.

Q9: Explain briefly the principles involved in frequency-shift keying (FSK). How is this signal detected?

A: Frequency-shift keying, known as F1 emission, provides keying of the radio-telegraph transmitter by changing the output frequency when the key is depressed in place of turning it off and on. By connecting a reactance modulator across the master oscillator and keying the former, the resonant frequency is changed (about 850 Hz) as the system is keyed. The oscillator operates at a low frequency, and because the frequency shift is small, good stability is maintained. Two levels are used, the *mark* at 425 Hz above the carrier, and at 425 Hz below is the *space*. The method of detection is similar to that required for regular FM—a ratio detector or discriminator tuned to the center frequency. A communications receiver could also be used for reception of FSK by tuning as for CW and adjusting the BFO so the audio note varies accordingly.

Q10: Under what conditions of maintenance and/or repair should a transmitter be retuned?

A: A transmitter should be retuned after any change has been made to the transmitter's circuitry, power supplies, or load. Replacing a tube in a high power RF amplifier usually requires retuning, because each tube has slightly different characteristics from all other tubes.

Q11: What might be the effect on the transmitted frequency if a tripler stage in an otherwise perfectly aligned FM transmitter became slightly detuned?

A: The output signal might be reduced in amplitude.

Q12: If an indirect FM transmitter without modulation is within carrier frequency tolerance but with modulation it is out of tolerance, what would be some possible causes?

A: The indirect (Armstrong) system of FM depends on a balanced modulator to generate sidebands, and if the transmitter is out of tolerance only during modulation, the modulator must be out of balance. This would cause unequal sidebands to be produced that in turn would cause the average frequency to be raised or lowered. The actual trouble could be a defective modulator device, capacitor, or excessive drive.

Q13: In an FM broadcasting transmitter, what would be the effect on antenna current if the grid bias on the final power amplifier tube were varied?

A: High-quality FM transmitters are operated class C, which means that the final power amplifier is biased well below the stage's cutoff point. If the grid bias were increased (operating point established at a more negative position), antenna current would decrease so long as the drive signal were held constant. In this case, distortion would increase, but circuit efficiency would be improved; a high-power signal would be required to drive the amplifier into its saturation region. If the grid bias decreases (operating point brought closer to tube cutoff), antenna current will increase if no change is made to the excitation signal, but the amplifier's efficiency will diminish and the tube will run much hotter than it should.

Q14: Draw a schematic diagram of each of the following stage of a superheterodyne FM receiver. Explain the principles of operation. Label adjacent stages.
  (a) Mixer with injected oscillator frequency.
  (b) IF amplifier.
  (c) Limiter.
  (d) Discriminator.

A: (a) Figure 5-3 is a mixer circuit with base injection. As the transistor operates nonlinearly, the two frequencies heterodyne to form the desired IF to the tuned converter plate circuit.

  (b) The IF amplifier circuit shown in Fig. 5-4 operates class A with the tuned circuits resonant at the selected IF frequency. The typical receiver has two or three stages of IF with a bandwidth of about 150 kHz.

  (c) A limiter stage is illustrated in Fig. 5-5. Its function is to remove amplitude variations from the IF signal prior to detection in the discriminator. This stage removes most of the noise (which is usually amplitude modulated) and, being a sharp cutoff type, operates with low collector voltage, which ensures limiting on even the weakest signals or noise.

  (d) A discriminator schematic appears in Fig. 5-6. A discriminator produces a conventional audio output from the frequency variations of the FM input. The discriminator also supplies AVC and AFC voltages. As the FM carrier deviates in one direction, the dc output voltage increases; it decreases as the carrier swings the other way. Because the broadcast FM wave deviates exactly with audio modulation, the dc variations from the

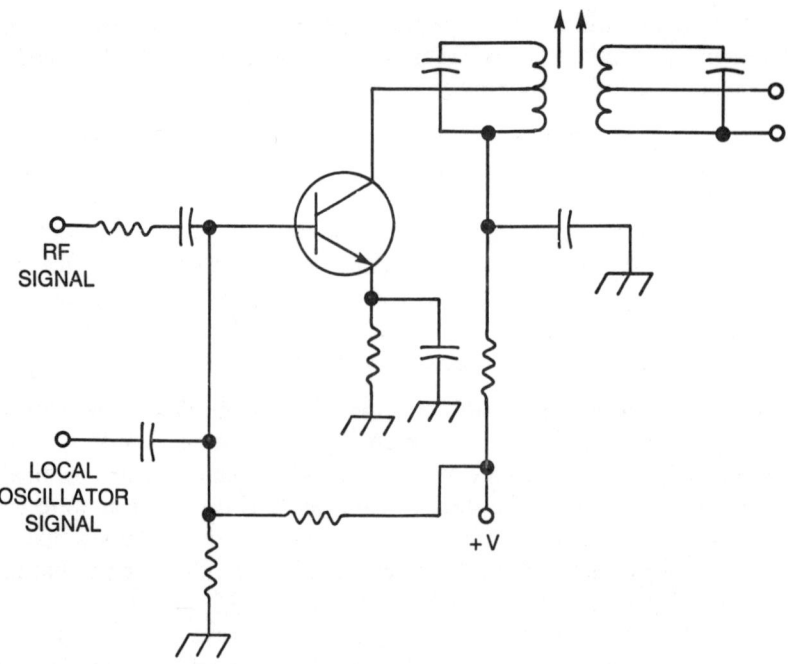

Fig. 5-3. Transistor mixer with base injection of LO signal.

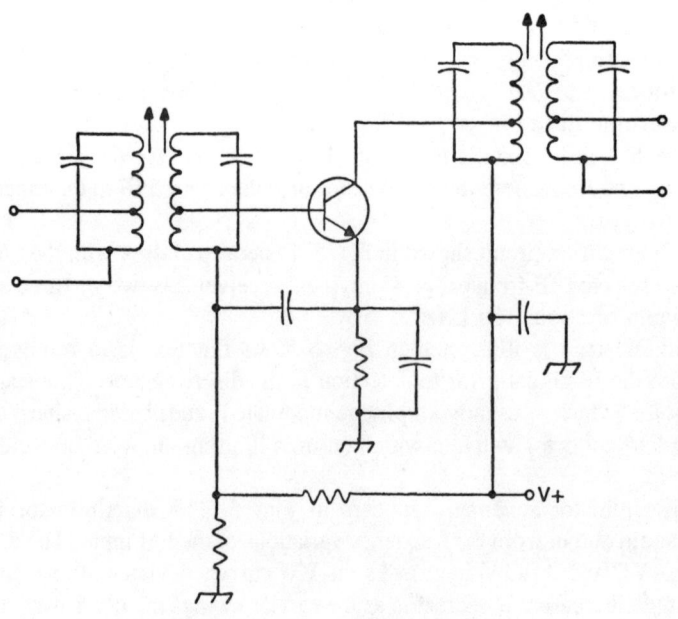

Fig. 5-4. IF amplifier circuit.

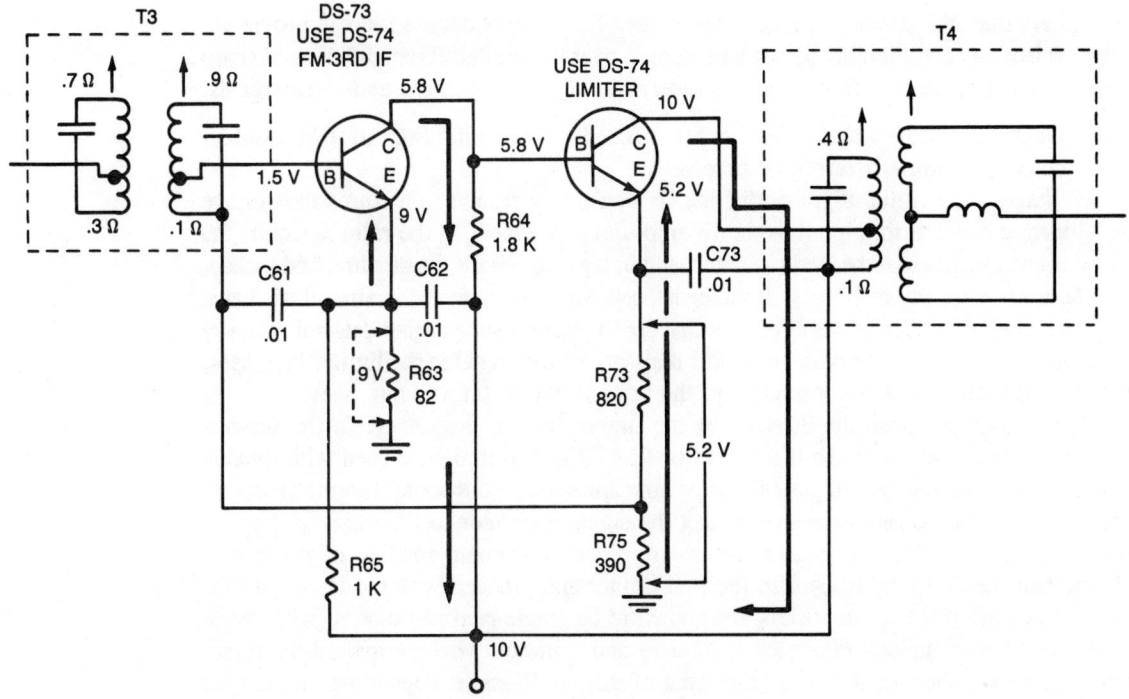

Fig. 5-5. Limiter circuit.

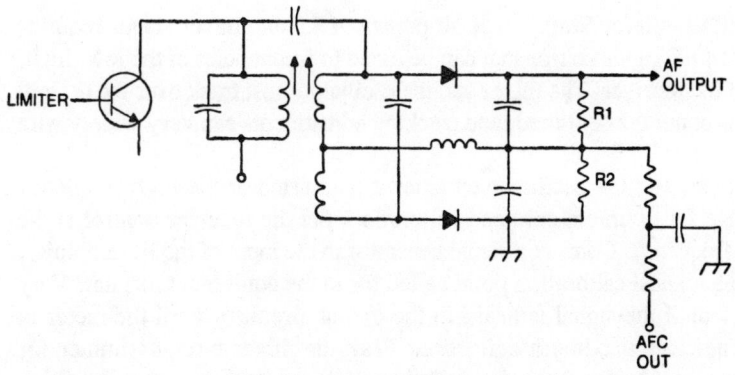

Fig. 5-6. Discriminator circuit.

discriminator reproduce the original modulation information. Because this stage lacks the ability to reject amplitude modulation (noise), one or two limiter stages must precede it to handle this problem.

Q15: Explain how spurious signals can be received or created in a receiver. How could this be reduced in sets having sealed untunable filters?

A: Spurious signals might be received from adjacent channels due to poor receiver selectivity, or such signals could be created by oscillating IF stages, local oscillators, or multipliers that generate heterodyne frequencies. The fixed bandpass filter improves receiver selectivity considerably by its flat response over the desired IF bandwidth and sharp drop-off on either side. This enables the receiver to reject most of the undesired signals.

Q16: Describe, step-by-step, the proper procedure for aligning an FM double-conversion superheterodyne receiver.

A: IF Stages. The meter alignment of the IF stages in a receiver measures the voltage developed at the detector circuit as the IF amplifiers are tuned. In the ratio detector, the meter is connected across the load capacitor, and in the gated-beam, quadrature, and locked-oscillator circuits, and audio output meter is used when the detector is altered so that it responds to AM. When aligning a receiver that uses a discriminator-limiter detector, connect the meter across the limiter base or emitter resistor. As the signal to the limiter increases, the current increases, and consequently, the voltage across the resistor rises.

Always work from the detector or the limiter toward the mixer. In the limiter-discriminator circuit, align the discriminator first. If a dual limiter is used with tunable coupling between the two stages, align the first limiter by connecting the meter across the resistor of the second limiter. Connect the signal generator to the input of the first limiter. Then tune the interstage coupling circuit for maximum reading of the meter. Connect the meter to the resistor of the first limiter and proceed with the IF alignment.

When all of the IF amplifiers are known to be single-peaked—that is, when none of them are overcoupled—tune each secondary and primary, working toward the mixer from the first-limiter input or from the input of the last IF stage, depending on the type of detector. Adjust each tuning control for maximum deflection of the meter. The signal generator must be unmodulated when the discriminator and ratio detector are aligned, and modulated for other detectors.

RF, Mixer, and Oscillator Stages. The alignment of RF and mixer stages requires an accurately calibrated RF signal source that can be tuned to frequencies in the low, high, and center portions of the receiver. The mixer and the oscillator must track over the desired range if the receiver is continuously tuned, and tracking adjustments can vary widely with the type of receiver.

The mixer trimmer and the oscillator trimmer are adjusted at the high frequency end of the tuning range for optimum gain and calibration. Set the receiver control at the designated alignment frequency. Connect a signal generator to the input of the RF amplifier. Set its frequency to the highest calibration point called for in the equipment manual. Vary the oscillator trimmer until the signal is heard in the output circuit or until the meter in the detector circuit indicates maximum deflection. Peak the mixer circuit trimmer for maximum output. Tune the receiver to the designated low-frequency alignment point. Tune the signal generator to the low end of the frequency range and set it at the low-frequency calibration point. Adjust the oscillator padder until the signal is a maximum at the detector. Then return the signal generator and receiver to the high-frequency calibration point. Retune the oscillator trimmer, if necessary, to bring the dial calibration of the receiver to the high-frequency calibration point. Retune the oscillator trimmer, if necessary, to bring the dial calibration of the receiver into correspondence with the frequency of the generator. Repeat these steps until both the low and the high frequencies are in calibration without the need for touching either the oscillator trimmer or the padder. Set the signal generator to a point midway in the frequency range, and check the calibration. When these steps have been carried out carefully, the calibration should be correct.

Alignment of the RF stage is done best with a noise generator. For the first rough initial alignment, antenna noise or a weak external signal supplied by the leakage from the signal generator or any other source will do. Peak the trimmer for maximum signal at the high end of the frequency range, and apply the noise-generator technique at the highest frequency to obtain the maximum signal-to-noise ratio.

Q17: Discuss the cause and prevention of interference to radio receivers installed in motor vehicles.

A: Most man-made noises fall into two general classifications: *impulse noise* and "hash". Impulse noise consists of sharp pulses of RF voltage that, when detected in a receiver, take the form of equally sharp pulses of audio voltage. They are often many hundreds of times greater in amplitude than the desired signal and make it impossible for the desired signal to be received. Perhaps the most common producers of impulse noise are the ignition systems of gasoline engines. Because many radio applications call for the installation of communication equipment in vehicles, impulse noise is a problem. Steps can be taken to eliminate much of this noise, but such elimination measures can never be perfect. There is always a residual component of the noise that can cause serious difficulties if the received signals are weak. The second kind of man-made noise, called pulse noise or "hash", is of a more continuous character. It appears as a broad band of many pulses that bear little or no relation to each other. Such noises are produced to a great extent by rotating electrical machinery, gas rectifiers, high-voltage transmission lines, and similar power devices. The noise from a small motor, although frequently weaker than the signal by a considerable amount, is capable of causing severe interference and possible interruption to reception.

After the source of the interference is determined, steps to eliminate it can then be initiated. The methods include *active* steps—that is, attempts to decrease the amplitude of the signals at the source of the noise—and *passive* steps—or attempts to combat the problem at the receiver. Active measures include the bypassing of radiating components, use of resistive spark-plug elements or resistive spark-plug wires, and insertion of choke coils in supply circuits. Passive measures include installation of noise limiters, clippers, and silencers at the receiver and shielding and improved grounding of receiver and power-supply circuits.

Q18: Explain the voltage and current relationship in a half-wavelength (dipole) antenna and a quarter-wavelength vertical antenna.

A: A half-wave length antenna is fed in the center, and is not grounded at any point. The current at the feed point is maximum, while the voltage is minimum, both being determined by the radiation resistance of the antenna and the applied power level. The current diminishes to minimum, and the voltage maximizes at the ends. The vertical antenna is a quarter wavelength long, and is fed at the base, which is insulated from ground in most cases (grounded vertical antennas also exist). The current and voltage distributions are similar to the dipole: the current is maximum at the feed point, and minimum at the top end; the voltage is minimum at the feed point and maximum at the top end.

Q19: What effect does the magnitude of the voltage and current, at a point on a half-wavelength antenna in "free space" (a dipole), have on the impedance at that point?

A: Antenna impedance at any point must equal the voltage-to-current ratio as Z equals E divided by I. Magnitude is of no importance except to establish the ratio; the greater the voltage or smaller the current, the greater the impedance will be.

Q20: How is the operating power of an AM transmitter determined by antenna resistance and antenna current?

A: By extension of Ohm's Law for power ($P = I^2R$); simply multiply antenna resistance by the square of antenna current.

Q21: What kinds of fields emanate from a transmitting antenna, and what relationships do they have with each other?

A: Radio waves in space have two kinds of fields—magnetic and electric. They are perpendicular to each other and both are at right angles to their direction of movement. Taken together, the electric and magnetic components form an "electromagnetic wave."

Q22: Can either of the two fields that emanate from an antenna produce an EMF in a receiving antenna? If so, how?

A: The fields radiated from the transmitting antenna induce an EMF in the receiving antenna as they pass, somewhat like the transformer where the fields around the primary "cut" the secondary windings and induce a voltage in that winding. It is true that the receiving antenna gets maximum pickup when parallel to the passing electromagnetic wave.

Q23: Draw a sketch and discuss the horizontal and vertical radiation patterns of a quarter-wave vertical antenna. Assume the center is at point "P".

A: Figure 5-7 illustrates the radiation pattern of a quarter-wave vertical antenna. The horizontal radiation (Fig. 5-7A) is equal in all directions around the antenna, while the vertical radiation pattern (Fig. 5-7B) forms an arc with a height of about 40 degrees at its maximum.

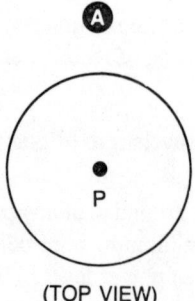
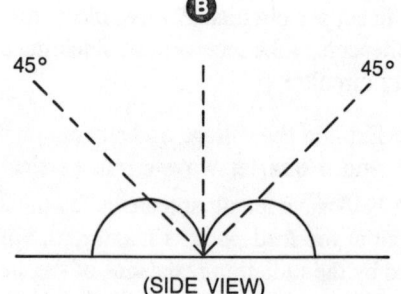

Fig. 5-7A. Horizontal radiation pattern of vertical antenna.

Fig. 5-7B. Vertical radiation pattern, quarter wave vertical antenna (side view).

Q24: Describe the directional characteristics, if any, of horizontal and vertical loop antennas.

A: The horizontal loop antenna is nondirectional with a minimum of radiation or reception capability vertically. Needless to say, it is seldom used in the horizontal plane as its primary purpose is not met. The vertical loop antenna has a bidirectional pattern horizontally, forming a figure 8 with the two maximum lobes in the plane of the loop and little or no radiation or pickup perpendicular to that plane. The vertical pattern of such a loop is nondirectional.

Q25: In speaking of radio transmissions, what bearing do the angle of radiation, density of the ionosphere, and frequency of emission have on the length of the skip zone?

A: The smaller the angle of radiation, the greater the length of the skip zone. If the angle between the direction of the wave and the earth's surface is too large, the wave will not return at all. The critical angle is the maximum angle at which the radiated wave will be reflected back to earth. As the density of the ionosphere increases, the greater the bending of the wave and the shorter the skip zone. This heavily ionized region extends from about 25 to several hundred miles above the earth's surface and exists in several layers. The "D" layer ranges to about 55 miles high and exists only in sunlight. The "E" layer at 65 miles is the lowest permanent layer and makes long-distance communication better, but the "F2" layer is most important in that regard. This layer ranges from 150 to 250 miles in height and is most consistent in density overall. Although high-frequency waves readily pass through the lower layers, the denser F2 layer might reflect or bend them back to earth.

The higher the frequency of emission below a critical value, the greater the skip zone as the degree of wave bending decreases. However, the frequency can be increased to a point where the wave is not bent enough to return, and no skip zone is possible. The energy of low-frequency sky waves is completely absorbed by the ionosphere, and skip above 50 MHz is not reliable. There is an area at the transmitting antenna where the sky wave is not returned due to the angle of radiation, and a skip zone exists between the limit of the ground wave and point where the sky wave might again be returned to earth as the radiation angle decreases.

Q26: Why is it possible for a sky wave to "meet" a ground wave 180 degrees out of phase?

A: When signals are radiated from an antenna, they radiate in many directions. Those signals radiated parallel to the earth are called ground waves, and those radiated into space are called sky waves. Sky waves strike atmospheric reflective layers of ionic material and reflect back to the earth. If a receiver is situated so as to receive both sky and ground waves from a single transmitter, signal cancellation could occur simply because the sky wave's path is longer than the ground wave. If the sky wave arrives later than the ground wave by a period equal to the duration of one-half cycle (or any multiple thereof), the two arriving signals will be 180 degrees out of phase, and cancellation results. In practice, the two waves are rarely exactly 180 degrees out of phase, so cancellation is not total.

The degree of cancellation depends on the relative strengths of the two signals. When both out-of-phase signals are of equal strength, there is "complete" cancellation; when one signal is stronger, only partial cancellation takes place.

Q27: What is the relationship between the operating frequency and ground-wave coverage?

A: The longer the wave, the more effective the ground-wave communications. Low-frequency waves actually follow the curvature of the earth. Conversely, very high-frequency signals tend to be attenuated during propagation and are thus not reliable for communications except on a line-of-sight basis. As a result of this phenomena, broadcast engineers prefer the low end of the AM band (550 to 1620 kHz) to the high end.

Q28: Explain the following terms with respect to antennas (transmission or reception):
- (a) Field strength
- (b) Power gain
- (c) Physical length
- (d) Electrical length
- (e) Polarization
- (f) Diversity reception
- (g) Corona discharge

A: (a) Field strength is signal strength induced in an antenna and measured in microvolts (μV) or millivolts (mV) per meter. If the field strength of a signal at a certain point from the transmitter measures 100 mV per meter, an induced voltage 100 mV would be measured in a conductor one meter in length.

(b) Power gain is a figure of merit for a directional antenna and is the ratio of power that must be supplied to a standard (dipole) antenna to reach a specific field strength figure at a certain distance as compared to the directional antenna power required to show the same field strength at the same distance. The same polarization must be used on each, and in order to figure the power gain of an antenna, the field gain figure is needed. Because the field gain is equivalent to the voltage induced in the directional antenna divided by the voltage induced in the comparison antenna, the power gain would be the square of that figure. Power is proportional to the square of the voltage.

(c) The physical length of an antenna is its actual length in feet or meters, and the practical physical length is shorter than the same wavelength in free space, for antennas close to earth's surface.

(d) The electrical length of an antenna is the physical length expressed in wavelength, radians, or degrees. It might be in degrees of a full hertz of the operating frequency, such as a resonant half-wave dipole (180 degrees). The electrical length can also be expressed in wavelengths in free space. A half-wave dipole operating on 20 meters would be one half of 20 meters in length at resonance or 10 meters long (39.37 inches per meter). The actual physical length of the dipole would be 5 percent shorter because it is not entirely free of surrounding forces that would decrease the velocity of a wave over that in free space. The length of a half-wave antenna in free space is 300,000,000 divided by 2 times the frequency in hertz which gives the frequency in meters. The resonant frequency is determined by the time required for a wave to travel the length of the antenna (180 degrees) and return (360 degrees). Because the usual approximation of the reduction in velocity is approximately 5 percent, simply multiply the electrical length by 0.95 for the physical length (as at least a first trial guess).

(e) Polarization is determined by the physical position of the antenna with regard to the earth and this indicates how the electric field propagates. Vertically polarized waves are generated by a vertical antenna, and the horizontal antenna radiates horizontally polarized waves.

(f) Space diversity reception is an efficient way of reducing fading by the use of more than one antenna. They are normally spaced several wavelengths (5 to 10) apart so that fading is not likely in all antennas at the same time. Each antenna is coupled to a separate amplifier with all feeding the same audio output stage. The use of a special AVC blocks all amplifiers except the one carrying the strongest signal, so the automatic use of the antenna receiving the strongest signal at the time reduces the fading effect to an absolute minimum. Diversity antenna systems are quite common in transoceanic communications. It is also possible to use frequency diversity and polarization diversity to help reduce fading.

(g) Corona discharge is an electrical discharge resulting from ionization of air, quite like a lightning discharge on a very small scale. Corona results from a voltage build-up around an HV conductor that tends to ionize the air near it. By rounding or balling all sharp or rough points, corona effect can be reduced considerably. Uninsulated or braided wire should never be used in high-voltage areas where corona discharge is likely to occur.

Q29: What would constitute the ground plane if a quarter-wave grounded (whip) antenna 1 meter in length were mounted on the metal roof of an automobile? Mounted near the rear bumper of an automobile?

A: An antenna mounted on the roof of an automobile has a good ground plane in the roof itself, which is a reasonable part of a wavelength in this case. Mounted near the rear bumper, a quarter-wave whip could use the bumper as a ground plane and possibly a portion of one fender. However, at this wavelength, the bumper would probably suffice.

Q30: Explain why a "loading coil" is sometimes associated with an antenna. Under this condition, would absence of the coil mean a capacitive antenna impedance?

A: A "loading coil" makes it possible to operate the antenna at a lower frequency than its actual length would permit. Short antennas (relative to wavelength) tend to be capacitive. The inductance of the coil in series with the antenna therefore makes it possible to resonate an antenna that is too short physically for the operating frequency and one that would look like a capacitive reactance without the "loading coil." This reactance could make it impossible to feed power into the antenna.

Q31: What radio frequencies are useful for long-distance communications requiring continuous operation?

A: Sky waves must rely on the presence of densely packaged atmospheric layers for reflection. Because these layers are inconsistent in their density and reflecting ability, ground waves are inherently more reliable than sky waves. Very short wavelength signals are readily absorbed by foliage and air and are thus dissipated more readily than longer wavelengths.

The U.S. Navy states that shore-based transmitters are able to transmit long-range ground-wave transmissions by using frequencies between 18 and 300 kHz with extremely high power. Because the electrical properties of the earth along with the paths the ground wave travels are relatively constant, the signal strength from a given station at a given point is nearly constant.

Q32: What type of modulation is largely contained in "static" and "lightning" radio waves?
A: Virtually all pulse-type interference takes the form of an amplitude-modulated signal—which gives FM one of its most striking advantages. Static, lightning discharges, ignition noises, and other impulse disturbances are effectively minimized or eliminated in FM receivers with adequate limiting circuitry.

Q33: Will the velocity of propagation differ in different materials? What effect, if any, would this difference have on the wavelength or frequency?
A: The velocity of propagation of radio waves differs according to the dielectric constant of the medium. It is always less than the speed of light by the factor K, which is 1 for air. It can be assumed that radio waves travel at maximum speed through air, atmosphere, or space (vacuum). The use of insulating material with a dielectric constant (K) greater than 1 decreases the speed of the wave and causes the physical length to be shorter than the electrical length. Velocity of propagation is lowered whenever the constant K is greater than unity (1).

Q34: Discuss series and shunt feeding of quarter-wave antennas with respect to impedance matching.
A: A series-fed quarter-wavelength antenna is ungrounded at the feed point, and the RF generator (i.e. transmitter) is therefore in series with the radiator element and ground. When the length is such that it is a quarter-wavelength at the operating frequency, the antenna will appear as a pure resistance at the point where energy is fed to the base of the antenna. The base of a series-fed quarter-wave antenna is insulated from ground, but a large metallic surface or ground plane must be in the immediate vicinity of the antenna's base for radiation to occur.

When the input impedance of the quarter-wave series-fed antenna is comprised primarily of radiation resistance, the efficiency of the antenna can be high. At frequencies above or below antenna resonance, inductive or capacitive reactances dominate the input impedance, and efficiency is low. When such reactances are unavoidable, they can be effectively canceled by introducing more inductance or capacitance in the series input circuit to cancel the effects. That is, series capacitance can be used to cancel an inductive reactance, and inductance can be added to tune out capacitive reactance.

The shunt-fed antenna is a grounded vertical quarter-wavelength radiator, often called a Marconi antenna. This system is also know by the generic term "image antenna." The ground is a fairly good conductor and acts as a large "mirror" for radiated energy. The ground surface reflects a large amount of energy that is radiated downward from an antenna mounted over it. It is just as though a mirror image of the antenna is produced; the image being located the same distance below the surface below the ground as the actual antenna is located above it. Even in the high-frequency range and higher, many ground reflections occur, especially if the antenna is erected over highly conducting earth, water, or a grounded screen.

Using this characteristic of the ground, a shunt-fed quarter-wave antenna can be made into the equivalent of a half-wave antenna. If such an antenna is erected vertically (its lower end is connected electrically to the ground), the quarter-wave behaves like a half-wave. Here, the ground takes the place of the missing quarter-wavelength, and the reflections supply that part of the radiated energy that normally would be supplied by the lower half of an ungrounded half-wave antenna.

The input impedance of a series-fed quarter-wave radiator varies, depending on the plane of the metallic ground surface. When the ground surface is perpendicular to the plane of the upright antenna, the impedance is precisely half that of the half-wave antenna, or 36-37 ohms. The grounded (shunt-fed) quarter-wave radiator cannot be fed at the base because the base itself is grounded directly. Because the impedance increases with the distance towards the end of the whip from ground, a suitable feed point can be located by attaching the feed line to a point somewhat above the ground surface. The point at which the antenna should be fed depends on the characteristic impedance of the feed line; the higher up the antenna the feed point is, the higher the feed-point impedance will be. Higher efficiency is observed when the feed point is less than halfway between the ground and the antenna's center.

Q35: Discuss the directivity and physical characteristics of the following types of antennas;
 (a) Single-loop
 (b) V-beam
 (c) Corner reflector
 (d) Parasitic array
 (e) Stacked array

A: (a) A single loop refers to the small loop antenna that consists of one or more turns of wire enclosed in an electrostatic shield. Generally circular, the loop acts as an inductance coil with a large diameter-to-length ratio and has maximum response in the direction of the plane of the loop. There is little or no pickup in a direction perpendicular to the plane.

(b) The V-beam consists of two radiating conductors arranged to form a V. It is directional in the open face. If each leg of the V is one wavelength long, the angle should be about 75 degrees. If unidirectional operation is desired, the ends of each leg should be terminated by load resistors to ground.

(c) The corner reflector has a driven element (which is a simple half-wave dipole) and two metal sheets forming a corner to act as a reflector. Its direction of radiation is principally away from the corner and can be further concentrated by decreasing the angle between the reflector sheets.

(d) The simplest form of parasitic array is a half-wave horizontal dipole acting as the driven element and a single parasitic element acting as a reflector. Placed at a distance of ¼ wavelength from the driven element, the parasitic element, in this case a reflector, should be about 5 percent longer than the driven dipole. Intercepting some of the energy from the driven element, the reflector re-radiates energy to combine with that radiated by the main element and concentrates it in that direction. The use of additional parasitic elements (with reflectors longer and behind the driven element and directors shorter and in front of the driven element) offers exceptional unidirectional characteristics and gain. This type of antenna is commonly known as a *yagi* and is extremely popular for a specific frequency or channel.

(e) The stacked array makes use of yagi elements mounted above each other (parasitic arrays) for exceptional gain. Aside from excellent directivity, stacking confines radiation to low angles, cutting ground losses and energy normally wasted in vertical radiation.

These directional characteristics in the horizontal plane offer improvement in field strength and range.

Q36: Draw a sketch of a coaxial vertical antenna, identify the positions, and discuss the purposes of the following components.
  (a) Radiator
  (b) Insulator
  (c) Skirt
  (d) Trap
  (e) Support mast
  (f) Coaxial line
  (g) Input connector

A: (a) Figure 5-8 shows a sketch of a coaxial vertical antenna. The radiator is the upper section and its length is about 95 percent of the free space quarter-length. It is an extension of the coaxial transmission line inner conductor.

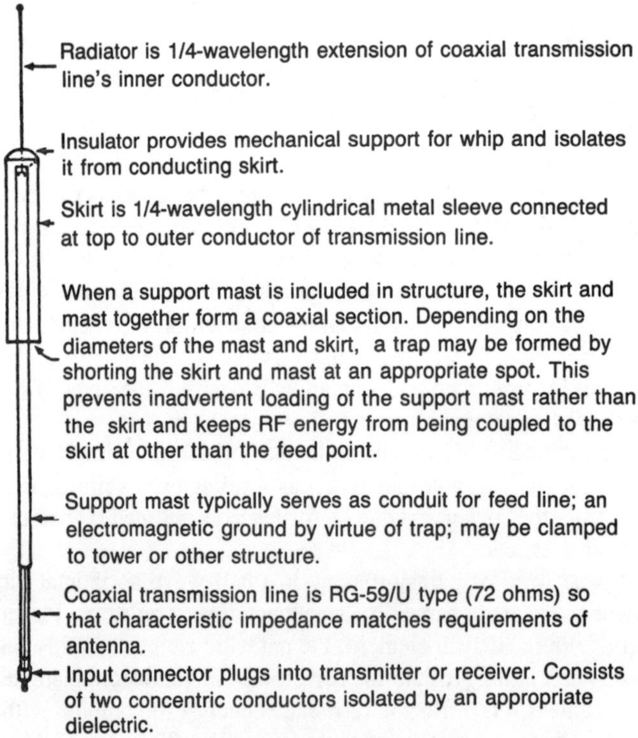

Radiator is 1/4-wavelength extension of coaxial transmission line's inner conductor.

Insulator provides mechanical support for whip and isolates it from conducting skirt.

Skirt is 1/4-wavelength cylindrical metal sleeve connected at top to outer conductor of transmission line.

When a support mast is included in structure, the skirt and mast together form a coaxial section. Depending on the diameters of the mast and skirt, a trap may be formed by shorting the skirt and mast at an appropriate spot. This prevents inadvertent loading of the support mast rather than the skirt and keeps RF energy from being coupled to the skirt at other than the feed point.

Support mast typically serves as conduit for feed line; an electromagnetic ground by virtue of trap; may be clamped to tower or other structure.

Coaxial transmission line is RG-59/U type (72 ohms) so that characteristic impedance matches requirements of antenna.

Input connector plugs into transmitter or receiver. Consists of two concentric conductors isolated by an appropriate dielectric.

Fig. 5-8. Coaxial vertical antenna.

(b) The insulator separates the upper and lower radiating sections and insulates the center conductor so that it will not short against the skirt.

(c) The metal cylinder mounted below the insulator is the skirt and it forms the lower radiating part of the antenna. It is also a quarter-wavelength and along with the whip is a half-wave dipole. A very high impedance is offered at the end near the insulator that minimizes current flow and prevents high vertical radiation.

(d) The trap is a portion of the skirt forming the outer conductor with part of the support mast forming the inner conductor of a quarter-wave coaxial section shorted at the bottom. The purpose of the trap is to eliminate current flow on the mast and the coax transmission line.

(e) The support mast is a good, strong pipe that holds up the entire antenna and also provides the inner conductor for the trap.

(f) The coaxial line is the transmission line used to carry energy from the transmitter to the antenna. Coaxial line, because of its low impedance (75 ohms) matches the input impedance or radiation resistance of the antenna, which is about 75 ohms.

(g) The input connector is a fitting for the coaxial cable to provide a convenient connection from transmitter to whip antenna.

The coaxial antenna is constructed to offer a handy, unbalanced feed arrangement and is a vertical half-wave type. Vertically polarized, it radiates at right angles to its plane and equally in all directions.

Q37: Why are insulators sometimes placed in antenna guy wires?
A: This makes the wires too short to resonant at the operating frequency by keeping their length below a half-wavelength, at which point they would absorb energy from the transmitting antenna and re-radiate, distorting the planned pattern. By using the familiar "egg type" insulators, the guy wire remains intact even if the insulator breaks.

Q38: What is meant by the characteristic impedance of a transmission line? On what physical factors is its value dependent?
A: When two wires are spaced proximally to each other, a capacitance exists between them. Wires also exhibit inductance. When an ac signal is fed to a capacitive, inductive circuit, the circuit represents a finite impedance. The characteristic impedance (also called "surge impedance") of a transmission line is the impedance it would have if it were infinite in length, and is equal to the square root of the L/C ratio. This impedance is considered to be purely resistive, and is constant for a given transmission line. The characteristic impedance is important in determining how well energy is transferred from the source of the signal to the load. For the infinitely long line, of course, all of the energy sent out on the line would appear to be dissipated in the line, and none would be reflected back to the source because of subsequent impedances down the line that might represent values other than the line's characteristic. Because lines are not infinitely long, it is important to terminate a transmission line in a load that represents the closest possible equivalent of the line's characteristic; otherwise, not all the energy on the line will be transferred to the load. In this case, some value of RF power would be reflected at the point of termination and would travel back down the line toward the source.

The physical factors that determine a line's characteristic impedance are: resistivity of the conductors, dielectric material separating the conductors (including the actual spacing between them), and the diameter of the conductors. For a given impedance value, the larger the diameter of the conductors, the more they must be spaced. The closer the spacing between conductors, the lower the characteristic impedance. The characteristic impedance of a line can be expressed mathematically by taking the square root of inductance divided by capacitance.

Q39: Why is the impedance of a transmission line an important consideration when matching a transmitter to an antenna?

A: Maximum power transfer occurs in any electrical system when the load impedance matches the source impedance. The characteristic impedance of the transmission line must therefore be equal to the antenna input impedance, or the input impedance at the transmitter will appear to vary with the length of the line, making maximum transfer of energy impossible.

When the line is exactly a quarter-wave or integer multiple of this length, the formula for the characteristic impedance of the line is:

$$Z_{line} = \sqrt{Z_{source}\, Z_{load}}$$

where Z (in) is the input impedance at the transmitter end, Z (load) is the antenna impedance and Z (line) is the characteristic impedance of the transmission line. If the transmitter output impedance is 140 ohms to a 40-ohm antenna impedance the characteristic impedance would be 80 ohms for the transmission line cut to an odd multiple of a quarter-wave by the above formula.

Q40: What is meant by standing waves; standing-wave ratio (SWR), and characteristic impedance as referred to transmission lines? How can standing waves be minimized?

A: When a transmission line is not terminated in its characteristic impedance, energy is reflected from the antenna (load) and combined with the power flowing out to the antenna to form standing waves. The standing wave ratio or SWR is the ratio of the maximum voltage to the minimum voltage along the line.

Standing waves can be eliminated by terminating the line in an impedance equivalent to the characteristic impedance of that line. In other words, the line must be matched to the antenna impedance.

The characteristic impedance of a transmission line is the input impedance of a line having infinite length, such as the input signal would "see" when flowing into a line of infinite length. Values of characteristic impedance actually vary from 25 to 600 ohms according to the type of line. Coaxial lines usually range from 25 to 90 ohms, parallel lines (twinlead, etc.) range from 100 to 600 ohms, with "open line" running to 600 ohms.

Q41: If standing waves are desirable on a transmitting antenna, why are they undesirable on a transmission line?

A: Standing waves in a conductor cause that wire or line to radiate energy, which is desirable in the antenna but definitely undesirable in a transmission line. The fact that standing waves exist in the transmission line indicates that a mismatch between the transmitter and antenna is causing a lower efficiency of energy transfer than is necessary or desirable.

Q42: What is stub tuning?

A: Stub tuning is a method of tuning a transmission line. A short length of transmission line is attached to the main line to eliminate standing waves. The stub effectively serves as an impedance-matching device and actually assists in matching the transmission line to the antenna. This reduces line losses and permits maximum power transfer from the transmitter to the antenna.

Q43: What would be the consideration in choosing a solid-dielectric cable over a hollow pressurized cable for use as a transmission line?

A: Solid cables use a resilient dielectric material such as polyethylene that is flexible and easy to install. It is lower in cost, has a higher loss tolerance and requires no special plumbing. The air-insulated cables are much more expensive but considerably more efficient. Some types are fully evacuated and filled with an inert gas under pressure and sealed to prevent moisture from accumulating within the cable, which results in losses. Seals are required at the ends and joints as well, and the mechanical faults along with bulky accessories restrict this type of transmission line to nonmovable transmitters.

Q44: Explain how a dip meter can be used to tune a vertical antenna used in the HF or AM broadcast band.

A: The dip meter, or "dipper", works on the principle that energy in a resonant tank circuit is reduced when the tank circuit is placed in close proximity to another resonant tank circuit that is tuned to the same frequency. The reduction of energy caused by the transfer to the other circuit results in a "dip" on the meter used to monitor the oscillator levels. A resonant vertical antenna is essentially a resonant tank circuit, so it can be tuned with a dipper. Figure 5-9 shows how this measurement is made. A small two- or three-turn "gimmick" link is connected to the input end of the coaxial transmission line, and the inductor from the dipper is brought in close proximity. As the dipper is tuned across the antenna resonant frequency, a brief dip is noted. The most common mistake made by users of the dipper is that they tune too fast and thereby miss the dip.

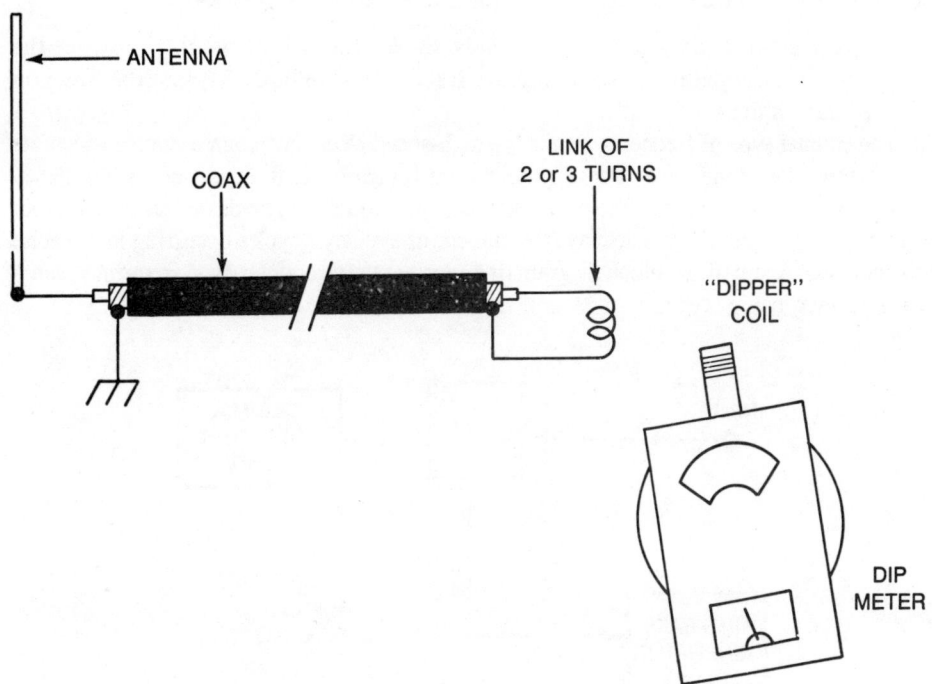

Fig. 5-9. Coupling dip meter to vertical antenna.

Q45: Draw a simplified circuit diagram of an absorption wavemeter (with galvanometer indicator). Explain the operation and give some possible applications.

A: The schematic in Fig. 5-10 is simply a calibrated LC circuit with an indicating device. The galvanometer and diode can be replaced with a simple flashlight bulb as an indicator at a slight sacrifice in accuracy.

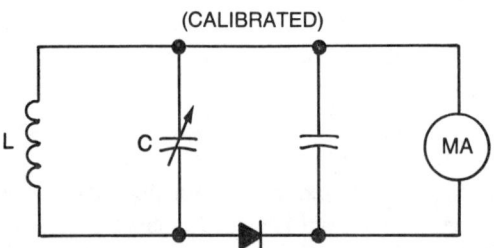

Fig. 5-10. Schematic of an absorption wavemeter.

Probably the major disadvantage of the absorption wavemeter is a tendency to detune the circuit under test as energy is absorbed from it. Nevertheless, this condition can be minimized by keeping the coupling between the two circuits as loose as possible. As resonance is approached, coupling should be reduced by moving the wavemeter farther away from the circuit being measured until the meter is just barely off zero. Tuning becomes much sharper as coupling is loosened, and when ready for final adjustment, the meter can be switched to maximum sensitivity for a more accurate indication.

Q46: Draw a block diagram showing only those stages that would illustrate the principle of operation of a secondary frequency standard. Explain the function of each stage.

A: The counter type of frequency meter is a high-speed electronic counter with an accurate crystal-controlled time base. These counters are typically used as secondary frequency standards. The combination time base and electronic counter provides a frequency meter that automatically counts and displays the number of events or cycles occurring in a precise time interval. A simplified block diagram of a representative counter-type frequency standard is shown in Fig. 5-11.

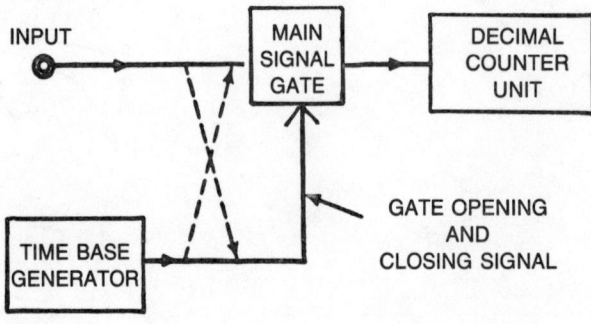

Fig. 5-11. Frequency counter, simplified block diagram.

In addition to making direct frequency measurements, the counter can measure periods, frequency ratios, and total events. A self-check feature enables an operator to verify instrument operation for most types of measurements. The internal oscillator is extremely stable. A secondary transfer standard is calibrated for accuracy against a primary standard (National Bureau of Standard's WWV broadcasts).

Q47: Draw a block diagram of a heterodyne frequency meter that includes the following stages: Crystal oscillator, crystal oscillator harmonic amplifier, variable frequency oscillator, mixer, detector and AF amplifier, AF modulator, and show the RF input and RF, AF, and calibration outputs.

Assume a bandswitching arrangement and a dial having arbitrary units, employing a vernier scale.
(a) Describe the operation of the meter.
(b) Describe, step-by-step, how the crystal should be checked against WWV, using a suitable receiver.
(c) Under what conditions would the AF modulator be used?
(d) Describe, step-by-step, how the unknown frequency of a transmitter could be determined by the use of headphones; by the use of a suitable receiver.
(e) What is meant by calibration check points, and when should they be used?
(f) If in measuring a frequency the tuning dial should show an indication between two dial-frequency relationships in the calibration book, how could the frequency value be determined?
(g) How could this meter be used as an RF generator?
(h) Under what conditions would it be necessary to recalibrate the crystal oscillator?

A: (a) The heterodyne frequency meter consists of a stable variable-frequency oscillator, a crystal oscillator (and harmonic amplifier), a mixer/detector, an audio amplifier, and sometimes an audio modulator, as shown in Fig. 5-12. The detector measures the difference frequency between the variable oscillator and the accurate standard crystal (or one of its harmonics). The resulting frequency difference is in the audio range, so the beat frequency is audible in headphones (or other suitable reproducer) after amplification. The variable-frequency oscillator is tuned by means of a corrector knob to get a zero beat; then the

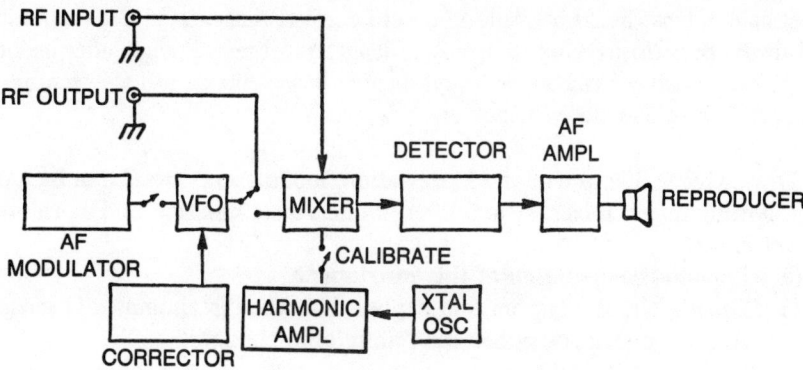

Fig. 5-12. Heterodyne frequency meter.

variable oscillator is known to be at the same frequency as the crystal. The RF input from the transmitter is then connected in place of the crystal oscillator. The variable oscillator is tuned to obtain a zero beat; this oscillator is now at the same frequency as the transmitter. The setting of the calibrated dial is read directly or converted frequency by means of a chart.

(b) To check the crystal frequency when one of the crystal's harmonics falls on the frequency of a primary frequency standard, simply tune a receiver to a WWV or WWVH channel and couple the RF signal from the receiver into the frequency meter's RF input. When the mixer is provided with a signal from the crystal oscillator that is exactly at the same frequency (or harmonic frequency) as WWV, a zero beat will be obtained in the headphones. When one of the crystal's harmonics does not fall on a WWV frequency, the VFO can be calibrated against WWV, then the VFO can serve as a standard for setting the frequency of the fixed crystal oscillator.

(c) The modulator makes it easier to keep track of the RF signal when the unknown cannot be found for comparison. The modulator also allows the VFO to be used as a signal generator for tuning and aligning receivers.

(d) See (a) above for explanation of how to measure unknown frequencies.

(e) Calibration check points provide an easy method for determining which harmonic of the crystal should be used, and for determining the proper range setting of the dials. When the frequency indication is not readable directly in kilohertz or megahertz, a numerical vernier indication can be compared against a like number in the calibration book, which lists the frequency that each numerical indication represents.

(f) Most frequency meters allow fairly easy interpolation of frequency readings because of the linearity between the dial readings and the frequency indications they represent. Thus, if a dial indicated a frequency to be between 4115 and 4116, the frequency would be proportionally between the two frequencies indicated. Assume that 4115 represents 152.75 MHz and 4116 represents 152.78 MHz. If the dial reading is two-thirds of the way from 4115 toward 4116, the frequency indicated will be proportionally the same between the two listed frequencies, or 152.77 MHz.

(g) When the switch in the output line from the VFO is placed in the position shown in the block diagram, the VFO portion of the frequency meter will serve as an accurate signal source and can be used in conjunction with the audio modulator for aligning receivers.

(h) To qualify as a secondary transfer standard, a frequency meter must be calibrated against a primary standard at certain specified intervals (normally six months). However, frequent calibrations should be made to ensure consistent accuracy. Calibration checks should always be performed following a significant temperature change, after making any change in the frequency meter's tube complement or power supply, and after transportation of the device from one place to another.

Q48: Draw a block diagram of an FM deviation (modulation) meter that includes the following stages: mixer, IF amplifier, limiter, discriminator, and a peak reading voltmeter.
    (a) Explain the operation of this instrument.
    (b) Draw a circuit diagram and explain how the discriminator is sensitive to frequency changes rather than amplitude changes.

A: (a) As can be seen by studying the block diagram Fig. 5-13, a deviation monitor is nothing more than a simple FM receiver without the audio or squelch circuits. Since

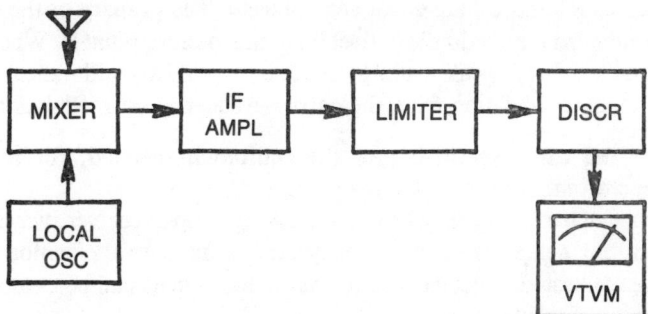

Fig. 5-13. FM deviation meter.

transmitter deviation is normally adjusted in the immediate vicinity of the instrument, the meter is not equipped with receiver circuits that provide a high degree of selectivity or sensitivity.

In operation, a signal from the transmitter under test is coupled into the mixer along with an appropriate-frequency signal from the monitor's local oscillator. The difference frequency is processed through the IF amplifier and limiter. (The limiter causes the IF signal to reach saturation so that AM components in the signal are minimized; that is, when an IF signal is amplified to saturation, there are no variations in signal amplitude and, consequently, little or no AM signal components.)

The discriminator recovers the audio from the IF signal and provides a varying dc output voltage that is directly proportional to frequency deviation. Because the signal has been amplified to saturation via the limiter stage, the discriminator's input signal is of varying frequency but constant amplitude.

(b) The discriminator circuit is shown in Fig. 5-14. Note that two resistors (R1 and R2) appear across the output. When no modulation is applied to the transmitter under test,

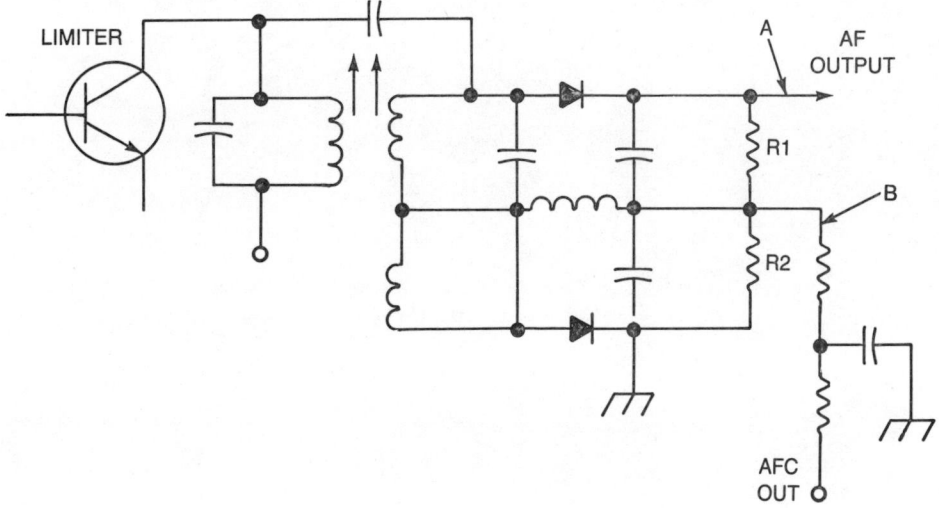

Fig. 5-14. Discriminator alignment connections.

205

the voltages across R1 and R2 are equal and opposite. The primary of the discriminator transformer is adjusted for maximum deflection of the meter at point B. When modulation is applied, the meter at A (a peak-reading voltmeter or VTVM) will indicate a dc voltage that is proportional to deviation from the carrier frequency on modulation peaks.

Q49: Describe the usual method, and the equipment needed, for measuring the harmonic attenuation of a transmitter.
A: The existence of harmonics may be observed using simple receiver, wavemeters, field-strength meters, dip meters, or spectrum analyzers. To measure attenuation, the detecting device must be calibrated in decibels so the harmonic signals can be compared with the strength of the fundamental signals.

The detecting device is placed at some fixed point in proximity to the transmitting antenna while the transmitter is operating. A calibrated attenuation control on the instrument is used to set the meter to 0 dB when it is tuned to the fundamental frequency. The device is then tuned to the desired harmonic and the negative-decibel reading noted. If the signal is not detectable, the attenuation is reduced incrementally until a reading is obtained. The attenuation is decibels will equal the meter reading plus the difference in decibels between the first setting and the second setting of the attenuation control. Each 10 dB represents an order of magnitude power change. Thus, if a 40 dB difference was noted between the fundamental and second harmonic, the transmitter's fundamental signal is 10,000 times more powerful than that of the harmonic.

Q50: Why is it important that transmitters remain on frequency and that harmonics be attenuated?
A: If transmitters drift off frequency or pass harmonics without insufficient attenuation, interference with other signals can hardly be avoided. All transmitters are required to remain on their assigned frequency and harmonics must be attenuated.

6

Basic Radiotelephone: Part IV

A battery is a simple source of electric power; whether it's a single cell or a combination of cells connected together in series for greater voltage, in parallel for greater current, or in series-parallel for greater voltage and current. A battery converts chemical energy into electrical energy and can be a primary (dry) cell (cannot be recharged and used again) or a secondary (wet) cell that (can be recharged and used over again many times).

The zinc-carbon dry cell consists of a zinc container forming the negative terminal with a paste-type electrolyte of sal ammoniac and manganese dioxide separating it from the carbon rod in the center, which is the positive terminal. The cell offers 1.5 volts, which drops as the zinc is decomposed by the chemical action of the electrolyte until it is of no further use and must be discarded. This type of cell is used for flashlights, small transistor radios, and other small devices requiring limited amounts of power.

The secondary cell, an example of which is a lead-acid storage cell or wet storage cell used in automobiles, has a normal output voltage of 2 volts per cell. The standard 12-volt auto battery consists of six of these cells connected in series for maximum voltage. With proper care, they often last for several years. The construction and care for these batteries is reviewed more completely in the Q & A part of this chapter.

MOTORS AND GENERATORS

A motor converts electrical energy into mechanical energy, and a generator does just the opposite—changes mechanical energy into electrical energy. If a current is applied to a wire or coil inside a magnetic field, the wire will move in a direction according to various factors involved. In order to allow continuous rotation, the current flowing through the coil must be reversed during each rotation, and this switching is provided by the commutator. A dc motor uses an electromagnet in the field winding that is connected in series with the armature coil to provide a series-wound motor. The field winding can be connected in parallel with the armature winding, in which case we have a dc shunt motor.

Although a dc motor can be used as a dc generator and the generator as a motor, best performance is realized only when the unit is specifically designed for the intended use. The shunt-wound motor offers reasonably constant speed under varying loads, but some slow-down might occur under heavier loads; torque is rather low, too. The voltage output from the shunt-wound generator is also fairly constant under various loads. A dc motor can be reversed by simply reversing the field or the armature winding connection. Because a running dc motor also generates an opposing voltage to that applied, counter EMF is only a little less than the applied voltage and it serves to reduce the current drawn by the motor. Actually, this current depends on the resistance of the motor, the voltage applied, and the counter EMF. The current drawn by the motor while running increases along with the load because that load increase slows the motor, causing the counter EMF to decrease; consequently, the current increases.

Most modern vehicles replace the generator with a rectified alternator. In most of these vehicles, a capacitor can be placed across the output line in a manner similar to the generator. *But beware*—this can ruin some alternators, especially if the wrong value capacitor is used. Consult the manual for the vehicle before proceeding. Another solution is to place a large value inductor in series with the power line to the affected radio communications or audio equipment.

MICROWAVE EQUIPMENT

Conventional transistors and other components become of little use at UHF frequencies, and they are of no use at all for microwave operation. The inductance in a fraction of an inch of wire or the capacitance in adjacent equally short wires or leads is sufficient to unbalance circuits and cause all sorts of problems in the microwave region. Hence, there is an entirely different concept circuit characteristics in the study of microwave communication equipment. Actually, the *klystron*, *magnetron*, and *traveling wave tubes* are much simpler than conventional pentodes and other multielement tubes if the principles of operation are understood.

Just about every loss conceivable increases with frequency in ordinary vacuum tube power amplifiers until, at microwave frequencies, they fail to show an output as great as their input signal. The inductance in the element connecting leads, grid-cathode capacitance, dielectric loss, and electron transit time are all major factors against them. Electron transit time (the time required for an electron to travel from cathode to plate) is comparatively short at low frequencies—about one-thousandth of a microsecond. This is about one-thousandth of a cycle at a frequency of 1 MHz, but at 1,000 MHz, the electron transit time amounts to one full cycle. Indications are that tube efficiency really drops excessively when transit time is greater than one tenth of a cycle, and if it reaches two or three times that figure, the tube becomes a big zero, because it will not amplify or even oscillate.

Using special designs eliminates or at least reduces some of the problems and extends the upper limit of the operating frequency of conventional tubes. The acorn, doorknob, lighthouse, and pencil types are a few of the original special designs for UHF operation;

many have been replaced with solid-state devices. However, there are definite limitations even today, and the newer principles of operation found in the klystron with its velocity modulation and the magnetron with a high-power output at super-high frequencies are explained a little later. In a magnetron, the transit time between cathode and anode determines the frequency of operation and it can be varied by changing the position of the external magnet. Klystron and traveling wave tubes also utilize transit time, but in a different way.

Certain properties of microwaves make them especially attractive for special situations. Probably the most important feature of microwave operation is the narrow beam of radiated energy produced with parabolic reflector antennas of modest size. This facilitates point-to-point transmission such as studio-transmitter links where the narrow beam makes it possible to focus most of the radiated power on the receiving antenna. Interference is easy to avoid and reliability is assured with surprisingly low power. Radar also makes good use of the concentrated beams offered in microwave operation, but due to other requirements such as sweeping beams and distances involved, extremely high power is used. Radar operations are discussed fully in Chapter 8. The width of a microwave carrier also makes it possible to comfortably accommodate many voice channels at one time with multiplexing.

TROUBLESHOOTING

Your exam might contain several questions on troubleshooting. The best suggestion is to use good common sense in answering them. Read the question over carefully and take your time answering. Don't look for something complicated; it's more likely to be something simple—but logical.

QUESTIONS AND ANSWERS

Q1: How does a primary cell differ from a secondary cell?
A: The primary cell must be discarded after use, because it cannot be recharged like a secondary cell. An example of the primary cell is the common flashlight battery (dry cell), while a typical secondary cell is the lead-acid cell (storage battery) used in automobiles.

A primary cell cannot be recharged, because one electrode has been partly destroyed by the chemical reaction. Recharging could not possibly restore it to its original condition. A secondary cell merely undergoes a chemical change when discharged, which the charging current reverses. Nothing has been dissolved or destroyed, so the battery is restored to its charged condition.

Q2: What is the chemical composition of the electrolyte of a lead-acid storage cell?
A: The electrolyte is a diluted solution of suphuric acid in distilled water with a specific gravity of 1.300 when completely charged. The acid forms 25 percent of the mixture by volume.

Q3: Describe the care that should be given a group of storage cells to maintain them in good operating condition.
A: Storage batteries require considerable care to keep them in top condition and to ensure maximum life. The water level is important and must not be too high or it will boil out while the cell is in use (either charging or discharging). The level should be maintained at about ¼ inch above the plates by adding pure distilled water only—never acid or

electrolyte unless some of the original has been spilled. Cells should be kept fully charged and never allowed to stand in partly or fully discharged condition because this results in *sulphation*. Overcharging slightly about once a month removes any sulphation. The charging rate should be low to eliminate excessive gassing or bubbling. Keep the battery terminals free of corrosion with a layer of petroleum jelly and wash away corrosion prior to this application with a solution of baking soda. Do not allow the solution to get into the battery cells. Adequate ventilation should be provided because hydrogen gas (plosive) is normally produced by a lead-acid battery that is operating improperly.

Q4: What causes sulphation of a lead-acid storage cell?
A: Sulphation can be caused by local action, particularly in batteries with a fairly high internal resistance. This type of problem can be prevented by making certain cell is charged fully before allowing it to stand idle for long periods. Another major cause of sulphation is operating the battery with insufficient electrolyte. When the level of the electrolyte drops below the surface of the tops of the plates, sulphation is almost surely the result. When excessive gassing occurs, sulphation is accelerated, too; to avoid this, be sure to avoid overcharging and overdischarging. Adding electrolyte to a fully charged cell can cause excessive gassing (the result of which is sulphation); this can be avoided by adding electrolyte only when some of the electrolyte in the battery has been lost by spillage. When the battery is fully charged or when the battery is about to be charged, distilled water can be added to bring the electrolyte to its proper level.

Q5: What is the result of discharging a lead-acid storage cell at an excessively high current rate?
A: The battery tends to overheat, thereby accelerating the discharge rate and perhaps damaging the battery permanently. The capacity of a battery is reduced during periods of excessive heating. Excessive heating causes sulphation, plate buckling, and electrolyte boil-off.

Q6: If the charging current through a storage battery is maintained at the normal rate but its polarity is reversed, what will be the result?
A: This would discharge the storage battery, but it would cause no damage unless permitted to continue. The battery would eventually be damaged if charged with reverse polarity for an extended period. Severe sulphation would also result, ruining the negative plates.

Q7: What is the approximate fully charged voltage of a lead-acid cell?
A: The fully charged voltage of a lead-acid cell is about 2.06 volts, and when fully discharged 1.75 volts. Actual voltage is dependent on temperature also, but it would be close to these figures. Measuring the terminal voltage is one method of determining its condition, and specific gravity of the electrolyte is the other. All cells are fully charged when the specific gravity reads 1.300; the battery should be recharged if it is below 1.140.

Q8: What steps can be taken to prevent corrosion of lead-acid storage cell terminals?
A: Clean the top of the battery and cell terminals with a baking soda solution, then coat the terminals with petroleum jelly. Exercise care with the baking soda solution—do not let it get into the cells. Connections to the terminals should be clean and tight before applying the lubricant.

Q9: How is the capacity of a battery rated?
A: The capacity of a battery is rated in ampere-hours, a multiple of current in amperes and time in hours. A fully charged battery rated at 100 ampere-hours should deliver 10 amperes continuously for 10 hours or 100 amperes for 1 hour, but the actual performance would be somewhat less as a result of hearing and chemical changes. The rate of discharge has considerable effect on the efficiency, as does the ambient temperature. Extremely high discharge rates or cold temperatures could reduce the ampere-hour capacity to less than half the actual rating.

Q10: What is power factor? Give an example of how it is calculated.
A: Power factor is a measure of the phase difference between the voltage and current or it can be expressed as the figure by which the product of E and I must be multiplied to secure the true power of a circuit. The power factor varies between zero and unity. When the phase angle between voltage and current is 90 degrees, the power factor is zero; and when voltage is exactly in phase with current, the power factor is unity. A high power factor is definitely desirable in lines carrying power, because circuit losses are greatly reduced and efficiency is better as a result. A low power factor is desirable in capacitors and inductors because the maximum phase angle reduces losses in these components. In order to find the true power of a circuit, the apparent power (E × I) must be corrected by the factor relating to the phase angle. Simply multiply EI times the cosine of the phase angle, which equals Z over R. If the resistance and reactance are known, impedance (Z) is equal to the square root of R squared plus X squared; then the power factor is equal to R over Z.

Q11: List the comparative advantages and disadvantages of motor-generator and transformer-rectifier power supplies.
A: The motor-generator can be operated independently from any ac supply. In mobile applications, the motor-generator (dynamotor) can be connected directly to the terminals of a battery. The efficiency of the dynamotor, however, is extremely low—on the order of 43 percent or so; thus, a great deal of power is required to operate such a unit. The dynamotor often proves to be the source of RF interference, too; the contacts make the break at an extremely high rate, and generate electrical impulses that can prove difficult to filter. On the other hand, dynamotors are capable of sustaining large current drains over long periods without overheating, so long as the battery supply is capable of delivering the required power. The dynamotor is typically characterized by a high-pitched whine (which can prove annoying). On the positive side of the ledger, the output of a dynamotor is a relatively ripple-free dc that does not itself require filtering other than with ordinary bypass capacitors.

A transformer-rectifier operating at a frequency of 60 Hz requires a conventional power-line source, so its portable applications are limited. There are transformer-rectifier combinations capable of operating at extremely high frequencies, but all such combinations do require a source of ac power, whether it be the power line, a vibrator supply, or a pair of switching transistors. A low-frequency combination requires considerable filtering to remove the ripple content and smooth the output to a usable dc for most receivers and transmitters. A high-frequency combination requires little filtering but could be the cause of inductively coupled whine (an audio signal at the switching rate) in the radio circuits it is powering. A low-frequency transformer-rectifier requires a transformer with a considerably heavy core mass—which means more cost, more weight, and more space consumption. A high-frequency combination means smaller transformer, less filtering, less weight, less space. At one time, the rectifiers were a major space consideration, too. Today, however, the space consumed by solid-state rectifiers—regardless of the circuit in which they are used—is negligible.

A transformer-rectifier combination is an extremely efficient energy converter. While a dynamotor can produce dc when connected to a battery, its efficiency is rarely more than 45 percent (and frequently less). A vibrator, transformer, and rectifier combination (assuming solid-state rectifiers are used) can convert battery voltage to B+ voltage with a total conversion efficiency of up to 70 percent. A solid-state switching circuit in concert with a transformer-rectifier circuit can change battery voltage to B+ with a conversion efficiency of up to 78 percent. Some switching circuits, which employ two transformers rather than one, can operate with conversion efficiencies as high as 98 percent.

The dynamotor has a relatively low reliability factor because it is a mechanical device with a moving assembly that can (and does) wear out. The transformer-rectifier has a high reliability factor because it has no moving parts. Even when a transformer-rectifier is used with a solid-state switching system to supply the necessary ac, the reliability can be quite high.

Q12: What determines the speed of a synchronous motor? An induction motor? A dc series motor?
A: The speed of a synchronous motor depends on the frequency of the supply voltage and the number of pairs of poles. The speed of the induction motor is related to the same factors plus the load, to some extent.

The speed of a dc series motor depends chiefly on the load, but to some extent on the voltage, type and number of turns in the armature, number of turns in the field and number of poles.

Q13: Describe the action and list the main characteristics of a shunt-wound dc motor.
A: Because the field winding and the armature winding of a shunt motor are in parallel, the field current and flux are independent of the armature current and of constant value. As usual, the armature current peaks with the load and is lowest with no load. Even though the shunt motor slows down noticeably with an increase in load, the variation in speed with load is limited enough for this type to be labeled the "constant-speed" motor. The starting torque of the shunt motor is less than satisfactory for heavy loads, but the speed is readily controlled by a rheostat in series with the field. As more resistance is added, the field current decreases and the motor speed increases.

Q14: Name the possible causes of excessive sparking at the brushes of a dc motor or generator.
A: Brushes not aligned, insulation between commutator segments too high, commutator dirty or rough, poles incorrectly connected, brushes binding in holders, incorrect tension, open or shorted armature coil, defective field coil, or excessive load on armature.

Q15: How can radio frequency interference, often caused by sparking at the brushes of a high-voltage generator, be minimized?
A: RF interference can be reduced by adding filter chokes in series with the brushes and bypass capacitors from the brushes to ground. Sparking interference often originates from

generator components that form tuned circuits, with the power leads radiating at the frequencies of those tuned circuits, and the spark energizes their oscillations. All radiating leads must be effectively bypassed as close to the generator as possible in order to provide effective relief from this source of interference.

Q16: How can the output voltage of a separately excited ac generator, at a constant output frequency, be varied?
A: The output voltage can be adjusted by means of a potentiometer connected in series with the field windings. The field windings carry dc to the generator from a separate voltage source. As the field voltage is decreased (by increasing the series resistance), the alternator's output voltage is reduced.

Q17: What is the purpose of a commutator on a dc motor? On a dc generator?
A: The commutator on a dc motor provides current in the required direction for each armature coil to cause a torque that acts to turn the armature. This switching action causes the armature current to reverse periodically and become ac, in effect, so that any given point on the armature leaving one field pole is repelled from it and attracted to the next pole. The commutator maintains contact between the armature and the supply voltage.

The commutator on a dc generator makes a direct current output possible by switching the ac generated by armature coils. As the brushes contact each armature segment, current flows in one direction in the output. The current in the armature windings is ac, and the output would also be ac except for the commutator action that switches in another coil just as the previous coil starts to reverse its current direction.

Q18: What can cause a motor-generator bearing to overheat?
A: Common causes are improper lubrication, poor alignment of the motor to the generator, broken bearings, bearings not aligned with the shaft, dirty bearings, overload, or a lack of ventilation. If a bearing overheats, the motor-generator should not be stopped; instead, remove the load and slow down the machine while making every effort to cool it by forced air and by applying oil and graphite. Continue running it slowly until the bearing cools to the normal temperature, then stop the motor and flush the bearing with light oil, followed by lubrication with regular weight oil. The bearing should still be in satisfactory condition if heating was not too severe. Check the machine for overload and proper condition.

Q19: What materials should be used to clean the commutator of a motor or generator?
A: If in reasonably good condition, the commutator can be cleaned with very fine sandpaper, light canvas material, or special commutator polishing paste. Rough or burned spots can be smoothed down with fine sandpaper. Emery paper must never be used because the metallic dust can very easily short out segments or even windings! The chocolate brown color indicates a commutator is in proper operating condition. Special care should always be exercised in servicing rotating electrical machinery to avoid possible injury or electrical shock.

Q20: If the field of a shunt-wound dc motor were opened while the machine is running under no load, what would be the probable result?
A: The speed of a shunt motor is dependent on the magnetic strength of the field: the higher the field strength, the slower the motor turns. An open field would cut off the field,

and the motor speed would increase beyond the design point. Such operation could seriously damage the motor.

Q21: Describe the physical structure of a klystron tube and explain how it operates as an oscillator.

A: The reflex klystron usually performs as an oscillator and has a cathode, resonator or anode, and repeller. Electrons from the cathode are drawn into a beam by the focusing electrode. The electron beam is attracted to the cavity resonator by its positive charge. Passing through the gap in the resonator, the beam interacts with the fields of oscillations in that cavity. When an electron passes through the gap in phase with the resonator oscillations, it passes on energy to the cavity fields and slows down. Passing through out of phase, the electron takes energy from the gap and picks up speed.

After passing through the gap, the electron beam is repelled by the strong negative charge of the repeller, which slows the beam to a stop and forces it to reverse direction. Passing through the gap again in the opposite direction, the electrons traveling at various speeds drift together in bunches. These bunches are in phase with cavity oscillations and give energy to the cavity to sustain oscillations. A layout sketch of the reflex klystron is shown in Fig. 6-1A.

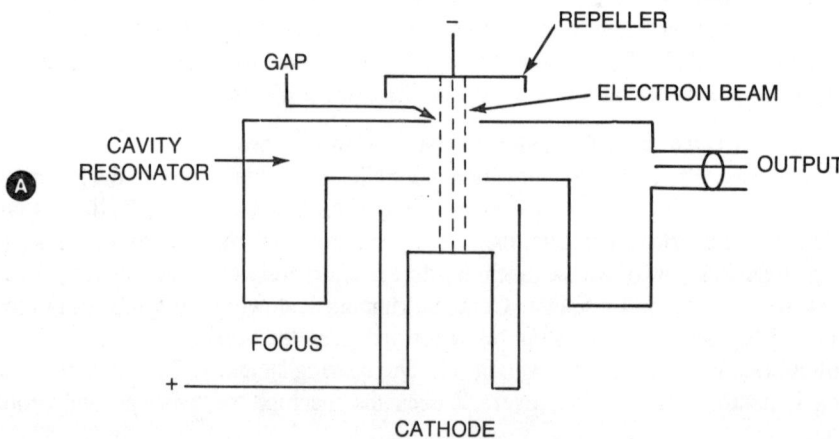

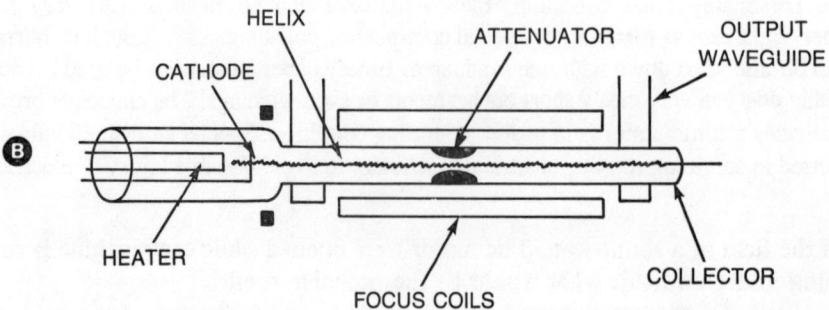

Fig. 6-1. Cross-sectional drawings (A) reflex klystron; (B) traveling wave tube (TWT).

Q22: Draw a diagram showing the construction and explain the principles of operation of a traveling-wave tube.

A: Acting as an RF amplifier at frequencies above 3000 MHz, the traveling-wave tube offers an exceptional bandwidth of about 1000 MHz. As noted in Fig. 6-1B, the tube has no resonant circuits. The cathode and gun anode produce a beam aimed through the long helix and received by the collector at anode potential. The input signal enters the waveguide at the helix end nearest the gun, traveling along the helix to the opposite end where it arrives as an amplified signal to be coupled to the load through the output waveguide. In order to be amplified, the signal must travel at a lower velocity than the electron beam in order to receive energy from that source. This causes the wave to increase in amplitude as it reaches the output end of the helix. The electron beam must be slowed down to about one tenth its normal velocity (speed of light), and this is accomplished by the wave traveling around the turns of the helix at the speed of light but moving forward according to the pitch and diameter of the turns. By satisfactory design of the helix, forward progress of the wave might be limited to the desired speed. Focusing coils restrict the beam diameter and steer it through the center of the helix.

Q23: Describe the physical structure and explain the operation of a multianode magnetron.

A: Figure 6-2 illustrates the structure of the multianode magnetron, which roughly consists of a rod-shaped cathode in a copper cylinder (anode). Several cavities in the anode form resonant tank circuits with the centrally located cathode an output coupling and a very strong permanent magnet. Instead of being drawn to the anode directly, electrons from the cathode are bent as they pass though a strong magnetic field and whirl around the cathode, traveling in the same direction as determined by the field according to Fleming's

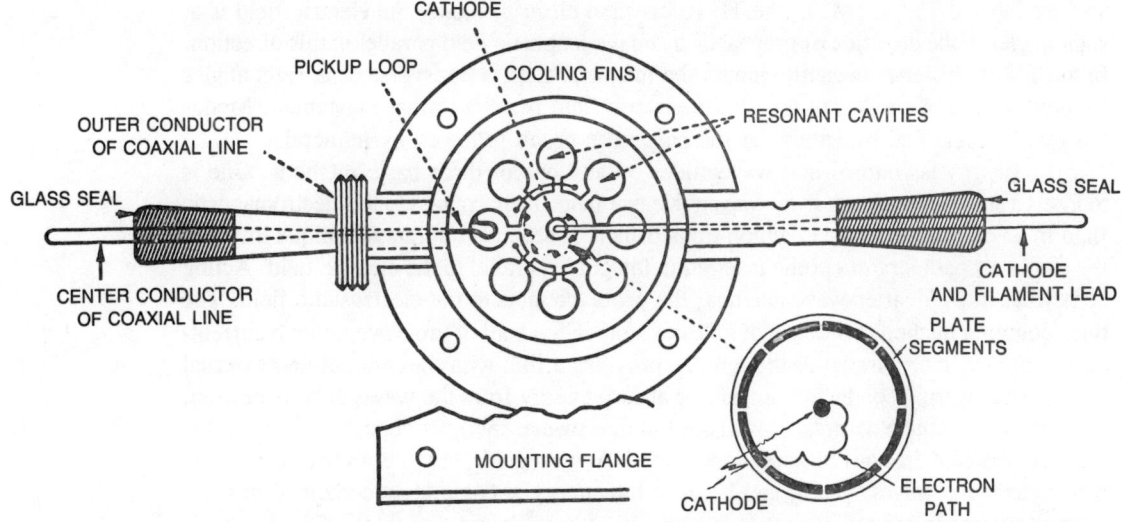

Fig. 6-2. Multinode magnetron construction.

rule. When the electrons pass the cavities or slots in the anode block, the cavities are shock-excited into oscillation and electrons passing the cavity slots out of phase take energy from the cavity, speed up, and being bent faster, return to the cathode. Electrons passing cavity slots in phase give energy to the cavity to sustain oscillations and slow down. Actually, most of the electrons contribute energy to the cavities and as a result of the short path of those that take away energy, the overall efficiency of the magnetron is not decreased to any great extent. In practice, operating efficiencies can be as high as 50 percent. Magnetrons are widely used in radar systems as pulsed power oscillators with peak power of 2 megawatts.

Q24: Discuss the following with respect to waveguides.
(a) Relationship between frequency and size.
(b) Modes of operation.
(c) Coupling of energy into waveguides.
(d) General principles of operation.

A: (a) Waveguides are used in place of transmission lines at UHF or microwave frequencies. The inside dimension of the waveguides varies in accordance with the design frequency band for which it is intended. Inside dimensions decrease with an increase in frequency, and the widest inside dimensions must equal at least one-half wavelength for the signal traveling through it. The narrow width of the rectangular waveguide is usually one-half the wider dimension. This is known as the "fundamental mode" (TE1,0) and is superior to all others because no power is wasted. The wider dimension must be greater than one-half wavelength but no greater than one wavelength and the narrow width less than one-half wavelength.

(b) The mode of operation is based on the configuration of the electric and magnetic fields inside the waveguide. The modes are determined by the shape of the waveguide and are labeled TE or TM. In the TE (transverse electric) mode, the electric field is at right angles to the direction of propagation and the magnetic field parallel in this direction. In the TM (transverse magnetic) mode, the magnetic field is transverse or at right angles to the direction of propagation with the electric field parallel to that movement. Modes are readily identified by letters for the particular group followed by numerals.

(c) Energy is coupled into waveguides by any one of three basic methods. One is to insert a small loop of wire in the waveguide that inductively couples to the electromagnetic field in a way similar to the common transformer. A second coupling technique involves the use of a small, straight probe inserted in the guide parallel to the electric field. Acting somewhat like a quarter-wave antenna, the probe couples to the electrostatic field. The third coupling method makes use of holes or slots in the wall of the waveguide; a current-carrying conductor parallel to these holes provides a link with the internal and external fields. The external conductor can add or absorb energy from the waveguide as desired.

(d) A waveguide permits transmission of microwave energy without the considerable loss encountered in conventional conductors. A waveguide has the ability to conduct electromagnetic waves within, and because the energy is completely contained in these waves, current-losses as encountered normally in wires is not a problem. The energy is in the electric and magnetic fields. Conventional electron flow as in a wire conductor is not required for waveguide transmission. Of course, the use of waveguides is restricted by practical dimensions, because the inside dimension of the guide must increase as the frequency of operation decreases. This makes their use practical for microwave frequencies only.

Q25: Explain briefly the construction and purpose of a waveguide. What precautions should be taken in the installation and maintenance of a waveguide to ensure proper operation?

A: A waveguide can consist of either round or rectangular pipes or channels for propagating microwave energy. Its purpose is clearly defined by the fact that it transmits high-frequency energy without excessive loss as would be unavoidable in the regular wire conductors. Precautions must be observed in installing sections of waveguides to be sure that all joints are strong mechanically and continuous electrically to reduce the possibility of loss. Tight, secure joints keep out moisture and dust particles that could adversely affect proper operation. Careful handling should eliminate bending, denting, or otherwise altering the designed shape of the waveguide and its resonant cavity.

Q26: Explain the operation of a cavity resonator.

A: The cavity resonator is a form of resonant circuit having an extremely high Q and intended for efficient operation in the microwave frequency range. Conventional LC resonant circuits are not practical due to the small physical size of inductance and capacity required in the microwave region. The resonant cavity is like a closed section of waveguide with the microwave energy therein reinforced at resonance, resulting in strong oscillations. This measured section of waveguide has a resonant frequency dependent on its dimensions. Cavities can be shaped in various ways—cylindrical, doughnut, rectangular, and spherical. They are energized in the same manner as waveguides.

Q27: How are cavities installed in vertical guides to prevent moisture from collecting? Why are long horizontal waveguides not desired?

A: Vertical sections should be installed with the end where the cavity is mounted through a choke joint at the bottom. Moisture-sealing gaskets should be used at each choke-coupling flange. Long horizontal waveguides have a tendency to collect moisture, making them undesirable. A small hole can be drilled at the lowest point in such existing installations to permit drainage of accumulated moisture.

TROUBLESHOOTING

Several troubleshooting questions of (multiple-choice type) are asked on most exams. The questions are accompanied by appropriate diagrams with numbered components. This section lists representative troubleshooting questions in the same manner as on the actual test.

It will be helpful for you to remember certain specifics about amplifiers. Many of the questions contain references to meter readings; if you have an idea as to the class of amplifier represented by the diagram and have a basic idea as to the normal readings, you will be in a better position to predict the results of any given malfunction. A class A amplifier draws current at all times (the constant current keeps the transistor or tube operating in the center of its linear region); a class B amplifier is biased near cutoff, which means that it will draw current only during the positive half-cycles of input signals. A class C amplifier will always contain tuned circuits—either in the input circuit or the output circuit, or both. Also, the class C amplifier should draw current for only a small portion of the input signal's cycle, because the device must be driven into conduction by some considerable portion of the input cycle before the device starts conducting.

It is also helpful to bear in mind the basic function of biasing: to control the flow of charge carriers. Without bias, the signal itself is the only control. Thus, if a class C amplifier loses its bias for some reason, current flows uncontrollably; and since class C amplifiers are designed to draw considerable current for short periods of each cycle, a continuous current almost certainly will destroy the device in short order. This is especially true in vacuum-tube, high-power RF amplifiers used in transmitters.

Q28: In the RF power amplifier circuit of Fig. 6-3, what would be the meter indication if capacitor C1 should short?
   (a) M1 would read higher
   (b) M1 would read lower
   (c) M2 would read lower
   (d) M2 would read zero
   (e) Both meters would read normal

A: If C1 shorts, there is no signal developed in the input tuned circuit, and consequently no signal applied to the grid. Under normal conditions, electrons flow from cathode to grid during positive half-cycles. The grid current charges the capacitor as it flows from the grid through the input tuned circuit. A portion of this charging current is diverted by the grid-leak resistor (R1), and so M1 will indicate a current flow. With C1 shorted, however, the input tuned circuit is effectively shunted to ground and can develop no signal. The grid, being at the same potential as the cathode, will have no curtailing effect on electron flow through the tube. The reading of M2 will increase, the reading of M1 will drop, and the circuit will cease to amplify.

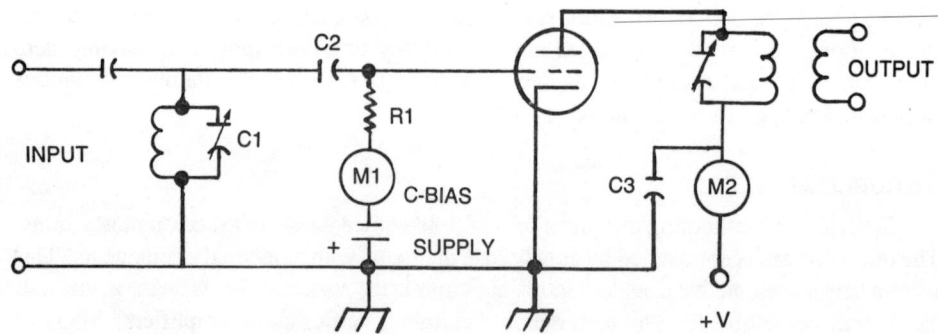

Fig. 6-3. RF power amplifier circuit.

Q29: If R1 should burn out (Fig. 6-3)
   (a) M1 would read zero
   (b) M2 would read lower
   (c) M1 would read higher
   (d) M2 would read higher
   (e) Both (a) and (b) above are true

A: When a resistor "burns out," it opens. With no resistor in the grid circuit, there is no discharge path for C2, so meter M1 would have to read zero. During the positive half of the grid signal, the grid becomes positive with respect to the cathode. The grid draws current and charges to the peak value of the grid signal so that plate of the capacitor connected to the grid becomes negative. The charge path is through the tube. During the negative half-cycle, the grid is driven negative with respect to the cathode. Ideally, C2 should discharge through R1 during negative half-cycles and be charged during positive half-cycles—the bias being the average pulsating dc value of the varying charge. With no discharge path, the capacitor remains fully charged. With no discharge current through the grid-leak resistor, there is no dc bias voltage. With no bias voltage, there is no control element that can be used to check the flow of electrons in the tube from cathode to plate (other than the signal itself); thus, the reading of M2 would increase and the tube could burn out.

Q30: If M1 reads normal while M2 is pinned against the maximum-current stop, there is an indication of:
  (a) Defective C1
  (b) R1 too low in value
  (c) Leaky C3
  (d) Open L2
  (e) Open L1

A: Plate voltage is applied to the oscillator tube though M2 and the output tank circuit. If M1 reads normal or low while M2 is pinned against its peg, there is an indication that excessive current is flowing in the plate circuit. If C1 is defective, plate current would drop. If R1 were too low in value, M1 would read higher than normal. An open L1 would remove the input signal and drop the reading of M1 to zero. If the bypass capacitor is leaky or shorted, however, the high voltage is effectively shorted to ground, and the reading of M2 would be excessive.

Q31: A resonant antenna operating on 1,580 kHz is delta-fed and has a feed point impedance of 19 ohms. Calculate the output power of the transmitter if there is 3 dB loss between the transmitter and antenna, and the antenna base RF ammeter reads 14.2 amperes.

A: The power applied to the base is found from:

$P = I^2 R$
$P = (14.2)^2 (19)$
$P = (201.6)(19) = 3830.4$ watts

Because 3 dB represents a 2:1 power ratio, the actual transmitter RF output power is twice the antenna base power, or $2 \times 3{,}830.4 = 7{,}660.8$ watts.

Q32: What is the frequency range associated with the following frequency designations:
  (a) VLF
  (b) LF
  (c) MF
  (d) HF
  (e) VHF

(f) UHF
(g) SHF
(h) EHF

A: (a) Very low frequency (VLF), below 30 kilohertz (kHz).
  (b) Low frequency (LF), 30 to 300 kHz.
  (c) Medium frequency (MF), 300 to 3,000 kHz.
  (d) High frequency (HF), 3,000 to 30,000 kHz.
  (e) Very high frequency (VHF), 30 to 300 megahertz (MHz).
  (f) Ultrahigh frequency (UHF), 300 to 3,000 MHz.
  (g) Super high frequency (SHF), 3,000 to 30,000 MHz.
  (h) Extremely high frequency (EHF), 30,000 to 300,000 MHz.

Q33: What is meant by the following emission designations: (a) A0, (b) A1, (c) A3, (d) A3A, (e) A5C, (f) F0, (g) F1, (h) F3, (i) F5, (j) P3D, (k) P3F?

A: Designations for various types of transmission are A (amplitude modulated), F (frequency or phase modulated), and P (pulse modulated):
  (a) A0 indicates continuous wave (CW) with no modulation.
  (b) A1 indicates CW telegraphy using on-off keying.
  (c) A3 indicates telephony with regular double sideband and carrier.
  (d) A3A indicates telephony, single sideband, with reduced carrier; also known as SSSC.
  (e) A5C indicates regular amplitude-modulated television (picture) emission with vestigial sideband.
  (f) F0 indicates FM carrier without modulation.
  (g) F1 indicates frequency-shift keying (no modulation).
  (h) F3 indicates regular frequency-modulated telephony such as used in FM broadcast stations.
  (i) F5 indicates frequency-modulated television.
  (j) P3D indicates pulse-modulated telephony with amplitude modulated pulses.
  (k) P3F indicates pulse-modulated telephony with phase-modulated pulses.

Q34: What is the basic difference between type approval and type acceptance of transmitting equipment?

A: Type approval is required for all transmitters in the commercial services and indicates testing by FCC personnel of submitted samples. If the transmitter meets frequency tolerance, harmonic suppression, and stability requirements, it is placed on the type-approved list, and you can then use it. Type acceptance is based solely on information submitted by the manufacturer or individual prospective licensee.

Q35: Define the following phrases:
  (a) Authorized bandwidth
  (b) Bandwidth occupied by an emission
  (c) Station authorization

A: (a) The authorized bandwidth is the maximum width of the band of frequencies, as specified in the authorization, to be occupied by an emission.

(b) The bandwidth occupied by an emission is the width of the frequency band (normally specified in kilohertz) containing those frequencies upon which a total of 99 percent of the radiated power appears, extended to include any discrete frequency upon which a power is at least 0.25 percent of the total radiated power.

(c) The station authorization is any construction permit, license, or special temporary authorization issued by the Commission.

Q36: Can stations in the public safety radio services be operated for short periods of time without a station authorization issued by the FCC?

A: No. No radio transmitter may be operated in the public safety service except under and in accordance with a proper station authorization granted by the Federal Communications Commission.

What notification must be forwarded to the engineer in charge of the Commission's district office prior to testing a new radio transmitter in the public safety radio service, which has been obtained under a construction permit issued by the FCC?

The date on which the transmitter will first be tested in such a manner as to produce radiation, giving the name of the permittee, station location, call sign, and frequencies on which tests are to be conducted. This notification shall be made in writing at least two days in advance of the test date. FCC Form 456 may be used for this purpose.

Q37: Where can standard forms applicable to the public safety radio services be obtained?

A: The standard forms can be obtained from the FCC office at 1919 M St. NW, Washington, D.C. 20054 or from any FCC field engineering office listed in this book.

Q38: In general, what type of changes in authorized stations must be approved by the FCC? What type does not require FCC approval?

A: Proposed changes that result in operation inconsistent with any of the terms of the current authorization require that an application for modification of construction permit and/or license be submitted to the Commission on FCC Form 400. Proposed changes that do not depart from any of the terms of the outstanding authorization for the station involved can be made without prior Commission approval. Included is the substitution of equipment on the Commission's "List of Equipment Acceptable for Licensing," and designated for use in the public safety, industrial, and land transportation radio services—provided that substitute equipment employs the same type of emission and does not exceed the power limitations as set forth in the station authorization.

Changes of name and mailing address do not require a formal application for modification of license (without changes in ownership, control, or corporate structure and without changes in authorized location of the base or fixed station or the area of operation of mobile stations); however, the licensee shall notify the Commission promptly of these changes.

Q39: The carrier frequency of a transmitter in the public safety radio service must be maintained within what percentage of the licensed value? Assume the station is operating at 160 MHz with a licensed power of 50 watts.

A: The carrier frequency of the transmitter must be within 0.0005 percent of its assigned frequency.

Q40: A VHF FM transmitter uses a single crystal oscillator followed by a chain of frequency multipliers that total a multiplication ratio of 24. What is the proper deviation of the reactance modulator following the crystal oscillator in order to achieve an FM deviation of 12 kHz?

A: To find the oscillator deviation, divide the required deviation by the frequency multiplication ratio: $DEV2 = DEV1/24 = 12/24 = 0.5$ kHz $= 500$ Hz.

Q41: What is the maximum percentage of modulation allowed by the FCC for stations in the public safety radio services that utilize amplitude modulation (AM)?

A: The modulation percentage must be sufficient to provide efficient communication and should normally be maintained about 70 percent on peaks but must not exceed 100 percent on negative peaks.

Q42: Define control point as the term refers to transmitters in the public safety radio service.

A: A control point is an operating position that meets all of the following conditions:
  (a) The position must be under the control and supervision of the licensee.
  (b) All monitoring facilities required by this section must be installed at this position.
  (c) The person immediately responsible for the operation of the transmitter is stationed at this position.

Q43: Outline the transmitter measurements required by the FCC for stations in the public safety radio service.

A: The licensee of each station having a transmitter with a plate input power to the final RF stage in excess of 3 watts must employ a suitable procedure to determine that the carrier frequency of each transmitter is maintained within the tolerance prescribed and the results entered in station records:
  (1) When the transmitter is initially installed.
  (2) When any change is made which may affect the carrier frequency or stability of the transmitter.

The licensee must employ a suitable procedure for determining the plate power input to the final RF stage of each base or fixed station transmitter over three watts to ensure operation within the maximum figure specified on the current station authorization. If direct measurement of plate current to the final stage is not practicable, the plate input power can be determined by a measurement of the cathode current in the final stage. In all such cases, the required entry should indicate clearly the quantities measured, the values thereof, and the method of determining the plate power input from those figures. These measurements and entries in the station records shall be made following the initial installation or after any change is made in the transmitter that could increase the power input, and at intervals that should not exceed one year.

The license of each station shall employ a suitable procedure to determine that the modulation of each transmitter, authorized to operate in excess of 3 watts input power does not exceed the limits specified in this part. This determination must be made and entered in the station records when the transmitter is installed initially and when any change is made in the transmitter that could affect the modulation characteristics. Intervals of such measurement should not exceed one year.

Q44: What are the general requirements for transmitting identification announcements for stations in the public safety radio service?

A: The required identification for stations in these services must be the assigned call signal at the end of each transmission or exchange of transmissions, or once for each 30 minutes of the operating period, as the licensee may prefer. However, a mobile station authorized to the licensee of the associated base station and transmits only on the transmitting frequency of the associated base station is not required to transmit any identification.

Q45: When a radio operator makes transmitter measurements required by the FCC rules for a station in the public safety radio service, what information should be transcribed into the station's records?

A: The results and dates of the transmitter measurements, as well as the name of the person making the measurements and expiration date of his license.

Q46: What are the FCC's general requirements regarding the records that must be kept by stations in the public safety radio service?

A: The records must be kept in an orderly manner and in such detail that the data required are readily available. Key letters or abbreviations can be used if proper explanation is given in the record. Each entry in the record must be signed by a person qualified to do so, one who has an actual knowledge of the facts to be recorded. No record or portion thereof shall be erased, obliterated, or willfully destroyed within the required retention period (one year). Any necessary correction can be made only by the persons originating the entry, who must strike the erroneous portion, initial the correction made, and indicate the date of the correction.

7

Sample Questions

If you feel you have made sufficient preparation, check your progress by using the 102 sample multiple choice questions in this chapter. Take a sheet of paper and mark down your choice (a), (b), (c), (d), or (e) beside each question number, but don't check your answers until you have completed the sample test. Then, by checking your answers against the list in the Appendix, you can easily grade your progress. Brush up on any specific areas where a need is shown, and check yourself on the sample test again.

NOTES ABOUT TWO-WAY RADIO

Most of the stations in the public safety radio service and land mobile radio service take advantage of FM's ability to suppress noise, a particularly important consideration in mobile operations. Automobile ignition noise may not be fully suppressed, but with FM it need be reduced only to a level slightly below that of the desired signal. As the noise generated by the car limits the range of the mobile unit, it must be reduced as much as possible in order to allow reliable communications.

Probably the best way to tackle the problem is to bypass or filter the major potential noise sources. These include the generator, voltage regulator, and ignition system. Because the noise can be considered as an ac signal, a large capacitor to ground will provide a low resistance path for the alternating component while leaving the dc in the circuit free to circulate. The capacitor should be as close to the noise source as possible, even mounted directly on the offending unit if possible. A series resistance (in the form of special spark plug wires) may be used in the high-voltage part of the ignition system, because the ignition coil pulses the spark plugs at 25,000 volts or more. The exceptionally high Q of the coil causes the duration of the pulse to be long, and by inserting a series resistance of about

10k, the duration is shortened with an appreciable reduction in total noise. Battery cables sometimes serve as noise radiators and should always be as short as practical. Resistors and capacitors used as noise suppressors in mobile installations must be special types to withstand harsh conditions. All capacitors have an inductance factor to a small extent, at least, and for the VHF operation, the coaxial capacitor is recommended for its high self-resonant frequency.

Ignition noise is a sharp, continuous, popping sound that varies in frequency with the speed of the motor, but it is stronger at idling speed. The installation of a suppressor between the coil and distributor will usually eliminate the problem. If further reduction is required, install a coaxial capacitor in the primary lead to the ignition coil (from the ignition switch) and resistor spark plugs, if special radio-resistance wire was not used in your system. Capacitors should not be used in high-voltage circuits, just as the use of resistors is not permissible in low-voltage circuits.

The alternator offers a whistling sound that increases in frequency as the motor speed increases. By installing a bypass capacitor on the alternator case (connect the capacitor lead to terminal A, the thick wire), the noise is eliminated at its source. On some vehicles, a parallel resonant filter or trap is used in series with the alternator output, rather than a bypass capacitor. Check vehicle manufacturer recommendations.

Instrument noise is difficult to track down, because it can be produced by the fuel, temperature and oil gauges and may be corrected only at the source—never at the dash. Connect a bypass directly to the sending unit, not at the dash. The symptom is irregular popping only when the car is moving.

Front-wheel static results from a build-up of a charge while the car is moving. Grease insulates the hub from the car body. Cure by installing grounding springs inside the front hub caps.

SAMPLE QUESTIONS

1. What other expression describes *difference of potential*?
   (a) Electromotive force.
   (b) IR drop.
   (c) Voltage drop.
   (d) EMF.
   (e) Any of the above.

2. What is the basic unit of EMF?
   (a) Coulomb
   (b) Ohm
   (c) Meter
   (d) Volt
   (e) Watt

3. What would be used to measure current in a circuit?
   (a) A voltmeter.
   (b) An ammeter.
   (c) An ohmmeter.

(d) A potentiometer.
   (e) None of the above.

4. What governs the ability of a material to conduct electricity?
   (a) The number of free electrons.
   (b) The type of insulation.
   (c) Its flexibility.
   (d) Its diameter.
   (e) Its heat resistance.

5. What is a hertz?
   (a) An ampere.
   (b) A megohm.
   (c) A cycle.
   (d) A kilovolt.
   (e) None of the above.

6. What is the value of a resistor colored brown, green, brown?
   (a) 50 ohms.
   (b) 100 ohms.
   (c) 150 ohms.
   (d) 151 ohms.
   (e) None of the above.

7. What is *skin effect*?
   (a) The resistance to high-frequency current in the center of a conductor.
   (b) The insulation losses in a conductor.
   (c) The average reactance between two inductors.
   (d) The resistance to high values of direct current in conductors.
   (e) The effect of a nonconductor on audio voltages.

8. How many degrees does 1 hertz represent?
   (a) 45
   (b) 90
   (c) zero
   (d) 270
   (e) 360

9. What should be used to replace a condenser?
   (a) A capacitor.
   (b) A resistor.
   (c) An inductor.
   (d) A battery.
   (e) None of the above.

10. What is impedance?
    (a) The resistance offered by a capacitor to current flow.
    (b) The resistance offered by an inductance to the flow of current.
    (c) The total opposition to current flow at a specific frequency.
    (d) The opposition of an inductor to changes in current.
    (e) The opposition of a voltage divider to current changes.

11. What is the impedance of an ideal parallel resonant circuit?
    (a) Zero.
    (b) Infinite.
    (c) Twice the value of a series resonant circuit.
    (d) One-half the value of a series resonant circuit.
    (e) The reciprocal of series resonant circuit.

12. A relay having a resistance of 1,000 ohms requires a current of 50 mA from a 100-volt source; what value of series resistance is needed?
    (a) 150 ohms
    (b) 1,500 ohms
    (c) 15,000 ohms
    (d) 100 ohms
    (e) 1,000 ohms

13. Why are interstage leads shielded?
    (a) To control stage gain.
    (b) To reduce distortion.
    (c) To increase magnetic coupling.
    (d) To reduce magnetic coupling.
    (e) None of the above.

14. What is the advantage of matching impedances?
    (a) Reduces harmonics.
    (b) Provides maximum power transfer.
    (c) Eliminates parasitic oscillations.
    (d) Reduces overmodulation.
    (e) Eliminates the need for neutralization.

15. What controls the amount of voltage induced in a conductor?
    (a) The strength of the magnetic field it cuts.
    (b) The length of the conductor cutting the lines of force.
    (c) The speed at which it cuts the lines of force.
    (d) The angle between the conductor and the lines of force.
    (e) Any of the above.

16. What is the total resistance of three 150-ohm resistors connected in parallel?
    (a) 450 ohms.
    (b) 300 ohms.

(c) 100 ohms.
   (d) 50 ohms.
   (e) 75 ohms.

17. What is the actual value of a 10 k resistor?
    (a) 10,000 ohms.
    (b) 10 ohms.
    (c) 100 ohms.
    (d) 1,000 ohms.
    (e) 10,000,000 ohms.

18. What is the total value of two 0.005 µF capacitors connected in series?
    (a) 0.010 µF
    (b) 0.0001 µF
    (c) 0.1 µF
    (d) 0.0025 µF
    (e) 0.00025 µF

19. To change the 0.005 µF capacitor to picofarards, how should the decimal point be moved?
    (a) Six places to the left.
    (b) Six places to the right.
    (c) Three places to the right.
    (d) Nine places to the left.
    (e) Nine places to the right.

20. What effect does increasing the Q have on a tuned circuit?
    (a) It increases the bandwidth.
    (b) It decreases the gain.
    (c) It decreases the bandwidth.
    (d) It causes distortion and radiation.
    (e) It inverts the output waveform.

21. What would cause the plates of a vacuum tube rectifier to become red hot?
    (a) Shorted filter capacitor.
    (b) Grounded filter choke.
    (c) Shorted bleeder resistor.
    (d) Short in the voltage divider or load.
    (e) All of the above.

22. Which of the following could prolong life of a vacuum tube?
    (a) Increased filament voltage.
    (b) Increased screen current.
    (c) Excessive plate current.
    (d) Insufficient grid bias.
    (e) Insufficient filament voltage.

23. When does the current lag the voltage by 45 degrees?
    (a) In a circuit with equal resistive and inductive reactance.
    (b) In a circuit with equal resistive and capacitive reactance.
    (c) A circuit with capacitive reactance and no resistive reactance.
    (d) A circuit with inductive reactance but no resistance.
    (e) In a circuit with equal capacitive and inductive reactance.

24. What is the main advantage of a tetrode over a triode?
    (a) Greater voltage gain.
    (b) Neutralization is not necessary.
    (c) Distortion factor is reduced.
    (d) Power output is considerably greater.
    (e) Any of the above.

25. If a coil with an inductance of 100 millihenrys (mH) and 500 turns has 200 turns added, what is the new value of inductance?
    (a) 1,500 mH
    (b) 1,200 mH
    (c) 196 mH
    (d) 120 mH
    (e) 140 mH

26. How could low-impedance headphones be properly connected to a high-impedance plate circuit?
    (a) With a matching stub.
    (b) With a low-loss transmission line.
    (c) By connecting an RFC in series.
    (d) Through a modulation transformer.
    (e) With an output transformer.

27. If a wattmeter read 220 watts, what power is used in 15 hours?
    (a) 330 watt-hours
    (b) 33 kilowatt-hours
    (c) 3,300 watts
    (d) 3.3 kilowatt-hours
    (e) 2,200 watt

28. If the prefix "meg" is 1,000,000, what is micro?
    (a) 0.000001
    (b) 0.00001
    (c) 0.0001
    (d) 0.001
    (e) 0.0000001

29. How can feedback be prevented in audio amplifiers with a common power supply?
    (a) By reducing the plate voltage.

(b) Increasing the grid bias.
(c) Reducing the screen resistor value.
(d) Increasing the value of the cathode bypass capacitor.
(e) By using decoupling resistors and capacitors.

30. What is the negative electrode of a lead-acid storage battery?
    (a) Monel metal.
    (b) Pure nickel.
    (c) Pure sponge lead.
    (d) Pure zinc)
    (e) Lead peroxide.

31. What is the electrolyte in a lead-acid storage battery?
    (a) Vinegar and water.
    (b) Sulphuric acid and water.
    (c) Nitric acid and distilled water.
    (d) Sodium hydroxide.
    (e) Acetic acid and carbon tetrachloride.

32. How can a storage battery be checked for condition of charge?
    (a) A hydrometer.
    (b) A VTVM.
    (c) A volt-ohmmeter.
    (d) An ammeter.
    (e) None of the above.

33. What is the value of a mica capacitor, reading from left to right top row (white, yellow, violet), second row (green, silver, brown)?
    (a) 0.047 µF, plus or minus 20 percent.
    (b) 947 pF, plus or minus 10 percent.
    (c) 470 pF, plus or minus 10 percent.
    (d) 940 pF, plus or minus 5 percent.
    (e) 9,400 pF, plus or minus 10 percent.

34. If a sine wave measures 150 volts peak, what is the effective voltage?
    (a) 135.0 volts.
    (b) 95.4 volts.
    (c) 212.1 volts.
    (d) 106.2 volts.
    (e) 75.1 volts.

35. What would replace the bleeder if more than one output is needed from the power supply?
    (a) A bridge rectifier.
    (b) A tapped filter choke.
    (c) A voltage divider.

(d) A swinging choke.
(e) A mercury-vapor rectifier.

36. What would result from an open bleeder resistor?
    (a) Improved regulation.
    (b) Choke overheating.
    (c) Rectifier damage.
    (d) Very little regulation.
    (e) Higher hum level.

37. What is the phase relationship between the grid and plate of a triode that is operated in a common-cathode circuit?
    (a) 0 degrees.
    (b) 360 degrees.
    (c) 180 degrees.
    (d) 270 degrees.
    (e) 90 degrees.

38. What is the main advantage of a class B vacuum tube amplifier over one operating class C?
    (a) Greater harmonic output.
    (b) More linear output.
    (c) Greater plate efficiency.
    (d) More economical.
    (e) Greater stability.

39. What is an advantage of the push-pull amplifier?
    (a) Even-order harmonics are cancelled.
    (b) Odd-order harmonics are cancelled.
    (c) Fundamental and other odd harmonics are cancelled.
    (d) High-order harmonics are cancelled.
    (e) Only the fundamental is left in the output.

40. What is the cause of frequency shift?
    (a) High plate voltage.
    (b) Overmodulation.
    (c) High screen voltage.
    (d) Insufficient regulation.
    (e) Excessive biasing.

41. Overmodulation usually causes:
    (a) Attenuation of the lower sideband.
    (b) Suppression of the carrier.
    (c) Increased bandwidth.
    (d) Improved signal-to-noise ratio.
    (e) Higher noise level.

42. What advantage does grid modulation have over plate modulation?
    (a) Offers better linearity.
    (b) Requires much less audio drive.
    (c) Much easier to control.
    (d) Provides better stability.
    (e) Overrides noise better.

43. What is the "flow of holes"?
    (a) Movement of positive carriers.
    (b) Movement of negative carriers.
    (c) Flow of electrons.
    (d) Junction current flow.
    (e) Reverse bias current.

44. What is the output power of a transmitter with 1.6 amps antenna current at a resistance of 50 ohms?
    (a) 80 watts
    (b) 128 watts
    (c) 4,000 watts
    (d) 96.4 watts
    (e) 91.2 watts

45. What is the purpose of diversity reception?
    (a) Practically eliminates fading.
    (b) Eliminates image frequencies.
    (c) Reduces electrical interference.
    (d) Improves quality.
    (e) Overrides weak signals.

46. What is the effect of an open cathode bypass capacitor on an audio amplifier?
    (a) Increased output, improved quality.
    (b) Less output, better quality.
    (c) Greater output, some distortion.
    (d) Practically no effect.
    (e) Tube would be biased to cutoff.

47. What effect would a shorted plate bypass capacitor have on an amplifier?
    (a) Output current would rise.
    (b) Output voltage would drop to zero.
    (c) Tube would be damaged.
    (d) Cathode resistor would burn out.
    (e) None of the above.

48. What is the mark of a Colpitts oscillator?
    (a) Split tank inductance.
    (b) Split tank capacitance.

(c) Isolation of the tank from the load.
  (d) Unusual feedback loop.
  (e) The quartz crystal.

49. What governs the speed of a dc series type motor?
  (a) The frequency of the source.
  (b) The strength of the field.
  (c) The load.
  (d) The applied voltage.
  (e) None of the above.

50. What is the function of the commutator in a dc generator?
  (a) Changes ac to dc.
  (b) Controls the speed.
  (c) Reduces the current on field windings.
  (d) Prevents armature coils from overheating.
  (e) Provides a smoothing action.

51. Where is the suppressor grid located?
  (a) Between the screen grid and plate.
  (b) Between the control grid and plate.
  (c) Between the control grid and screen grid.
  (d) Between the cathode and control grid.
  (e) Between the cathode and filament.

52. What potential is supplied to the suppressor grid of a vacuum tube?
  (a) Same as the screen grid.
  (b) Slightly higher than the screen grid.
  (c) Lower than the plate voltage.
  (d) Higher than the control grid.
  (e) Same as the cathode.

53. Which audio stage supplies information to the final RF amplifier?
  (a) Damper
  (b) Modulator
  (c) Limiter
  (d) Oscillator
  (e) Mixer

54. What may cause excessive grid current fluctuating during plate tuning?
  (a) Insufficient grid bias.
  (b) Poor voltage regulation.
  (c) Improper grid drive.
  (d) Improper neutralization.
  (e) None of the above.

55. What is "type approval" equipment?
    (a) Manufacturer's data submitted to FCC.
    (b) Tested by a licensed operator.
    (c) Tests by FCC personnel.
    (d) Test approved by an installation agent.
    (e) Approved and tested by the licensee.

56. How often should the carrier frequency of a 3 kW crystal-controlled transmitter be checked?
    (a) At least once a year.
    (b) At least once a month.
    (c) Twice a year.
    (d) Once a week.
    (e) Every 24 hours.

57. What is the maximum allowable bandwidth for an AM broadcast station?
    (a) 2 kHz
    (b) 4 kHz
    (c) 8 kHz
    (d) 10 kHz
    (e) 20 kHz

58. What is required to prevent frequency drift in the crystal oscillator?
    (a) Low temperature operation.
    (b) Constant temperature.
    (c) Maximum crystal current.
    (d) Excessive feedback.
    (e) Any of the above.

59. What happens when excitation is lost in a class C amplifier with grid leak bias?
    (a) The output increases.
    (b) Plate current is excessive.
    (c) Plate current drops to a low level.
    (d) Self-oscillation takes place.
    (e) Output is distorted.

60. What causes self-oscillation in an audio amplifier?
    (a) Open grid resistor.
    (b) Defective bypass capacitor.
    (c) Leaky filter capacitor.
    (d) Gassy tube.
    (e) Any of the above.

61. What is the conduction angle of a class A amplifier?
    (a) 120 degrees.
    (b) 180 degrees.

(c) 270 degrees.
(d) 360 degrees.
(e) 25 degrees.

62. What class of amplifier is most efficient?
    (a) Class A
    (b) Class AB
    (c) Class B
    (d) Class C
    (e) None of the above.

63. What is the function of the screen grid?
    (a) Decreases secondary emission.
    (b) Reduces interelectrode inductance.
    (c) Eliminates harmonic generation.
    (d) Increases secondary emission.
    (e) Lowers interelectrode capacitance.

64. What is the result of plate current reaching its maximum level?
    (a) Plate saturation.
    (b) Plate resonance.
    (c) Demodulation.
    (d) Parasitic oscillations.
    (e) Plate cutoff.

65. What is the approximate efficiency of a class A amplifier?
    (a) 100 percent.
    (b) 95 percent.
    (c) 70 percent.
    (d) 50 percent.
    (e) 20 percent.

66. What is the output voltage of the usual primary cell?
    (a) 4.5 volts.
    (b) 1.5 volts.
    (c) 2.1 volts.
    (d) 9.0 volts.
    (e) 22.5 volts

67. What is a common type meter movement?
    (a) Faraday
    (b) Edison
    (c) D'Arsonval
    (d) Miller
    (e) Marconi

68. Which of the following describes a primary cell?
    (a) Lead-acid cell.
    (b) A rechargeable cell.
    (c) A wet cell.
    (d) Large current cell.
    (e) None of the above.

69. Lightning tends to cause _____ _____ of a carrier.
    (a) Pulse modulation.
    (b) Phase modulation.
    (c) Frequency modulation.
    (d) Amplitude modulation.
    (e) Harmonic radiation.

70. How is field strength normally expressed?
    (a) Joules per square meter.
    (b) Microvolts per meter.
    (c) Microvolts per centimeter.
    (d) Kilowatts per mile.
    (e) Reactance per mile.

71. What is the bias level of a class C amplifier?
    (a) Slightly positive.
    (b) Slightly negative.
    (c) Well below cutoff.
    (d) The cutoff value.
    (e) Zero.

72. What frequencies are covered by the high-frequency (HF) band?
    (a) 30 to 300 MHz.
    (b) 3 to 30 MHz.
    (c) 30 to 300 kHz.
    (d) 3 to 30 kHz.
    (e) 300 to 3,000 kHz.

73. What is referred to as an A5 transmission?
    (a) Television.
    (b) FM telephony.
    (c) AM telephony.
    (d) AM telegraphy.
    (e) Unmodulated AM.

74. What is the meter reading shown in Fig. 7-1?
    (a) 0.0258 volt
    (b) 2580 volts
    (c) 25,800 volts

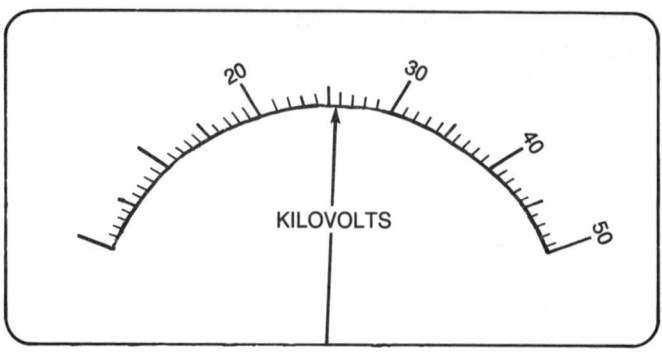

Fig. 7-1. High-voltage meter.

    (d) 258,000 volts
    (e) 2,580,000 volts

75. What is the effect of a decoupling circuit?
    (a) Introduces oscillations.
    (b) Improves regulation.
    (c) Provides better transfer of power.
    (d) Provides interstage shielding.
    (e) Isolates stages from each other.

76. What is the purpose of the limiter in an FM receiver?
    (a) Limits intermediate frequency bandwidth.
    (b) Removes amplitude variations.
    (c) Establishes AVC level.
    (d) Controls audio peaks.
    (e) Prevents overmodulation distortion.

77. What is the function of a traveling-wave tube?
    (a) UHF amplifier/oscillator.
    (b) VHF oscillator.
    (c) FM modulator.
    (d) AM buffer amplifier.
    (e) LF intermediate-frequency amplifier.

78. What problem results from using long, horizontal waveguide sections?
    (a) Modes overlap.
    (b) Reflections are caused.
    (c) Moisture accumulates.
    (d) Difficult to match impedances.
    (e) Signal losses are increased.

79. What is gained by the use of a capacitor input filter in place of the choke input type?
    (a) Higher dc peak voltage.
    (b) Much improved regulation.
    (c) Lower ripple content.
    (d) Easier to maintain.
    (e) Less filtering required.

80. If a transformer has more turns on the primary than the secondary, it would be called a:
    (a) Push-pull transformer.
    (b) Step-down transformer.
    (c) Step-up transformer.
    (d) Modulation transformer.
    (e) Power transformer.

81. Which type microphone requires a dc source?
    (a) Crystal microphone.
    (b) Dynamic microphone.
    (c) Carbon microphone.
    (d) Ribbon microphone.
    (e) Ceramic microphone.

82. What is a desirable quality of the ribbon type microphone?
    (a) Has a wide frequency range.
    (b) Rugged construction.
    (c) High output level.
    (d) Low-impedance output.
    (e) All of the above.

83. What components form an RC circuit?
    (a) Resistor and a coil.
    (b) Resistor and a choke.
    (c) Resistor and a capacitor.
    (d) Relay and capacitor.
    (e) Relay and a coil.

84. What is pre-emphasis?
    (a) Low audio frequencies amplified more.
    (b) High audio frequencies amplified more.
    (c) High audio frequencies attenuated.
    (d) Low audio frequencies attenuated.
    (e) Lows attenuated, high frequencies boosted.

85. What is "high-level" modulation?
    (a) Modulation level over 100 percent.
    (b) Modulation input to the plate circuit of the final amplifier.

(c) A modulation percentage below 85.
(d) When two modulators are used in parallel.
(e) When modulation is introduced in the grid of the final.

86. When an AM transmitter is modulated 100 percent with a sinusoidal tone, what is the percentage increase in antenna current?
    (a) 22.5 percent.
    (b) 25.5 percent.
    (c) 60 percent.
    (d) 85 percent.
    (e) 100 percent.

87. What characteristics of an audio tone determine the modulation percentage in an FM transmitter?
    (a) Amplitude and phase of the tone.
    (b) Amplitude and frequency of the tone.
    (c) Only the frequency of the modulating tone.
    (d) Only the phase of the modulating tone.
    (e) The pulse interval of the tone.

88. In a quarter-wave transmission line, which of the following characteristics is not true?
    (a) When shorting the far end, the input impedance is infinite.
    (b) When the far end is terminated in an open circuit, the input is extremely low.
    (c) A quarter-wave section always repeats its load.
    (d) In the quarter-wave section, the input impedance reverses the output impedance.
    (e) The greater the value of the terminating resistance, the lower the input impedance will be.

89. What is the advantage of terminating a transmission line in an impedance value equal to its characteristic impedance?
    (a) Offers maximum power transfer.
    (b) Greatly reduces line radiation.
    (c) Provides a 1:1 current ratio without losses.
    (d) Provides an excellent standing wave ratio.
    (e) All of the above.

90. What is the current in a 72-ohm transmission line with an input of 10,000 watts? Assume that the line is terminated in a 72-ohm resistive load.
    (a) 12.38 amperes.
    (b) 23.56 amperes.
    (c) 11.78 amperes.
    (d) 13.88 amperes.
    (e) 138.8 amperes.

91. What is the purpose of using top loading in a standard broadcast antenna?
    (a) Controls the power output.

(b) Increases the effective height of the antenna.
(c) Decreases the effective height of the antenna.
(d) Improves the vertical field intensity.
(e) Produces a directional radiation.

92. Which efficiency factor would be used when the authorized night-time power of a standard broadcast station is different from daytime power and the operating power is computed by the indirect method?
    (a) Efficiency factor for minimum rated carrier power.
    (b) Efficiency factor for maximum rated carrier power.
    (c) Efficiency factor for regular daytime power.
    (d) Efficiency factor for average rated carrier power.
    (e) Efficiency factor for regular night-time power.

93. What is the center frequency of an FM broadcast station?
    (a) The instantaneous output frequency.
    (b) The frequency of the unmodulated carrier.
    (c) The average between maximum and minimum peak values.
    (d) The average difference between maximum and minimum excursions.
    (e) The output frequency of the FM oscillator.

94. What method must be used to determine the operating power of an FM broadcast station?
    (a) The Armstrong method.
    (b) The electron coupled method.
    (c) The indirect method.
    (d) The direct method.
    (e) E_p times I_p.

95. What is the function of the reactance tube in an FM broadcast transmitter?
    (a) Prevents overheating the swinging choke.
    (b) Provides better stability with plate modulation.
    (c) Modulates the master oscillator.
    (d) Improves power supply regulation.
    (e) Stabilizes the modulator balancing.

96. What frequency tolerance is an international broadcast station allowed?
    (a) 0.00005 percent.
    (b) 0.0005 percent.
    (c) 0.005 percent.
    (d) 0.003 percent.
    (e) 0.001 percent.

97. What is the necessary bandwidth of the IF strip in an FM receiver for acceptable quality?
    (a) 25 kHz

(b) 50 kHz
  (c) 75 kHz
  (d) 100 kHz
  (e) 150 kHz

98. What is the speed of a 220-volt 60 Hz, 4-pole, 3-phase induction motor?
    (a) 2150 RPM.
    (b) 1250 RPM.
    (c) 1750 RPM.
    (d) 2250 RPM.
    (e) 2500 RPM.

99. If a voltmeter with a 0 to 250 microampere scale and a suitable series resistor offers a full-scale reading of 500 volts, what is the ohms-per-volt rating of the meter?
    (a) 4,000 ohms per volt.
    (b) 2,000 ohms per volt.
    (c) 1,000 ohms per volt.
    (d) 5,000 ohms per volt.
    (e) 10,000 ohms per volt.

100. What accuracy at full scales is required of meters used to measure voltage and current in the final stage of a standard broadcast station?
    (a) 10 percent.
    (b) 5 percent.
    (c) 2 percent.
    (d) 1 percent.
    (e) 0.02 percent.

101. What is the purpose of the variable attenuator for a speech input system?
    (a) Attenuates undesired background.
    (b) Provides control of the voltage gain.
    (c) Attenuates undesired frequencies.
    (d) Serves as an impedance match between the signal and the amplifier.
    (e) Adjusts the amplifier frequency response.

102. What is the purpose of a line equalizer?
    (a) Balances the input and output circuits.
    (b) Uniform frequency response at all times.
    (c) Increased power transfer.
    (d) Eliminates standing waves.
    (e) Ensures correct impedance matching.

8

Radar Endorsement

The expansion of radar during recent years into the small boating and other fields has resulted in additional opportunities for the licensed technician. Kits and other inexpensive radar type systems are available, but the installation, maintenance, and repair of such equipment requires a radiotelephone license with the radar endorsement. Small radar units are also being used in light planes and some guidance systems. Only a brief coverage of the subject is attempted here, but for those with a greater interest, many comprehensive books are currently available.

Radar is actually an acronym of *r*adio *d*etection *a*nd *r*anging, a method of detecting objects by means of radio waves (it is now also accepted as a noun—not an acronym). A radar system radiates microwave energy in pulses that bounce off objects in their path. The reflections return to the system receiver and are displayed on the oscilloscope screen. The interval between transmission and reception of the pulses is timed to compute their distance away. Although there are other forms of radar transmission, we are interested in pulse transmission, which is the type commonly used in search and direction-determining equipment.

RULES AND REGULATIONS

Before proceeding with the actual radar equipment involved, a study of the Commission's Rules and Regulations governing operation, installation, and maintenance is covered in the first part of the question and answer section.

So that the operation of each system is clear and questions regarding their purpose can readily be answered, each unit is individually reviewed in the question and answer section.

The *echo box* is a device used for periodic system checks. It utilizes a high-Q resonant cavity for a reference standard. The cavity can be varied in length to change the frequency as desired. Pulsing the transmitter shock-excites the cavity resonator into oscillation, which continues for a short time after the transmitter pulse. A signal is returned to the receiver and appears as an artificial pattern on the CRT indicator. By measuring the length of the spokes or size of the intensified area on the screen when the system is in good working condition, a standard can be set for future checks, because a decrease in spoke length or distortion in shape would indicate improper performance.

QUESTIONS AND ANSWERS

Q1:   Within what frequency bands do ship radar transmitters operate?
A: The following bands are authorized: 2900 to 3100 MHz; 5460 to 5650 MHz; 9320 to 9500 MHz.

Q2:   What are the FCC license requirements for the operator who is responsible for the installation, servicing and maintenance of ship radar equipment?
A: Although fuses can be replaced without a license, the operator responsible for the installation, maintenance and servicing of the ship radar equipment must hold a first or second class radiotelephone or radiotelegraph license as well as the ship radar endorsement.

Q3:   Who may operate radar equipment in the ship service?
A: The master of the ship may designate any crew member to operate a ship radar station during the course of normal duty.

Q4:   Under what conditions may a person who does not hold a radio operator license operate a radar station in the ship service.
A: Any crew member designated by the master of the ship may operate the radar station so long as the station does not possess external controls that affect the tuning, frequency, or other operating parameter of the equipment.

Q5:   Who may make entries in the installation and maintenance record of a ship radar station?
A: Entries in the maintenance and installation log may be made by the station's licensee or any authorized licensee who performs a maintenance adjustment or service of the equipment.

Q6:   What entries are required in the installation and maintenance record of a ship radar station?
A: The following entries must be made:
  (a) Date and place of initial installation.
  (b) Any required steps taken to eliminate any interference found to exist at the time of such installation.
  (c) The nature of any complaint, including interference to radio communication arising following the initial installation and the date of same.
  (d) Reason for complaint or trouble leading to same and the component or part responsible.
  (e) Corrective measures taken and date.

(f) Name, license number and date of radar endorsement on the operator license of the responsible operator supervising or engaged in the installation, service or maintenance.

Q7: Draw a block diagram of a radar system, labeling the antenna, duplexer, transmitter, receiver, timer, modulator, and indicator.

A: See the sketch in Fig. 8-1.

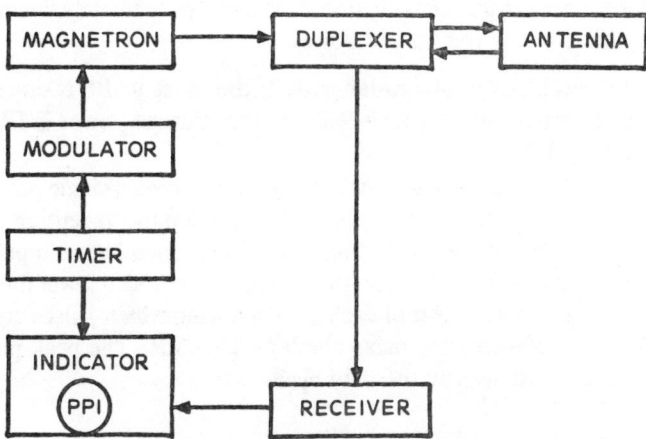

Fig. 8-1. Block diagram of a radar system.

Q8: Explain briefly the principle of operation of a radar system

A: A radar system transmits high-power RF pulses of very short duration but at regular intervals from a direction antenna. A portion of the energy from the pulse is reflected back from the target to the receiver in the system, with the direction determined by the position of the antenna and the distance of the target by the elapsed time required for the return of the reflected signal. The modulator or pulser in Fig. 8-1 controls the pulsing of the magnetron oscillator to permit extremely short duration pulses of high power to be generated and fed through a waveguide to a horn-type antenna that radiates them in a narrow, searchlight type beam. The antenna rotates continuously to permit scanning of the desired area.

The reflected signal picked up by the receiver is amplified and displayed on a cathode ray tube known as a *Plan Position Indicator* (PPI). The duplexer makes it possible to use the same antenna for transmitting the pulse and receiving the reflected echo by disconnecting the receiver during transmission and the transmitter between pulses to allow the echo to be received. The timer provides synchronization between the scope (PPI) and the pulse transmission to ensure accurate measurement of the elapsed time from pulse transmission to echo reception as shown by position of the PPI screen.

Q9: What component determines pulse repetition rate (PRR) in radar?

A: The timer or synchronizer is responsible for determining the pulse repetition rate of a radar system.

Q10: What was the purpose of the rotary spark gap used in some older radar sets?
A: Although not used in modern equipment, the rotary spark gap was a mechanical substitute for the modulator tube and offered a mechanical method for discharging the pulse-forming network and forming the output pulse by modulating the magnetron directly at a high level.

Q11: What is the purpose of an artificial transmission line in a radar system?
A: Its function is to form the shape of the modulation pulses with an LC network that actually resembles a regular transmission line. This network accurately controls the length, shape and magnitude of the transmitted pulse.

Q12: What is the peak power of a radar pulse if the pulse width is one microsecond, the pulse repetition rate (PRR) is 900 and the average power is 18 watts? What is the duty cycle?
A: Peak power in radar is the average power transmitted during a single pulse and average power is the average of the transmitted power during the pulse repetition period, or the start of a pu'_ to the start of the succeeding pulse. The period between pulses is always many times the pulse duration which makes the average power much lower than peak power. The duty cycle is that fractional part of each second during which pulses are transmitted, and is the result of multiplying the pulse width by the PRR. The peak power is found by dividing the average power by the duty cycle.

$$\text{Peak power} = \frac{\text{average power}}{\text{duty cycle}} = \frac{18}{0.0009} = 20,000 \text{ watts}$$

Q13: Explain briefly why radar interference to a radiotelephone receiver is frequently characterized by a steady tone in the radio speaker?
A: The steady tone is the PRR of the radio transmitter or a harmonic thereof and it is detected by the receiver. The signal may be picked up through local power lines or by reception of the radiation. As the radar pulses are rich in harmonics as well as strong in peak power, they are not tunable.

Q14: Describe how various types of interference from a radar installation can be apparent to a person when listening to a communications receiver.
A: When listening to radar interference with an AM or FM communications receiver, the sound is a characteristic musical buzzing, similar to ignition interference but at a much higher repetition rate. Often, such interference varies in intensity proportional to the radar's sweep. Most squelch-operated receivers "see" radar signals as noise; thus, the interference is not noticeable except when an accompanying communications signal defeats the squelch and is detected. In such cases, the radar pulses tend to desensitize the receiver, rendering it incapable of receiving faint signals that would otherwise be receivable.

Q15: How are various types of radar interference recognized in auto-alarm equipment? In direction-finding equipment?
A: The characteristic audio indication of radar interference may be observed by listening to the auto-alarm or direction-finding equipment with an amplifier-speaker or a set of headphones. Owing to the nature of radar interference (its variation in intensity with range sweeping), it is often difficult to locate with direction-finding equipment. The most

satisfactory method is to observe a signal strength indicator while monitoring during periods of peak disturbance, and take readings from several locations.

Q16: On what frequencies should the radar serviceman look for radar interference to communication receivers on ships equipped with radar?

A: The frequencies susceptible to interference would depend on the type of interference (mixing, or intermod, for example, can occur on any frequency); however, UHF equipment in the 450 MHz region would be most vulnerable.

Q17: Why is it important that all units of a radar installation be thoroughly bonded to the ship's electrical ground?

A: Grounding is important as a safety measure to eliminate the danger of electrical shock and to prevent radar interference to other electronic units.

Q18: Would there be danger in testing or operating radar equipment aboard ship when explosive or inflammable cargo is being handled?

A: Definitely, radar equipment should never be operated or tested while explosives or inflammable cargo is being handled. The high-frequency radar pulses could dielectrically produce enough heat in the material to cause it to ignite, and possible arcing in the system would pose a threat as well.

Q19: What precaution should a radar serviceman observe when making repairs or adjustments to a radar set to prevent personal injury to himself or other persons?

A: Power should be removed from the equipment before making adjustments. Electrolytics and power supply filters should be discharged by shorting to ground for 30 seconds. The magnetron and other heat-producing elements should be allowed to cool in order to prevent possible burns.

Q20: In checking a direction finder for interference from radar equipment, would it be advisable to rotate the D-F loop while checking?

A: Radar interference tends to be characterized by a constantly changing intensity, particularly where the interference signal is being radiated by the radar's revolving antenna. This lack of a continuous and stable source renders a direction finder useless unless it is held in a fixed direction during each radar sweep. The best approach is to take independent readings from a variety of directions without moving the loop during each reading. The readings can be compared following the taking of sample signals.

Q21: Two small-boat targets are side-by-side. Their dimensions together are $\frac{2}{3}$ the cross-range beam dimension of the radar. What does the operator of the radar see on the scope?

A: A single target. The radar cannot discriminate two targets that are less than the 3 dB beamwidth of the radar antenna.

Q22: Would a radar installation be likely to cause interference to radio receivers if long connecting lines were used between the radar transmitter and modulator?

A: Yes.

Q23: What steps can be taken by the radar serviceman to eliminate a steady-tone of interference to radio communication receivers, or interference to loran evidenced by "spikes"?

A: The serviceman should check to see that appropriate filters are installed in the receiving equipment and that the reciprocal filter types are installed and properly deployed on the radar equipment. Where the interference is attributable to leakage through the power lines, the serviceman should make certain that appropriate RF filter chokes have been used to isolate the radar gear from the lines.

Q24: What determines the azimuth and range resolutions of a shipboard radar unit?
A: The azimuth resolution is set by, and specified with, the -3 dB beamwidth of the radar antenna. For parabolic "dish" antennas, the principal dimension is the diameter of the antenna. The range resolution is set by the pulse width.

Q25: Name at least four pieces of radio or electronic equipment aboard ship that might suffer interference from the radar installation.
A: ADF equipment, loran gear, auto-alarm equipment, radiotelephone installations and other communications equipment.

Q26: What can cause bright flashing pie sections to appear on a radar PPI scope?
A: The flashing pie sections, sometimes referred to as "spoking," as displayed on the plan position indicator, normally signifies failure of the AFC circuit in the receiver. The cause is often a defective crystal in the AFC, but other possibilities include an improperly keyed magnetron or even failure of the magnetron. Irregular rotation of the deflection coil or yoke could also cause a similar indication and would be the result of a defective servo amplifier or mechanical binding in the assembly.

Q27: What symptoms on a radar scope would indicate that the radar receiver mixer crystal is defective?
A: A defective crystal mixer is indicated by the presence of excessive noise. While a single-ended or unbalanced crystal mixer will not exhibit the noise-reduction and noise cancellation capability of a balanced or hybrid mixer, there is a noticeable difference in noise content on a radar signal when the mixer crystal begins to deteriorate. Similarly, crystal defects can take the form of phase-shifting of the displayed signal as well as a general diminishing of displayed-signal quality (amplitude and intensity).

Q28: What tests do radar servicemen make to determine whether or not the radar receiver mixer crystal is defective?
A: The crystal in a radar receiver's mixer stage is a high-quality germanium diode. Normally, the crystal is easily accessible, as it plugs directly into a waveguide section. To check the crystal element, it is only necessary to check the rectification capability of the device with an ohmmeter. When forward-biased, the device should indicate low resistance; when reverse-biased, the resistance across the terminals should be quite high. The ratio of resistance between forward- and reverse-biasing should be at least 200 to 1. If the reverse-bias resistance reading is not on the order of hundreds of thousands of ohms, the crystal junction has probably deteriorated and excessive leakage is indicated.

Q29: In a radar set, what are the indications of (1) a defective magnetron, (2) a weak magnet in the magnetron, (3) a defective crystal in the receiver converter stage?
A: When a magnetron is supplied with a rectangular plate-voltage pulse of proper magnitude and duration and is itself operating as it should, the ultrahigh-frequency output waveform should be a uniform wavetrain. If this output voltage is rectified and filtered,

it appears as a square wave whose pulse width is equal to the duration of the wavetrain. If a nonrectangular pulse is supplied to the magnetron, or if the magnetron itself is not operating properly, the pulse envelope is not a square wave. Hence, observation of the pulse envelope from a transmitter gives an indication of the magnetron performance. In general, if the frequency or plate voltage of a magnetron changes, the power output changes, which also makes a change in the observed pulse-envelope height. As a further check on the operation of magnetrons, coaxial wavemeters are sometimes used to measure the wavelength of oscillation or to check the relative power outputs between different tubes.

Under certain conditions, the frequency of the magnetron is very sensitive to the RF tuning. In these cases, the mere variation in the impedance of the line resulting from faulty rotating joints or from other causes can shift the magnetron frequency by several megahertz.

A magnetron must not have its plate voltage applied when the magnetic field is not present. If this is done, the plate could be bombarded and the tube destroyed almost at once.

Another precaution to observe concerns the powerful permanent magnets that are frequently used to supply the magnetic field. Striking or jarring these magnets or touching them with a magnetic material such as a screwdriver greatly lowers their field strength.

Q30: What is the purpose of a klystron in a radar set?
A: A klystron can be used as an amplifier, oscillator, or mixer. It usually has a low power-output capability, and is typically used in the local oscillator in the receiver portion of a radar set. The klystron's output is heterodyned with the RF input to produce an intermediate frequency. Some modern klystrons are high power, and are used to replace magnetrons in radar transmitters.

Q31: Explain briefly the principle of operation of the reflex klystron.
A: A klystron tube using cavity resonators is very critical to adjust, because the tuning and spacing of the cavities are interdependent. Therefore, when it is desirable to use a velocity-modulated tube as an oscillator only, a simplified form called the reflex klystron is used. This tube is so named because the same set of grids is used for both bunching and catching, and a negative repeller plate is provided to force the electrons to retrace their paths after their first passage through the grids.

By proper adjustment of the negative voltage on the repeller plate, the electrons that have passed the bunching field might be made to pass through the resonator again at the proper time to deliver energy to this circuit. Thus, the feedback needed to produce oscillations is obtained and the tube construction is simplified. Spent electrons are removed from the tube by the positive accelerator grid or by the grids of the resonator. Energy is coupled out of the cavity by a one-turn coupling loop. The operating frequency can be varied over a small range by changing the negative potential of the repeller. This potential determines the transit time of the electrons between their first and second passages through the resonator. The output of the oscillator is affected considerably more than the frequency by changes in the repeller voltage. This is because the output depends upon the fact that the electrons go back through the resonator just at the time when they are bunched and at exactly the decelerating half-cycle of oscillating resonator grid voltage. The volume of the resonant cavity is changed to change the oscillator frequency. The repeller voltage may be varied over a narrow range to provide minor adjustments of frequency.

Three typical tubes of the reflex klystron type are the McNally tube, using an external cavity with screw plugs for tuning; the Pierce (or Shepard) tube, using a cavity sealed in the tube and tuned by means of flexing the tube envelope to vary the grid spacing; and the klystron 417, a large tube whose cavity can be tuned by flexing the sides of the cavity. These reflex klystrons are by far the most widely used types of local or beating oscillators in ultrahigh-frequency receivers.

Q32: Can a simple magnetron generator be used in a coherent radar?
A: No. The coherent radar requires the pulse-to-pulse phase relationship to be maintained, which a magnetron cannot do. Generally, klystrons or traveling wave tubes are used in coherent radars.

Q33: Draw a simple block diagram of a radar receiver, and label the signal crystal, local oscillator, AFC crystal stage, IF amplifier, and discriminator.
A: See Fig. 8-2 for a block diagram of a radar receiver.

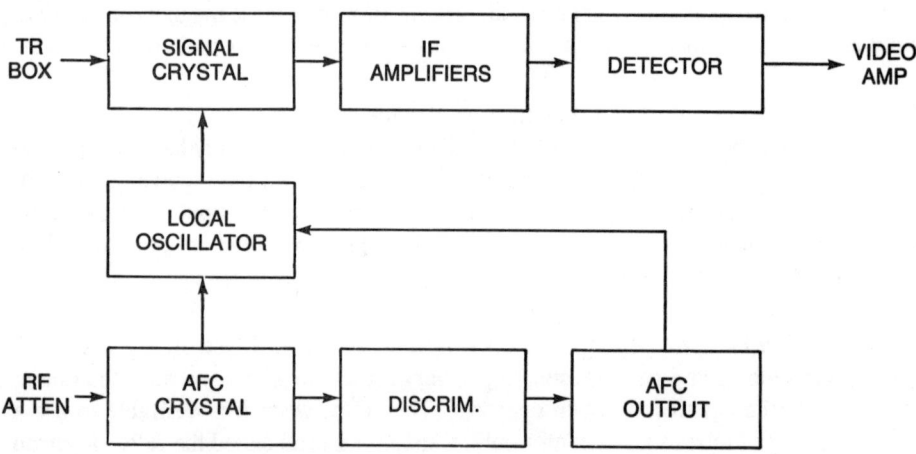

Fig. 8-2. Block diagram of a radar receiver.

Q34: What type of detector is used frequently in radar receivers?
A: The usual type of detector (mixer) is a crystal diode.

Q35: What care should be taken when handling silicon crystal rectifier cartridges for replacement in radar superheterodyne receivers?
A: Silicon crystals are extremely sensitive to the electrical field and static charges and should be wrapped in lead foil or kept in a lead container when not in use. The technician should touch a convenient ground to discharge static charges from his/her body before handling and inserting the exposed crystal in the set. Mechanical shock and strong electrical fields must be avoided while handling this type of diode.

Q36: What nominal intermediate frequencies are commonly used in radar receivers?
A: The commonly used intermediate frequencies are 30 and 80 MHz.

Q37: What is "sea return" on a radar scope?
A: Sea return is the reception and radar scope display of signals that strike ocean wave crests. Since the reflection of such signals depends on the angle of the wave surface at any given instant, the interference is not normally a long-term problem.

Q38: Explain briefly the purpose of the sensitivity time control circuit in a radar set.
A: Receiver gain is reduced automatically for closer targets by the sensitivity time control, reducing sea return interference and overload of the receiver as a result of the strong reflections from nearby targets. The STC control should be carefully adjusted until the solid sea return pattern becomes weaker and the closer ship targets are readily observed.

Q39: What is the purpose of the discriminator stage in a radar superheterodyne?
A: It serves as part of the automatic frequency control (AFC) circuit that prevents drift in the frequency of the local klystron oscillator. Any drift in the klystron frequency causes the intermediate frequency of the receiver to drift, developing a dc voltage across the discriminator output. After amplification, the dc voltage is impressed on the klystron repeller grid, causing it to return to the correct frequency.

Q40: Draw a diagram of a cathode ray tube used in radar, showing the principal electrodes in the tube and the path of the electron beam.
A: See Fig. 8-3.

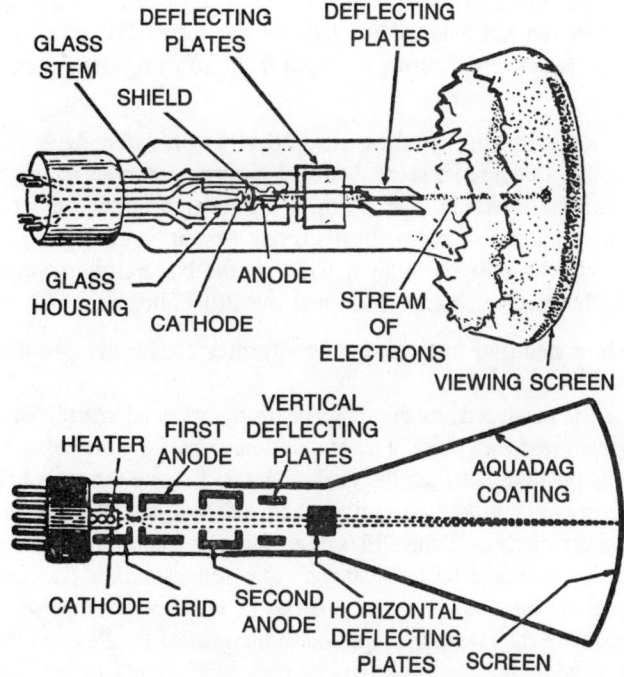

Fig. 8-3. Cathode ray tube construction.

Q41: What is the distance in nautical miles to a target if it takes 12.3 microseconds for a radar pulse to travel from the radar antenna to the target, back to the antenna and be displayed on the PPI scope?

A: The distance to the target would be one nautical mile. A radar pulse travels 1 nautical mile and returns in 12.3 microseconds.

Q42: Explain the principle of operation of the cathode ray PPI tube and the function of each electrode.

A: Referring to Fig. 8-3, electrons are emitted by the cathode and accelerated by the first anode toward the face of the tube. The focusing coil confines the electron stream to a narrow beam approaching a point at the face of the CRT. The Aquadag coating on the bell of the tube is the second anode, which, due to its high potential, accelerates the beam to a high velocity. Beam brightness is regulated by the amount of negative bias on the grid electrode, and the fluorescent material on the screen or face of the tube causes light wherever the beam strikes it. The deflection coils cause the beam to sweep across the screen according to the timing of the current flowing through them. In sets using PPI (plan-position-indicator), the deflection coils rotate around the neck of the tube in step with the rotation of the radar antenna. This means that the sweep line on the screen rotates radially around the center in sync with the antenna.

Q43: What is the purpose of the Aquadag coating on radar cathode ray tubes?

A: Aquadag is a conductive material of graphite particles used to paint the inside of cathode ray tubes to form a second anode. The entire painted surface assumes the high-voltage potential applied at the terminal on the side of the tube. The coating also forms an electrostatic shield to prevent external voltages from affecting the sweep of the electron beam.

Q44: What is meant by the "bearing resolution" of a radar set?

A: This is the ability of the radar set to distinguish between targets having the same range but differing azimuth directions. The width of the radar beam is the determining factor in bearing resolution; a narrow beam affords better separation of targets at the same radial distance than would be possible with a wide beam. Naturally, resolution is affected considerably by the receiver components and the PPI indicator scope.

Q45: Explain how heading flash and range-marker circles are produced on a radar PPI scope.

A: Heading flash is produced, as the radar beam points dead ahead, by a switch in the antenna system that provides a pulse of short duration to intensify the radial line representing the heading. This positive pulse on the grid of the PPI scope causes a bright sweep line and enables the operator to know exactly when the dead-ahead position is reached.

Range-marker circles on the PPI screen indicate range distances and enable the operator to quickly estimate target distance. The range-marker oscillator, along with squaring and peaking circuits, produces a series of short positive pulses or "pips" that are synchronized with the sweep and applied to the grid of the PPI. As the beam sweeps out from center to edge, the accurately spaced pips appear and form range-marker circles as the sweep rotates.

Q46: What precautions should the service and maintenance operator observe when replacing the cathode ray tube in a radar set?

A: Turn off the power supply system and completely discharge all high-voltage capacitors with a well-insulated screwdriver or similar tool. Extreme care must be taken when removing and handling the tube to avoid implosion of the glass envelope and possible serious injury to anyone in the vicinity.

Q47: Draw a simple sketch showing a synchro generator located in the radar antenna assembly connected to a synchro motor located in the indicator to drive the deflection coils. Show the proper designation of all leads and where ac voltages (if needed) are applied.

A: See the sketch in Fig. 8-4. Some ship radar systems employ additional PPI indicators and it is necessary to synchronize the rotation of the sweep line on the CRT with the rotation of the radar antenna. This is normally handled by a servo system.

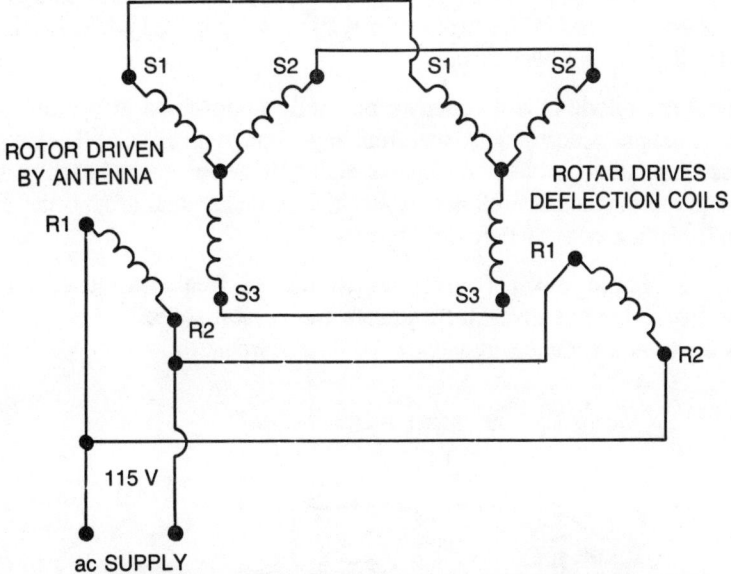

Fig. 8-4. Diagram of a synchro-generator circuit.

Q48: In what range of frequencies do magnetron and klystron oscillators find application?

A: The frequency range normally covered by a magnetron oscillator is 600 to 30,000 MHz, while klystron oscillators operate in a range from 3,000 to 30,000 MHz.

Q49: Draw a simple cross-section sketch of a magnetron showing the anode, cathode, and direction of electron movement under the influence of a strong magnetic field.

A: Refer back to Fig. 6-2 for this drawing in its complete form. The insert of this drawing, showing electron movement, is shown in Fig. 8-5.

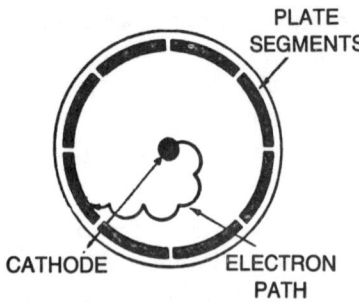

Fig. 8-5. Cross section of a magnetron showing plate segments.

Q50: Explain briefly the principle of operation of the magnetron.
A: The plate element of the multianode magnetron is made up of cavity resonators that receive energy from the movement of electrons outside each cavity opening. If the electron is accelerated by the cavity field, energy is taken from the cavity, but if on the other hand the electron is slowed by the cavity field (as is usually the case) then energy is given, which sustains oscillations. Magnetrons usually function as pulsed power oscillators with an extremely high output peak power.

Q51: Why is the anode of a magnetron normally maintained at ground potential?
A: This is for the protection of personnel from high-voltage shock as well as simplification of the chassis insulating problem. A negative high-voltage pulse is fed to the cathode that is centered in the magnetron well out of reach. The metal shell around the magnetron is grounded, making construction simple and safe.

Q52: Draw a simple mixer (converter) circuit as frequently used in a radar superheterodyne receiver and indicate the crystal stage.
A: Fig. 8-6 shows a common frequency converter circuit.

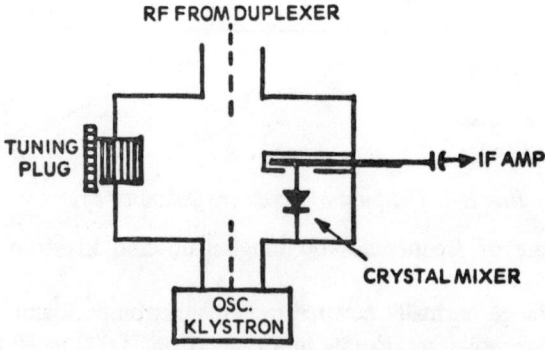

Fig. 8-6. Simple mixer or converter circuit.

Q53: Describe briefly the construction and operation of radar TR and anti-TR boxes. What is the purpose of a "keep-alive" voltage?
A: Ship radar systems use a common antenna for transmitting and receiving, which makes it necessary to protect the receiver from the very high-powered pulse of the transmitter. On the other hand, the transmitter would absorb too much power from the reflected signal

between pulses during reception. The duplexer is a type of switching arrangement made up of specific lengths of waveguide plus two spark-gap tubes. During transmission, both spark gaps are fired and the gap resistance becomes very low. The waveguide input to the quarter-wave lines out to the gaps now becomes high. This makes it possible for most of the pulse energy to pass directly to the antenna with very little getting into the TR or anti-TR boxes with the very high input impedance as seen from the waveguide. During reception of the reflected echoes, the spark gaps do not operate because the received signal voltage is far too weak to break down the air gaps. The anti-TR box is now a half-wave transmission line shorted at the far end which makes its input impedance zero at the waveguide. The quarter-wave length between the waveguide entrance to the TR-box and the entrance to the anti-TR box is thus terminated in zero impedance which makes its input impedance very high. The received signal, therefore, takes the lower impedance path to the receiver.

The TR box electrically isolates the receiver from the transmitter during pulse transmission to prevent damage to the receiver. The anti-TR box electrically isolates the transmitter from the receiver during reception of the reflected echo to avoid a loss in signal voltage.

The "keep-alive" voltage is a constant negative (1000 volts) applied to a third electrode inside one of the main electrodes. This keeps the gas and vapor slightly ionized to permit easier arcover during pulse transmission, and it also protects the receiver crystal.

Q54: Describe briefly the construction of a waveguide. Why should the interior of the waveguide be clean, smooth and dry?

A: Waveguides are normally made of hollow metal stock with either a rectangular or circular cross section. In some instances, weight considerations dictate the use of aluminum and other times brass is used. Plating the inside with silver improves conductivity to a great extent and a gold or rhodium protective flashing prevents or retards corrosion. Smooth interiors prevent troublesome reflections and dirt must be avoided to eliminate transmission loss. Moisture also contributes to arcing and transmission problems.

Q55: Why are waveguides used in preference to coaxial lines for the transmission of microwave energy in most shipboard radar installations?

A: Coaxial line losses are so great at the microwave frequencies required for radar that their use for long runs is not practical. When properly designed, waveguide losses are extremely low at microwave frequencies. The waveguide has neither the dielectric loss or the copper loss of the conventional coaxial line, since the waveguide has air for a dielectric and eliminates the thin inner conductor where most of the copper loss occurs. Thus, a waveguide having the same diameter as the coaxial line can carry much more power.

Q56: Why are rectangular waveguides generally used in preference to circular cross-sectional waveguides?

A: The use of circular waveguides makes polarization of the wave more difficult to control and for this reason they are only occasionally used in radar. Since the electric field has a tendency to change direction at bends in circular waveguides, the polarization changes. With the rectangular waveguide, though, the desired polarization is readily maintained. A rotating joint permits free movement of the antenna with respect to the fixed waveguide; it must be circular, while the waveguide feeding the joint is rectangular. The frequency range at the dominant mode is limited to a great extent in the circuit waveguide than in the rectangular.

Q57: Describe how the waveguide is terminated at the radar antenna reflector.
A: There are variations of the horn radiator which point into the parabolic reflector and form the energy into a narrow beam. The horn radiator must be large, compared to the operating wavelength, which is quite practical at these frequencies. The electromagnetic horn operates like an acoustic horn by matching the impedance of the waveguide to the impedance of free space. The parabolic reflector directs the energy into a narrow beam for accurate tracking.

Termination of the waveguide at the antenna may be achieved by means of a polystyrene window with the correct physical dimensions to provide the impedance match required. The window is placed at the focal point of the parabolic reflector and acts as a matching device between the waveguide, reflector and free space.

Q58: What precautions should be taken when installing vertical sections of waveguides with choke-coupling flanges to prevent moisture from entering the waveguide?
A: Each guide section should have the end with the choke at the bottom to avoid collection of moisture in the choke joint. By using a gasket at each choke-coupling flange, the flange bolts may be tightened enough to ensure a rain-tight joint.

Q59: Draw a longitudinal section of a waveguide choke joint and explain briefly its principle of operation.
A: See Fig. 8-7. The choke joint includes a slot-type groove having a depth of a quarter wavelength, making the input impedance across the circular groove infinite. By acting as a resonant element, the choke groove transfers energy across the junction without electrical contact. The distance from the groove to the waveguide is also a quarter-wave and it feeds into the infinite impedance at the input to the choke flange. Because this is a quarter-wave open line, its input impedance at the guide is zero and energy passes freely

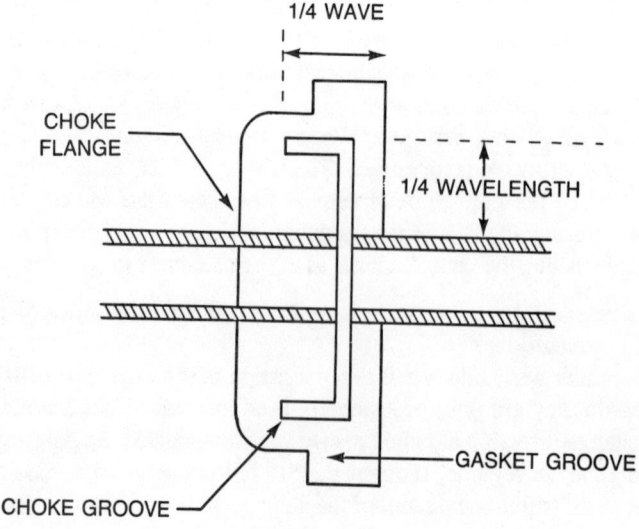

Fig. 8-7. Sketch of a longitudinal waveguide choke filter.

across the break at the flange without loss. Choke joints provided low-loss connections between parts of a system, provide mechanical isolation against vibration, and permit removal of sections for easy repair or replacements.

Q60: Why are choke joints often used in preference to flange joints to join sections of waveguide together?

A: The cost of choke coupling is much less than direct contact flange-joint coupling, which requires precision machining to ensure perfect contact all around and smooth, continuous inside surfaces. Perfect alignment is also important to avoid reflections and other losses, and this adds considerably to the cost of the flange-joint. This precision work is not required with choke joint coupling, because electrical contact around the outside of the choke groove is desirable. Of course, the choke joint is somewhat inferior to the precision-machined direct-contact flange joint from a loss standpoint, but the difference in signal losses certainly does not justify the much greater cost in most applications.

Q61: When installing waveguides, why should long, perfectly level sections be avoided? Why is a small hole about one-eighth inch in diameter sometimes drilled on the underside of an elbow in a waveguide near the point where it enters the radar transmitter?

A: A slight slant in long horizontal sections of the waveguide permits condensed moisture to drain out through the small hole at the lowest point that has been drilled on the underside for this purpose.

Q62: Describe how a radar beam is formed by the parabolic reflector.

A: The narrow beam of RF energy reflected by the parabolic reflector compares with the reflection of light by a parabolic reflector in an ordinary searchlight. Feeding the energy into the reflector at its focal point causes most of the RF to be focused into a narrow beam that is reflected by the "dish." The greater the diameter of the parabolic reflector, the narrower that beam (and greater the gain) will be.

Q63: What effect if any does the accumulation of soot or dirt on the antenna reflector have on the operation of a ship radar?

A: A thin layer of soot, dirt, or paint on the reflector has little or no effect on the operation of the ship radar unit, because microwaves are able to penetrate an average accumulation with very little loss. An excessive amount of foreign material on the reflector surface, however, will decrease the efficiency of the system on weak targets; therefore, such matter should be cleaned off. Any accumulation of dirt on the plastic window will cause considerable loss and must not be permitted.

Q64: What considerations should be taken into account when selecting the location of the radar antenna assembly aboard ship?

A: Obstructions must be avoided as much as possible when locating a radar antenna, and the scanning area around the ship should be reasonably clear of any objects that would interfere. It is most important that the forward or bow area be completely clear and the location of the antenna must not require a longer waveguide section from the transmitter than is necessary. It also must be accessible for routine maintenance.

Q65: What is the purpose of an echo box in a radar system? Explain the principle of operation of the echo box. What indications can be expected on a radarscope

when using an echo box and the radar set is operating properly? When the set is not operating properly?

A: The purpose of the echo box is to offer a phantom or artificial target for tuning the receiver and indicate or test the overall performance of a radar system.

An echo box is resonant cavity with a very high Q, which is shock excited into oscillation by the transmitter pulse. The cavity rings or oscillates for several microseconds after the transmitted pulse ends, as a result of its high Q, and its radiation is received and displayed on the PPI scope. This appears as lines or spokes extending out from the hub, in the case of a motor tuned box, or as an intensified area or large spot in a box set at resonance.

When the radar system is operating properly, the spokes or intensified area extend out quite a bit from the center of the display on the scope. If the radar is not functioning properly, the artificial target would be much smaller or even not visible at all. The echo box actually provides a useful reference signal for evaluating the performance of the radar system. Normal radar target signals do not furnish a satisfactory means of checking the system because of the many variables involved, such as atmospheric conditions, character of different signals, and the lack of adequate reference material.

Q66: Draw a simple diagram of an artificial transmission line showing inductance and capacitance, source of power, the load, and electronic switch.

A: To avoid the bulk of an actual line, an artificial line can be built of coils and capacitors which has approximately the same characteristics as the line but occupies a smaller space. The circuit for an artificial line is shown in Fig. 8-8. The distributed inductance and resistance of the line are lumped in several choke coils, while the distributed capacitance can

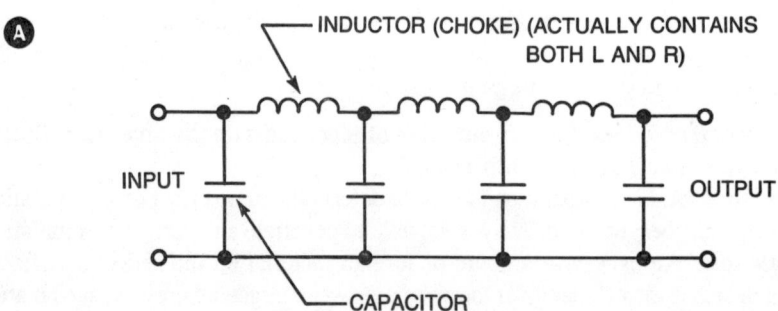

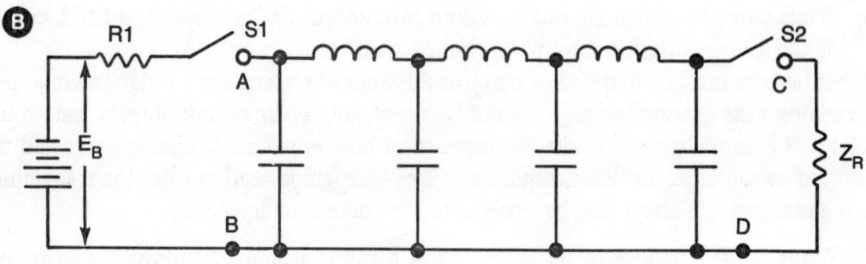

Fig. 8-8. Artificial transmission line.

258

be represented by capacitors, and the conductance is omitted entirely as it is too small even to consider. If the action of the actual transmission line with its evenly distributed R, L, and C is to be closely approximated, the sections must be small and numerous. In most cases, however, from three to eight sections produce as much of the required transmission line action as is needed for use in radar time-delay circuits.

Artificial transmission lines also may perform other duties in addition to introducing time delay or phase shift. These duties include action as filters to block or pass certain frequencies, and as models for laboratory demonstration of transmission-line action.

One important use in radar is the storage of energy and the subsequent delivery of the same energy at a predetermined rate to form pulses. In Fig. 8-8B, if switch S1 is closed S2 remaining open, the artificial transmission line will charge through R to the voltage of E_D, the charge being retained on the capacitors as electrostatic field energy. In application, operating current may be obtained from the cathode circuit of an appropriate stage.

Q67: Who has the responsibility for making entries in the installation and maintenance record of a ship-radar station?
A: The licensed operator doing the work and the station licensee are jointly responsible for making the proper entries in the installation and maintenance record.

Q68: May fuses be replaced in ship-radar equipment by a person whose operator license does not contain a ship-radar endorsement?
A: Yes, fuses may be replaced by unlicensed operators, but all other repairs, tests, and installations must be performed by a properly licensed technician.

Q69: What precautions should a radar serviceman take when working with or handling a magnetron to prevent weakening or damaging it?
A: A magnetron should be handled with the care and respect due any precision device; shocks and blows must be avoided. Steel tools or parts cannot be in the proximity of the magnet, and extreme heat can damage or weaken it.

Q70: Draw a simple block diagram of a radar duplexer system; label the waveguide, TR box, anti-TR box, receiver, and transmitter.
A: A block diagram of the duplexer is shown in Fig. 8-9.

Q71: What is required to operate a ship-radar station?
A: Authorization by the ship's master permits any person so designated to operate the radar station.

Q72: How does the "keep-alive" potential low the arc resistance in a TR box?
A: The constant "keep alive" voltage slightly ionizes the gas and vapor in the tube and lowers the breakdown resistance of the gap. A negative potential of about one thousand volts is used for this purpose.

Q73: What causes receiver paralysis in radar and how may it be avoided?
A: Although TR and anti-TR boxes normally protect the receiver from paralysis or blocking, when a strong signal does get through, the amplifiers are overdriven and blocking results from the residual charges. The blocking or paralysis of the receiver can be avoided

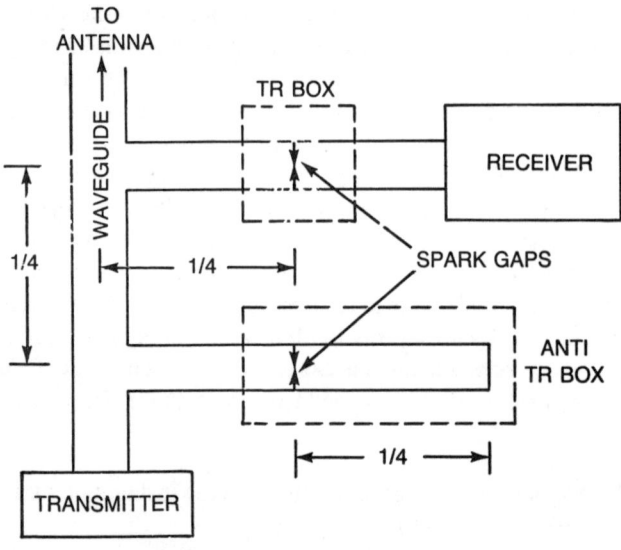

Fig. 8-9. Block diagram of a radar duplexer.

or greatly reduced by the application of a disabling rectangular pulse to one or more stages at the critical moment.

Q74: What kind of RF amplifiers are employed in some radar receivers?
A: Conventional RF amplifiers are just not practical at the microwave frequencies normally used for radar systems. Noise is high, gain is low and degenerative feedback is excessive as a result of the inductive reactance in the cathode leads. However, at microwave frequencies parametric amplifiers, some bipolar transistor amplifiers and especially GaAs FET amplifiers are used, with GaAs FETs, low-noise figures (e.g., 1.5 dB) are available.

Q75: What limits the number of RF amplifier stages in a radar receiver?
A: Noise figure requirements and maximum permissible mixer input signal level.

Q76: Why is stagger-tuning commonly employed?
A: Stagger tuning broadens the bandpass of the IF amplifiers and either single or double-tuned coupling may be used.

SAMPLE QUESTIONS

1. Within what frequency bands do ship radar transmitters operate?
    (a) 5460 to 5650 kHz
    (b) 9100 to 9320 MHz
    (c) 9320 to 9500 MHz
    (d) 3340 to 3760 MHz
    (e) 5120 to 5420 kHz

2. What are the FCC license requirements for the operator who is responsible for the installation, servicing, and maintenance of ship radar equipment?
    (a) First or second class radiotelegraph license.

(b) Second class radiotelephone license.
(c) First class radiotelegraph license.
(d) General radiotelephone or radiotelegraph license with radar endorsement.
(e) Marine radiotelephone permit with radar endorsement.

3. What component determines the pulse repetition rate in a radar system?
   (a) Marker generator
   (b) Timer
   (c) Magnetron
   (d) Artificial transmission line
   (e) Echo box

4. What was the purpose of the rotary spark gap used in older radar?
   (a) Pulses the synchronizer.
   (b) Eliminates the pulse-forming network.
   (c) Modulates the magnetron directly.
   (d) Controls the oscillator frequency.
   (e) Protects the deflection coils.

5. What precautions should a radar serviceman observe when making repairs or adjustments to a radar set to prevent personal injury to himself or others?
   (a) Check all protective devices.
   (b) Turn the power off and post a danger sign.
   (c) Wear gloves, goggles and other safety equipment.
   (d) Jumper interlock switches only after posting a warning.
   (e) Shut off the power, discharge the high-voltage capacitors with an insulated tool, and handle the CRT with extreme care.

6. Why is it important that all units of the radar system be thoroughly bonded to the ship's electrical ground?
   (a) Prevents accidental shock and interference to other equipment.
   (b) Provides a high-impedance path for RF energy.
   (c) Makes it easier to make electrical measurements.
   (d) Provides a reference point for accurate comparison.
   (e) Helps to lower the static charge peak.

7. What is the purpose of an artificial transmission line in a radar system?
   (a) Determines the operating frequency.
   (b) Provides trigger pulses for proper sweep.
   (c) Determines the output of the video detector.
   (d) Couples the flared waveguide to the reflector.
   (e) Determines the shape and duration of the transmitted pulse.

8. What is the usual way of terminating a waveguide at the radar antenna reflector?
   (a) Feeding it into a resonant cavity.
   (b) By matching it with a quarter-wave stub.

(c) Connecting it to a choke joint.
   (d) Forming a horn by flaring out the end.
   (e) Matching it with a loading coil.

9. What is the peak power of a radar pulse if the pulse width is 2 microseconds, pulse repetition rate (PRR) is 700 and the average power is 14 watts?
   (a) 10,000 watts
   (b) 100 kilowatts
   (c) 14,000 watts
   (d) 70 kilowatts
   (e) 1,000 kilowatts

10. What is the duty cycle of the radar transmitter if the pulse width is 1 microsecond and the pulse repetition rate (PRR) is 900?
    (a) 0.009
    (b) 0.00009
    (c) 0.0009
    (d) 0.000009
    (e) 0.00001

11. If we multiply the pulse width of a radar transmitter by its PRR, what would the product indicate?
    (a) Average power
    (b) Duty cycle
    (c) Peak power
    (d) Efficiency factor
    (e) Bandwidth

12. What type of detector is used frequently in radar receivers?
    (a) Ratio detector
    (b) Selenium diode
    (c) Grid leak
    (d) Horizontal diode
    (e) Silicon crystal diode

13. What are the indications of a defective crystal in the receiver converter (mixer) stage of a radar system?
    (a) Low front-to-back ratio.
    (b) Weak echo and no "grass" on the PPI scope.
    (c) High reading on the crystal current meter.
    (d) Low reading on the crystal current meter.
    (e) High "grass" level and weak targets.

14. What is the purpose of the klystron tube in a radar set?
    (a) Local oscillator in the receiver.
    (b) Pulse amplifier.

(c) Drives the modulator grid.
   (d) Determines the transmitter frequency.
   (e) Controls the AFC circuit.

15. Weak echo signals might be received if the radar receiver has a:
    (a) Low signal-no-noise ratio.
    (b) Good RF amplifier.
    (c) Klystron mixer.
    (d) High signal-to-noise ratio.
    (e) Narrow bandpass.

16. What determines the frequency of the radiated energy in a radar transmitter employing a magnetron?
    (a) The timer circuit.
    (b) The magnetron circuit.
    (c) The reflex klystron circuit.
    (d) The choke joint circuit.
    (e) The parabolic reflector.

17. What are the indications of a weak magnet in the magnetron in a radar unit?
    (a) AFC good and no change in oscillations.
    (b) Oscillation amplitude less and frequency steady.
    (c) Increased magnetron current and oscillations cease.
    (d) Decrease in magnetron current and oscillations stop.
    (e) Decrease in place voltage and oscillations.

18. What symptoms on a radar scope would indicate a defective crystal in the receiver converter stage?
    (a) Bright target with excessive noise.
    (b) Pattern drifts across the PPI.
    (c) Noise level below average.
    (d) No target or "grass" with advanced gain.
    (e) Weak or no target; "grass" high.

19. When checking the front-to-back ratio of the radar receiver mixer crystal with a good ohmmeter, what would be a reasonable figure for a crystal in good operating condition?
    (a) 5 to 1
    (b) 1.5 to 1
    (c) 8 to 1
    (d) 10 to 1
    (e) 20 to 1

20. In a radar set, what are the indications of a defective magnetron?
    (a) Good AFC, no range marks.
    (b) Poor AFC, fuzzy targets.

(c) No targets but stable frequency.
(d) Weak targets, current normal.
(e) High current, good AFC.

21. What care should be taken when handling silicon crystal rectifier cartridges for replacement in radar superheterodyne receivers?
    (a) Discharge static body charges by touching the nearest ground.
    (b) Wrap carefully in soft paper when storing.
    (c) Grasp firmly when removing.
    (d) Store near the transmitter for handy use.
    (e) Wipe off dust particles when dropped.

22. What intermediate frequencies are common in radar receivers?
    (a) 220 and 440 MHz
    (b) 47.5 and 95 MHz
    (c) 455 and 675 kHz
    (d) 30 and 60 MHz
    (e) 10.7 and 41.5 MHz

23. What is the purpose of the sensitivity time control circuit in a radar set?
    (a) Balances the echo box.
    (b) Regulates the pulse duration.
    (c) Reduces the gain on nearby targets.
    (d) Synchronizes the sweep with the transmitter.
    (e) Raises the gain for sea return.

24. What is the purpose of the discriminator stage in a radar receiver?
    (a) Part of the AVC circuit.
    (b) Causes the IF amplifier to drift and develop a dc voltage.
    (c) Supplies control voltage to the klystron for frequency correction.
    (d) Eliminates unwanted targets or echoes.
    (e) Applies a frequency correction potential to the magnetron.

25. What is the purpose of Aquadag coatings on radar cathode ray tubes?
    (a) Bunches electrons.
    (b) Forms the second anode.
    (c) Acts as a suppressor.
    (d) Protects against dangerous shock.
    (e) Prevents cathode blocking.

26. What is "sea return" on a radar scope?
    (a) Radar signals returned over water.
    (b) Echo from a distant target.
    (c) Noise from an electronic buoy.
    (d) Signals from a marine repeater.
    (e) Reflection of the signal from sea waves.

27. What might produce bright flashing pie sections on a radar PPI scope?
    (a) Defective TR box.
    (b) Defective deflection coil.
    (c) Defective parabolic reflector.
    (d) Defective AFC circuit.
    (e) Defective crystal holder.

28. How long does it take a radar pulse to reach a target one nautical mile away and return?
    (a) 12.3 microseconds
    (b) 1.23 microseconds
    (c) 6.15 microseconds
    (d) 24.6 milliseconds
    (e) 1.235 seconds

29. What is indicated by the "bearing resolution" of a radar set?
    (a) Consistent reception of long-range targets.
    (b) The error factor between sea return and true target.
    (c) Proper focusing of targets on the PPI screen.
    (d) Ability to distinguish between targets with the same range but differing azimuth directions.
    (e) Ability to distinguish between targets with the same azimuth directions but differing ranges.

30. If the elapsed time is 40 microseconds between transmission and reception of the radar signal, how far away is the target?
    (a) 3,280 yards
    (b) 6,460 yards
    (c) 5,280 yards
    (d) 1,640 yards
    (e) 6,560 yards

31. Who may operate a ship radar station?
    (a) Anyone holding a third-class radiotelephone permit.
    (b) Anyone holding a second-class radiotelegraph license.
    (c) The ship's master or anyone so authorized by him or her.
    (d) Only the holder of a valid FCC license or permit.
    (e) Only the ship's radio operator.

32. What license is required for the installation, servicing, and maintenance of ship radar?
    (a) First class radiotelegraph license.
    (b) First or second class radiotelephone or telegraph.
    (c) General radiotelephone or telegraph with radar endorsement.
    (d) Marine radiotelephone permit.

33. What type of radar interference would be apparent to anyone listening to a communications receiver?
    (a) A strong clicking like a rotating device.
    (b) Intermittent buzzing and popping noise.
    (c) A steady tone and hash from the radar motor-generator.
    (d) Receiver paralysis or blocking.
    (e) A series of popping sounds like ignition noise.

34. On what frequencies would the radar serviceman look for interference to communication receivers from a radar station?
    (a) 2,900 to 3,100 MHz
    (b) 9,320 to 9,600 MHz
    (c) 30 to 60 MHz
    (d) 5,460 to 5,650 MHz
    (e) Any communications frequency

35. How would radar interference to auto-alarm equipment be recognized?
    (a) By sparking of the relay contacts.
    (b) Checking the tubes for internal arcing.
    (c) Plugging phones into the alarm and listening for identifying sounds.
    (d) Checking the supply voltage on the alarm.
    (e) By shutting off the radar unit.

36. Why must the interior of waveguides be kept clean, smooth and free of moisture?
    (a) To prevent shock hazard.
    (b) To keep losses at a minimum.
    (c) To prevent polarization.
    (d) To eliminate serious rusting.
    (e) To maintain resonance.

37. Why are waveguides used in preference to coaxial lines in most shipboard radar installations?
    (a) Lower loss.
    (b) Less expensive.
    (c) Longer life.
    (d) Neater appearance.
    (e) Easier to terminate.

38. Why are rectangular cross-section waveguides usually preferred over circular cross-section waveguides?
    (a) Rotating joints are easier to install.
    (b) Electric field has greater tendency to change.
    (c) Less frequency range at the dominant mode.
    (d) Desired polarization easily maintained.
    (e) Better harmonic attenuation.

39. What precautions should be taken when installing vertical sections of waveguides with choke-coupling flanges to prevent moisture from entering the waveguide?
    (a) Use good gasket cement.
    (b) Bend the flange to allow moisture to run off.
    (c) Tape the outside with electrical tape.
    (d) Use a gasket at each flange and tighten securely.
    (e) Drill a hole in the bottom for the moisture to escape.

40. Why must long, perfectly level sections of waveguides be avoided?
    (a) Prevent the accumulation of condensed moisture inside.
    (b) This unbalances the magnetic fields.
    (c) Prevents overloading the antenna.
    (d) Makes impedance matching more difficult.
    (e) Prevents too much loss in the TR box.

41. Why are choke joints generally used in preference to flange joints to join sections of waveguides?
    (a) Electrical contact is better.
    (b) Lower cost due to wider tolerances.
    (c) Less signal loss.
    (d) Easier to maintain.
    (e) Overall efficiency is greater.

42. What is the purpose of an echo box in a radar system?
    (a) Low Q cavity to absorb sea return interference.
    (b) Provides a phantom target for tuning and evaluating performance.
    (c) Controls and regulates the echo timing interval.
    (d) Provides range indication by measuring the echo.
    (e) Amplifies the returning echoes from weak targets.

43. Who may make entries in the installation and maintenance record of a ship radar station?
    (a) Ship's master or person authorized by him.
    (b) Any person with a valid FCC license.
    (c) Only the station licensee.
    (d) Licensed operator responsible for work or under his supervision.
    (e) The ship's radio operator only.

44. Why would it be dangerous to operate or test a radar system aboard ship when explosive or inflammable cargo was being handled?
    (a) Vibration from the scanning system.
    (b) Reflections from targets.
    (c) Friction produced in the waveguides.
    (d) Static picked up by the receiver.
    (e) Possible arcing in the radar set or RF arcing from the beam.

45. What indication on a loran scope would result from radar interference?
    (a) Spikes moving across the screen and "grass" near the scanning lines.
    (b) Spikes near the scanning lines not moving.
    (c) Fuzzy detail on the screen.
    (d) Flashing pie sections extremely bright.
    (e) Radial spokes from the center of the scope.

46. When the radar set is operating properly, which pattern would appear on the PPI screen as a result of the echo box?
    (a) Radial spoke pattern or intensified area.
    (b) A non-sinusoidal wave.
    (c) A series of spikes and some "grass."
    (d) A damped RF pattern.
    (e) A sine wave of varying intensity.

47. Who has the responsibility for making proper entries in the installation and maintenance record of a ship radar station?
    (a) Only that person authorized to operate the station.
    (b) The radio operator and the ship electrician.
    (c) The person assisting in the work.
    (d) Licensed operator doing the work and the station licensee jointly.
    (e) The ship's master and his first officer jointly.

48. How are heading flashes produced on a radar PPI scope?
    (a) The action of the antenna waveguide.
    (b) By the local oscillator klystron.
    (c) These are a function of the echo producing box.
    (d) This is produced by the marker generator.
    (e) Cam-operated microswitch closed when the antenna is dead ahead.

49. How are range-marker circles produced on the radar PPI scope?
    (a) Intermittent shorting of the echo box.
    (b) Blanking of the AFC circuit.
    (c) Synchronized, short, positive pulses applied to the PPI grid.
    (d) Continuous impulse applied to the PPI for ranging.
    (e) Special CRT face-plate.

50. What is the purpose of the "keep alive" electrode in TR and anti-TR boxes?
    (a) Eliminates overloading and damaging the main spark gap.
    (b) Prevents an arc at the main gap when the transmitter is pulsed.
    (c) Increases the spark gap impedance during conduction.
    (d) Provides the proper delay in flash-over after pulsing.
    (e) Permits easier arc-over during pulse transmission.

9

Technician Certification Programs

A certification program is operated by one or more private organizations for the purpose of providing a baseline for judging individual technicians' knowledge. As is also true with the FCC licenses, the certificate does not in and of itself prove that the technician is, indeed, the best qualified for any given job. What the certificate does, however, is provide the potential employer or client with an indication that the person has met the minimum requirements of knowledge and skills required to perform duties in that particular area.

When the Federal Communications Commission began to deregulate much of the broadcasting and communications industries, they turned to these private certification organizations to provide employers with the credentialing schemes needed. Some of these are very new and were formed only in response to the new needs of deregulation. Others, such as ISCET, have been around for two decades or more. Some of these organizations with their addresses are listed below.

Association for the Advancement of Medical Instrumentation (AAMI)
1900 Fort Myer Drive
Arlington, VA 22209
Certifies clinical engineers and biomedical equipment technicians.

DeVry Industrial Training Division
2201 W. Howard Street
Evanston, IL 60602
(800) 323-4253

Electronic Technicians Association
Indianapolis, IN
Attn: Dick Glass

International Society of Certified Electronic Technicians (ISCET)
2708 Westbury, Suite 8
Fort Worth, TX 76109
(817) 921-9101

National Association of Radio and Telecommunications Engineers
(NARTE)
P.O.B. 15029
Salem, OR 97309
(503) 581-3336

National Association of Business and Educational Radio (NABER)
P.O. Box 19164
Washington, D.C., 20036
(202) 887-0920

National Institute for Certification in Engineering Technology (NICET)
2029 K Street, N.W.
Washington, D.C., 20006
(202) 463-2335

National Radio Institute (NRI)
McGraw Hill Continuing Education
3939 Wisconsin Avenue N.W.
Washington, D.C., 20007
(202) 244-1600

Society of Broadcast Engineers (SBE)
P.O. Box 50844
Indianapolis, IN 46250
(317) 842-0836

These organizations might offer one or several examinations. Those organizations that offer more than one exam might offer exams in several different fields (or areas of concentration) or several levels of examination in the same field. The ISCET series of examinations is an example of the former, while the SBE is an example of the latter.

The possession of a certification or FCC license is clear evidence of a certain calibrated minimum knowledge in certain specified areas of electronics. The certificate is a banner of competence to prospective employers and lets them know that you possess at least the minimum credentials to work in their field. From a pragmatic point of view, the "cert" is useful in several ways. It gives you job mobility by establishing your credentials with employers in distant cities. While the "cert" does not allow the employer to assess all of your qualifications, they can at least establish your minimum knowledge level—a fact that could cause him or her to hire you over a noncertified applicant.

Some of the available certifications are also useful for establishing credentials with state and local licensing officials. Technicians in many parts of the country are required to be licensed before practicing. The certification is a good "trial run" for some localities and a mandatory requirement in others.

ABOUT STUDY GUIDES

Neither this nor any other book does you a favor by sneaking a look at the FCC or certification examination questions. The FCC does, however, publish a syllabus from which the questions are taken. For most two-way radio, communications, or broadcasting technicians or engineers, the questions in this book also address the preparation you need to undertake to successfully pass the certification examinations. Besides the "philosophical" reasons for not publishing exact verbatim questions, there is also a practical matter: examinations can change in a heartbeat, but study guides take time to catch up. The best preparation for any examination is to thoroughly study the subject matter so you can answer any question they toss at you.

GOOD STUDY HABITS

It is probably begging the question to mention that good study habits will bring the applicant to an exam-passing competence faster than poor study habits. However, it is still useful to review some of the basics. Some educational experts recognize the "four R's of good studying," or *read, reread, review* and *regurgitate* (or *remember*). The technique is to read the material and slowly turn it over in your mind. Then reread it to answer any questions that might have popped up or to refresh your memory about details.

If there are any problems to solve or questions to answer in the textbook material, then don't skip over them just because they weren't assigned. One author (JJC) once took a course where the professor routinely took all but one of his test questions from the back of chapter questions in the assigned textbook. If you really studied, then there was no excuse for less than an "A".

After reading and rereading, it is necessary to review the material in your mind. Rehash it mentally or verbally (as is best suited for you) to see if you are comfortable with the material. In this stage, it is sometimes useful to study with another person (or a small group—emphasis on the word "small"). The process is to use each other as sounding boards of your own level of knowledge.

The fourth "R" (regurgitation) automatically follows the first three if they are done properly. Do not be discouraged by seemingly poor early results. Studying is merely a skill, and skills can be learned by doing—just like soldering or roller skating or riding a motorcycle. Repetition and hard work sharpen your skill so that you eventually reap a high harvest yield per hour of study time.

Concerning study time, schedule time whenever and wherever you can. Try to schedule a specific time each week, or make a daily schedule if possible. Whatever the schedule, making it routine and periodic is a good tactic.

TEST TAKING

Many tests and examinations mark mileposts in our personal development and achievement. All through school and into later life, tests and written examinations become necessary to qualify you for one thing or another. Beyond class examinations in your technical institute, you might want to take the Federal Communications Commission examinations or one of the certification examinations. While the purpose of the examinations are to identify qualified people, flunking the examination does not automatically mean that the person is not qualified! There are some people who simply have trouble with examinations no matter how well they study or how well they know the material.

There are two facets to successfully taking a test: *preparation* and *strategy*. Preparation is what this book is all about. It is your study guide for the FCC and most other certification examinations. But strategy is also an important (and often overlooked) element of taking a test, yet so few people really understand its meaning.

Consider the way most people attack a written examination. They start with question number one and work linearly through the examination to the last question. Unless the instructions for the exam specifically require this approach (and some do), it is all wrong—especially if there is a time limit to the exam.

Proper testing technique calls for you to *go through the test several times*. On the first pass through the exam, answer all of those questions that you know off the top of your head. On the second pass, answer those questions that require little thought or a simple mathematical calculation. On the third pass, answer even harder questions or tougher arithmetic calculations. For example, on the second pass you might answer a question on the physical length of a half-wave dipole antenna, while on a subsequent pass answer a question that requires calculation using a complex network theorem. Finally, as you come to the end of the test, work on the really tough questions.

When trying to work out the tougher questions, it often helps to work from a position of eliminating the obviously wrong answers. This tactic improves your chances of guessing the correct answer if you don't know it.

Understand the grading structure of the examination. If all problems or questions are given equal weight, then it is best to use the strategy above exactly. But if there are different weights for different problems, then it might pay to redirect your attentions. For example, if one problem is 30 points out of 100, then it is best to work on it first, right? . . . not necessarily! If all of the other lower-point questions are easy, then do those first and you will have a 70-percent score without even touching the big one.

Also, look at the instructors or examination coordinators grading habits. For example, one professor I studied under in engineering school used to give 90 percent of the credit for each problem for the mathematical set-up (which proves you understand the problem and can solve it) and only 10 percent for the numerical answer ("turning the crank" on your calculator, as he put it). I once got an "A" on a final examination with a numerical score of 80, and I had not provided a single numerical answer! I just set up all of those problems and did not have the time to go back and "turn the crank" for all of them. Such is a test taker's life.

It is important that you be thoroughly familiar with the operation of the calculator that you take into the examination. You don't want to waste valuable time or get the wrong answer just because you are unfamiliar with the machine.

Be sure you arrive at the test site rested, fresh, and with an empty bladder. Try not to arrive too soon after a large meal: it can be disastrous to work a test when you are uncomfortable, drowsy, or must answer the call of nature. Also, don't arrive too hungry. Bring an adequate supply of sharp #2 lead pencils. Don't ever work a test in ink unless required to do so by the instructions. I recall a high school electronics shop teacher who used to have a motto: *A sharp pencil means a sharp worker*. Trite, yes, but true.

Obtain as much "intelligence" information as you can find on the test. Because of the constantly changing nature of such examinations, it is a great idea to interview recent takers of the test to help you study.

APPENDIX

ANSWERS TO SAMPLE QUESTIONS

Basic Law

1. (b)	6. (b)	11. (d)	16. (d)
2. (c)	7. (b)	12. (e)	17. (c)
3. (e)	8. (c)	13. (a)	18. (c)
4. (d)	9. (b)	14. (b)	19. (d)
5. (a)	10. (d)	15. (c)	

Series O: Basic Operating Practice

1. (a)	6. (a)	11. (b)	16. (e)
2. (a)	7. (c)	12. (d)	17. (c)
3. (d)	8. (c)	13. (d)	18. (e)
4. (c)	9. (d)	14. (a)	19. (b)
5. (a)	10. (c)	15. (b)	20. (a)

Series M: Maritime Operating Practice

21. (c)	26. (c)	31. (b)	36. (c)
22. (a)	27. (e)	32. (a)	37. (a)
23. (c)	28. (b)	33. (a)	38. (a)
24. (a)	29. (d)	34. (e)	39. (a)
25. (b)	30. (a)	35. (c)	40. (b)

Basic Radiotelephone

1. (e)
2. (d)
3. (b)
4. (a)
5. (c)
6. (c)
7. (a)
8. (e)
9. (a)
10. (c)
11. (b)
12. (e)
13. (d)
14. (b)
15. (e)
16. (d)
17. (a)
18. (d)
19. (b)
20. (c)
21. (e)
22. (e)
23. (a)
24. (b)
25. (c)
26. (e)
27. (d)
28. (a)
29. (e)
30. (c)
31. (b)
32. (a)
33. (c)
34. (d)
35. (c)
36. (d)
37. (c)
38. (b)
39. (a)
40. (d)
41. (c)
42. (b)
43. (a)
44. (b)
45. (a)
46. (b)
47. (b)
48. (b)
49. (c)
50. (a)
51. (a)
52. (e)
53. (b)
54. (d)
55. (c)
56. (a)
57. (c)
58. (b)
59. (b)
60. (e)
61. (d)
62. (d)
63. (e)
64. (a)
65. (e)
66. (b)
67. (c)
68. (e)
69. (d)
70. (b)
71. (c)
72. (b)
73. (a)
74. (c)
75. (e)
76. (b)
77. (a)
78. (c)
79. (a)
80. (b)
81. (c)
82. (e)
83. (c)
84. (b)
85. (b)
86. (a)
87. (b)
88. (c)
89. (e)
90. (c)
91. (b)
92. (b)
93. (b)
94. (c)
95. (c)
96. (d)
97. (e)
98. (c)
99. (a)
100. (c)
101. (b)
102. (b)

Radar Endorsement

1. (c)
2. (d)
3. (b)
4. (c)
5. (e)
6. (a)
7. (e)
8. (d)
9. (a)
10. (c)
11. (b)
12. (e)
13. (d)
14. (a)
15. (d)
16. (b)
17. (c)
18. (e)
19. (e)
20. (b)
21. (a)
22. (d)
23. (c)
24. (c)
25. (b)
26. (e)
27. (d)
28. (a)
29. (d)
30. (e)
31. (c)
32. (c)
33. (c)
34. (e)
35. (c)
36. (b)
37. (a)
38. (d)
39. (d)
40. (a)
41. (b)
42. (b)
43. (d)
44. (e)
45. (a)
46. (a)
47. (d)
48. (e)
49. (c)
50. (e)

INDEX

A
absorption wavemeter, 202
ADF equipment, 248
air-insulated cable, 201
alpha cutoff, 84
alpha gain, 83, 84
alternators, 208, 226
AM modulation, 138
AM radio, bandwidth for, 235
AM television, 220
AM transmitter, 171-173
ambient temperature, 64, 65
ammeter, 142
amperes, 25
amplification, 81
amplification factor, 57
amplifiers, 50, 56-59, 66, 89-99, 137, 230, 232, 233, 234, 235
   audio, 137
   bipolar transistor power, 114-130
   bridge audio, 130
   Class A, 116-121
   Class AB, 123-128
   Class B, 121-123
   classifications of, 89
   common element classification of, 89
   complementary symmetry, 128
   conduction angle classification of, 92
   coupling methods, 98
   crossover distortion in, 123
   Darlington pairs and, 128
   efficiency of, 92-94, 117
   feedback vs. nonfeedback classification of, 95
   frequency response classification of, 97
   modulated, 173
   power, 131-134, 168
   push-pull, 52, 121, 125
   quasi-complementary symmetry, 128
   radio frequency, 138
   totem pole, 126
   transfer function classification of, 94
   transistor, 99
   transistor RF power, 131-134
   VFET, 166, 167
   voltage, 168
amplitude, 40
anode, 34
antenna towers, marking and lighting, 9, 19
antennas, 180, 192, 219, 240
   automobile, 195
   coaxial, 198, 199
   corner reflector, 197
   dipole, 191
   impedance in, 192
   loading coils and, 195
   location of, 257
   loop, 193
   parasitic array, 197
   properties of, 194
   quarter-wave, 191, 196
   radiation patterns for, 192
   reflectors for, 256
   single-loop, 197
   skip zone in, 193
   stacked array, 197
   transmitting, 200
   V-beam, 197
anti-TR boxes, 254, 255, 259, 268
Aquadag coating, 252, 264
Armstrong oscillators, 144, 145, 187
artificial transmission lines, 246, 258, 259, 261
attenuation, 206
attenuator, 242
audio amplifiers, 97, 137, 233
authorized bandwidth, 220, 221
auto-alarm equipment, 246, 248
automatic frequency control (AFC), 184, 251
azimuth, 248

B
bandpass filters, 33
bandpass switch, 178
bandstop filters, 33, 46
bandwidth, 171
   AM radio, 235
   authorized, 220, 221
bandwidth of emission, 182, 183, 220-221
bass control circuit, 165
batteries, 64, 207, 209, 236, 237
   capacity of, 211
   charge-discharge rate in, 210
   corrosion in, 210
   sulphation and overcharge in, 210
battery bias circuit, 156
beam power, 46, 49
bearing resolution, 252, 265
beat frequency oscillator (BFO), 176
beta cutoff, 84
beta gain, 83, 84, 128
bias, 34, 35, 58, 63, 65, 77-80, 86-88, 112, 156, 158, 237
   current-cutoff voltage, 55
   improper, 61, 62
bias network, 85
bias resistors, oscillators, 153

bias voltage, 156
bipolar transistor power amplifiers, 114-130
bleeder resistors, 68, 69, 231, 232
breakdown, 78
bridge audio amplifiers, 130
bridge rectifiers, 66, 68
buffer amplifiers, 91, 153
bypass capacitor, 153

C

cables, 201
call signs, 9, 11, 13, 17, 19
calling frequency, 10, 17, 21-23
capacitance, 29, 30, 31, 40, 42, 50, 60
capacitive reactance, 32
capacitor coupled amplifiers, 99, 100
capacitors, 30-32, 42, 43, 45, 229, 231, 233
carbon microphone, 166, 167
carrier frequency, 171, 221, 235
carrier shift, 173
cascaded amplifiers, 160, 163
cathode, 34, 61
cathode ray tubes, 251, 252, 253
cathode resistors, 168
cavity resonator, 217
center frequency, 241
center-tap power supply, 66, 67
characteristic impedance, 199, 200
charge carriers, 75
charge-discharge rate, 231
charges, 17, 22
chirp, 91
choke joints, 256, 257, 267
choke-input filters, 68
chokes, 68, 239
circuit components and properties, 33-37
Class A amplifiers, 116-121, 160, 235, 236
Class AB amplifiers, 123-128
Class B amplifiers, 121-123, 159, 232
Class C amplifiers, 169, 187, 217, 235, 237
coast-to-land radio contact, 23
coaxial antennas, 198, 199
coaxial line, 198, 199, 255
coefficient of coupling, 42
coefficient of crystal, 152
coherent radar, 250
coils, 41, 230

collector current, 64, 65, 100
collector-bias audio preamplifier, 158
collector-to-base voltage, 64, 65
collector-to-emitter voltage, 64, 65
Colpitts oscillators, 144, 147, 148, 233
common base/collector/emitter amplifiers, 66, 89-91
commutator, 213, 234
complementary symmetry amplifier, 128
components, properties of, 26-33
condensers, 227
conduction angle, 92, 235
conductors, 37, 40, 73, 227, 228
constant-speed motor, 212
continuous wave (CW), 220
control point, 222
control-grid voltage, 61
control-grid-to-plate capacitance, 60
copper-oxide diode, 69, 70
corner reflector antennas, 197
corona discharge, 194, 195
counter-EMF, 208
coupling, 42, 98, 153, 172
covalent bonds, 74
crossover distortion, 123
crystal microphone, 167, 168
crystal oscillator, 222, 235
crystal-controlled transmitters, 235
crystals, 138, 153
  coefficient of, 152
  frequency of, 152
current, 25, 37, 226, 230
current-cutoff bias voltage, 55
cutoff, 46, 50, 66

D

D'Arsonval movements, 135, 140, 236
damped sine wave, 136
Darlington pair, 128
dc amplifiers, 97
dc circuit, power measurement in, 142
dc operating voltage, 61, 62
dc series motor, 212, 234
de-emphasis, 184
decoupling circuits, 126, 155, 238
depletion MOSFET, 107
depletion zone, 77, 105
detectors, 250, 262
diaphragm, 167

dielectric, 30
difference of potential, 226
differential phase inverter, 162
differentiator circuit, 39
diodes, 34, 35, 46, 47, 71
dip, 173
dip meter, 201
dipole antennas, 191
direct coupled amplifiers, 98
direct current, 25
direction-finding signals, 4
discriminator, 187, 189, 205, 264
displaying licenses, 17, 19, 23
dissipation, 61, 64, 65
distortion, 9, 96, 123, 137, 155
distress calls-messages-signals, 4, 7, 10-15, 20-23
diversity reception, 233
DMM, 135
doping, 35, 74
downward modulation, 173
drift, 235
dry cells, 207
duplexer systems, radar, 259, 260
duty cycle, radar, 262
DVM, 135
dynamic instability, 173
dynamotor, 212

E

echo box, 244, 257, 258, 267, 268
eddy currents, 43
Edison Effect, 47
effective off-resistance, 110
efficiency, amplifier, 59, 92-94, 117
efficiency factor, 241
EHF, 220
electrneutrality, 72
electrodynamometer, 141
electrolyte, battery, 209, 231
electrolytic capacitors, 32, 43
electromagnetic fields, 38, 39, 208
electromagnetic wave, 192
electromagnetism, 41
electron theory, 25-26, 37, 72
electrons, 72
electrostatic fields, 38
emergencies, 10, 19
EMF, 37, 40, 226
emission designations, 220, 237
emitter-follower, 91, 160, 161
emitter-to-base voltage, 64, 65
energy, 143, 144
enhanced MOSFETs, 108, 109,

110
equalizers, 242

F
fall, 37
family of curves, 100
FCC rules and regulations, 1-23
feedback, 230
feedback amplifiers, 95
feedback oscillators, 136, 151
field strength, 194
field-effect transistors, 104-114
filter chokes, 212
filters, 33, 44, 46, 68, 189
FLF, 219
flow of holes, 233
flywheel effect, 136
FM deviation meter, 185, 204
FM radio/television, 220
  center frequency, 241
forward-bias, 63, 77, 78, 81
free electrons, 73
free space, 192
frequencies
  calling, 17, 22, 23
  carrier, 171, 221, 235
  designations of, 219
  intermediate, radar, 264
  measurement of, 180
  shared, 22
  sideband, 171
  vacuum tubes and, 62
  working, 17
frequency counter, 202
frequency drift, 235
frequency meter, 203
frequency modulation systems, 179
frequency multiplier, 169-171
frequency response, 84, 97
frequency shift, 173, 232
frequency-modulated oscillator, 181
frequency-shift keying (FSK), 186, 220
front-to-back ratio, 263
full-wave power supply, 66, 67
fundamental mode, 216
fusistor, 119

G
gain, 58, 83, 84
gain-bandwidth product, 84, 85
galena, 72
General Radiotelephone Permit, 1
general station operating procedures, 9-10
generators, 207-208 211, 213, 234, 253
germanium, 73, 74, 77
government radiotelegrams, 4
grid current, 234
grid modulation, 172, 233
grid-leak biasing, 158
grids, 34, 47, 48, 232, 234
ground wave, 193, 194
grounded element, 89
guy wires, 199

H
half-wave power supply, 66, 67
harmful interference, 5, 8
harmonic attenuation, 206
harmonic generator, 170, 171
harmonics, 169, 186
Hartley oscillators, 144, 146, 147, 181
hash noise, 191
Hazeltine method, 169
heading flashes, 268, 252
headphones, 166, 230
heater voltage, 60, 61, 62
heterodyne frequency meters, 180, 203
HF, 219, 237
high-level modulation, 239
high-pass filters, 33, 46
high-vacuum diodes, 69, 70
hole flow, 76
holes, 72, 75, 76, 233
hysteresis loss, 43

I
identification announcements, 223
IF amplifier, 187, 188
IGFET, 107, 111
ignition noise, 226
image antenna, 196
image frequency, 176
impedance, 26, 29, 38, 44, 192, 228
  characteristic, 199, 200
impedance coupling, 153
impedance matching, 38, 200, 228
impedance transformation, 91
impulse noise, 191
indicating instruments, 135-136, 140
inductance, 29, 30, 40, 41, 230
induction motor, 212, 242
inductors, 45

input capacitance, 60
input connector, 198, 199
inspection
  radio station, 4, 6
  U.S. vessel by foreign power, 16
instrument noise, 226
insulators, 30, 37, 43, 73, 198, 199
integrator circuit, 39
intercepted messages, 8
interference, 9
  harmful, 5, 8
  motor-caused, 212
  radar, 246, 247, 266, 268
intermediate frequencies, 264
intermodulation distortion, 137
international broadcast stations, 241
international distress frequency, 10
interstage leads, 228
interstage transformer, 125, 159
inverters, 162
IR drop, 37

J
JFETs, 51, 71, 89, 104-106, 111, 113, 160, 161
junction potential, 123
junction voltage, 78

K
keep-alive voltage, 254, 255, 259, 268
klystrons, 208, 209, 214, 249, 253, 262

L
lead-acid storage battery, 207, 209, 231
leakage current, 109
left-hand rule, 41
LF, 219
licenses, 3-5, 8, 10, 18
lightning, 196, 237
limiter, 187, 189, 238
line equalizer, 242
linearity, 137
load, 69
load lines, transistors, 100
loading, 152
loading coils, 195
logbooks, 4-7, 223, 244, 259, 267, 268
long-distance communications, 195
loop antennas, 193
LORAN, 247, 248, 268
lost licenses, 3

loudspeakers, 38
low-pass filters, 33, 44, 46

M

magnetic dipoles, 41
magnetism, 41
magnetrons, 208, 209, 248-254, 259, 263
Marconi antenna, 196
Marine Radiotelephone Operators Permit, 1
maritime service operating procedures, 10-17
Mayday, 11-15
mercury-vapor rectifier, 69, 70
meters, 135, 140, 180, 185, 201, 204, 230, 236, 237, 242
MF, 219
microphones, 9, 17, 137-138, 166-168, 239
microwave equipment, 208-209
Miller oscillators, 144, 150
milliammeter, 143
mixer, 187, 188, 190
modulated amplifiers, 173
modulation, 172
  AM, 138
  high-level, 239
modulation meter, 185, 204
molecular alignment, 41
monolithic microwave integrated circuits (MMIC), 98
MOSFETs, 89, 104, 106, 111
  depletion, 107
  enhanced, 108
motor-generator power supply, 211, 213
motors, 207-208, 211, 242
  interference caused by, 212
  sparking in, 212
  speed, 212
multimeters, 135
multipliers, 138, 169, 170
multivibrator, 144, 147, 148
mutual inductance, 41

N

n-type semiconductors, 75
narrowband, 186
navigation signals, 4
negative bias, 34
negative feedback, 95, 154
negative resistance, 48
neutralization of RF stages, 168, 169

neutrons, 72
noise, 155, 191, 226
nonfeedback amplifiers, 95
npn transistor, 80

O

off-resistance, 110
Ohm's Law, 26
on-off keying, 220
op amp, 54
open cathode bypass capacitor, 233
oscillator frequency, 176
oscillator stage, 190
oscillators, 136-138, 150, 165, 233
  crystal, 222, 235
  feedback and, 151
  frequency-modulated, 181
  parasitic, 151, 169, 170
  Q, 153
  shielding, 153
  stability factors in, 152
output capacitance, 60
overcharging, battery, 210
overmodulation, 9, 184, 232
overtone crystal, 152

P

p-type semiconductors, 75
pan, 14, 21
parabolic reflectors, 257
parallel configuration, 26, 44
parasitic array antennas, 197
parasitic oscillator, 151, 169, 170
peak current, 143, 144
peak inverse voltage, 69, 78
peak positive pulse voltage, 60, 61
peak reverse voltage, 78
pentavalent, 74
pentodes, 46, 48, 49, 60, 208
phase inverters, 160, 162
phase modulator, 184, 185
phase splitter, JFET, 53
phonetic alphabet, 10, 20
pi network, 39
Pierce oscillators, 144
piezoelectric bimorph cells, 138
piezoelectric effect, 167
pitch, 179
plan position indicator (PPI), 245
plate current, 236
plate dissipation, 61
plate modulation, 233
plate resistance, 57
plate voltage, 60, 61

pn junction, 76-79
pn junction diodes, 71, 72
pnp transistor, 80
polarization, 194
posting licenses, 10, 19, 23
potential, 37, 226
potentiometers, 213
power, 25, 27, 37, 43, 143, 144
power amplifier, 134, 168
power dissipation, 61, 62
power factor, 33, 142, 211
power supplies, 36, 66-69, 157, 211, 230
power transistor, 115, 132
pre-emphasis, 182, 184, 239
preamplifier, two-stage bipolar transistor, 53
preamplifier circuit, JFET, 51
primary cells, 207, 209, 236, 237
priority messages, 11-14, 21
propagation of radio signals, 193, 196
protons, 72
public safety radio, 221-223
pull, 91
pulse repetition rate (PRR), 245, 246, 261, 262
pulse voltage, 60, 61
pulse-modulation, television, 220
punch through region, 78
push-pull amplifiers, 52, 121, 125, 138, 159, 160, 164, 232
push-pull frequency multiplier, 169, 170

Q

Q value, 153, 229
quarter-wave antennas, 191, 296
quasi-complementary symmetry amplifiers, 128
quiescent point, 85, 101

R

radar, 243-268
  anti-TR boxes for, 254, 255, 259, 268
  artificial transmission lines, 246
  authorization for, 259
  auto-alarm equipment, 246
  automatic frequency control in, 251
  azimuth of, 248
  bearing resolution, 252, 265
  block diagram of, 245
  cathode ray tubes in, 251

coaxial line, 255
coherent, 250
crystal mixer in, 248
defective crystals in, 262, 263
detectors, 250, 262
discriminator, 264
duty cycles in, 262
echo signals in, 263
explosive and inflammable cargo and, 247, 267
frequency bands of, 260
front-to-back ratio in, 263
grounding for, 247, 261
heading flash, 252, 268
installation and maintenance, 244, 259, 261, 265, 267
interference and, 246, 247, 266, 268
intermediate frequencies for, 264
keep-alive voltage in, 254, 255, 259, 268
klystrons and, 249
licensing requirements, 244, 260
logbooks for, 244, 259, 267, 268
magnetron and, 248
operation of, 245, 265
plan position indicator, 245
pulse repetition rate, 245
range resolution of, 248
range-marker circles, 252, 268
receivers for, 250
repairs and adjustments, 247
RF amplifiers for, 260
rotary spark gap, 246
rules and regulations, 243-244
scoping in, 265
sea return in, 251, 264
sensitivity time control, 251, 264
ship transmitters, 244
spiking, 247
spoking, 248
stagger-tuning, 260
superheterodyne receivers for, 254
targeting with, 247, 252, 265
TR boxes for, 254, 255, 259, 268
unlicensed operator and, 244
waveguides for, 255, 266
radar duplexer system, 259, 260
radiation patterns, antenna, 192
radiators, 198, 199
radio frequency amplifiers, 138
radio stations
  AM, 222
  inspection of, 4, 6

logbooks for, 223
random noise, 154
range resolution, 248
range-marker circles, 252, 268
RC circuits, 39, 239
RC networks, 40
reactance, 29, 32, 181
reactance modulator, 179, 184, 185
reactance tube, 241
rebroadcast, 4, 7
receivers
  AM, alignment in, 177
  AM, superheterodyne, 175
  automobile, interference in, 191
  bandpass switch for, 178
  FM, 241
  FM, alignment of, 190
  FM, superheterodyne, 187
  radar, 250, 254
  selectivity and sensitivity in, 178
  spurious signal reception in, 189
rectifiers, 36, 37, 66, 68, 69, 70, 229
reflectors, 256-257, 261
reflex klystron, 214, 249
regeneration, 145
relays, 28, 29, 37, 44, 228
reluctance, 40
remote cutoff, 46, 50
renewing licenses, 4, 6
residual magnetism, 40
resistance, 26, 37-41, 48, 228
resistance coupling, 153
resistors, 27, 38, 39, 68, 168, 228, 229
  color codes for, 28, 44, 227
resonance, 32, 44, 61, 62
resonant circuits, 228
resonant frequency, 181
Restricted Radiotelephone Permit, 1
reverse current, 77
reverse-bias, 63, 77, 78
revoked license, 7
RF amplifiers, 131-134, 234, 260
RF choke, 168
RF power amplifier, 168
  30MHz circuit using, 133
  troubleshooting questions on, 218-219
RF power transistors, 132
RF stage, 190
  neutralization of, 168
RF voltage amplifiers, 168
ribbon microphones, 239
ripple voltage, 69

RL networks, 41
rms voltage, 143, 144
rotary spark gap, 246, 261

S
safety message, 12
safety signal, 4, 14, 15, 22
saturation, 55, 155
scoping, 265
screen grid, 61, 236
sea return, 251, 264
Second Class Radiotelephone Permit, 1
secondary cells, 207, 209
secondary emission, 57
secrecy provisions, 6
security, 14, 15, 21, 23
selectivity, 178
selenium diode, 69, 70, 72
self-bias, 65, 112
self-inductance, 41
self-oscillation, 165, 235
semiconductors, 35, 63, 71-76
sensitivity, 135, 178
sensitivity time control, 251, 264
series configuration, 26, 44
series-fed quarter wave antenna, 196
shared frequencies, 22
shells, 73
SHF, 220
shielding, 153, 228
shunt-fed quarter wave antenna, 196
shunt-wound dc motor, 212, 213
sidebands, 171, 173, 182, 183
signal frequency, 176
silicon crystals, 250
silicon diode, 69, 70, 73, 74
sine waves, 231
single sideband, reduced carrier, 220
single-ended transistor amplifier, 116
single-loop antennas, 197
single-sideband suppressed-carrier (SSBSC), 139, 174, 175
sinusoidal signals, 181
skin effect, 38, 227
skip zone, 193
sky wave, 193, 195
soldering, 45
solid-state electronics, 71-134
SOS, 11-15
sound levels, 9, 17

281

source bias, 157
space charge, 57
space diversity, 194, 195
sparking, motors, 212
spiking, 247
split secondary transformer, 126
split-load inverter, 162
spoking, 248
stabilization, 66
stacked array antennas, 197
stages, 234
stagger-tuning, 260
standing waves, 200
standing-wave ratio (SWR), 200
static, 196, 226
station authorization, 220, 221
stray capacitance, 152
stub matching, 230
stub tuning, 200
sulphation, 210
superheterodyne receivers, 175, 176, 187
support mast, 198, 199
suppressor grid, 234
surge impedance, 199
suspended license, 5, 7, 8
swinging chokes, 68
synchro-generator circuit, 253
synchronous motor, 212

T
30 MHz transistor power amplifier, 133
300-watt HF power amplifier, 134
tank circuits, 136
technician certification programs, 269-273
temperature-compensated crystal oscillators (TCXO), 138
term of license, 4, 6
testing, 10, 15, 19, 20, 22
tetravalent, 74
tetrodes, 34, 46, 50, 230
thermistors, 66
Third Class Radiotelephone Permit, 1
third-mode crystal, 152
tickler, 145
tolerance, 27
tone-control circuits, 165
top loading, 240
total harmonic distortion, 96, 137
totem pole amplifiers, 126
TR boxes, 254, 255, 259, 268
transconductance amplifiers, 94, 95, 107
transducer, 167
transfer function, 94
transformer coupled amplifiers, 99
transformer-rectifier power supply, 211
transformers, 43, 126, 166, 239
transistor amplifiers, 99-104
transistor dissipation, 64, 65
transistor RF power amplifiers, 131-134
transistors, 35, 36, 63, 64, 71, 72, 208
 amplification, 81
 basic theory for, 79-88
 bias voltage for, 156
 biasing, 80, 86, 87, 88
 field-effect, 104
 forward-biased, 81
 frequency response of, 84
 gain, 83
 load lines in, 100
 ratings for, 100
 RF power, 132
 thermal stability in, 86
transmission lines, 180, 240
 artificial, 246, 258, 259, 261
 characteristic impedance, 199
 impedance matching, 200
transmitters, 169, 233
 AM, 138, 171, 173, 240
 AM, FM reception in, 173
 buffer amplifiers in, 153
 crystal-controlled, 235
 FM, 187, 240
 harmonic attenuation in, 206
 measurements for, 223
 placement of, 8
 retuning, 186
 SSBSC, 174
 testing, 20
 type-approval and type-acceptance, 220
 unattended, 9, 17, 18
transresistance amplifiers, 94, 95
trap, 198, 199
traveling wave tube, 208, 209, 214, 215, 238
treble control circuit, 165
triodes, 34, 46, 47, 50, 56, 57, 59, 230, 232
tripler stage, 187
tuned-in tuned-out oscillators, 144, 145
TVM, 135

two-stage amplifier, 153
two-way radio, 225-226
type approval, 235

U
UHF, 62, 63, 208, 220
unlicensed persons, 10, 20
untuned lines, 180
upgrading licenses, 3
urgency signal, 4, 6, 14

V
V-beam antennas, 197
vacuum tubes, 34, 36, 46, 51-62, 71, 154, 208, 229, 234
valence electrons, 73
varactor, 181
variable attenuation, 242
variable frequency oscillator (VFO), 180
vectors, 33, 45
vernier scale, 180
VFET amplifier, 166, 167
VHF, 62, 63, 219
video amplifiers, 97
violation of FCC rules, 5, 7, 8
voltage, 25, 37
voltage amplifier, 82, 168
voltage divider bias circuit, 157
voltage drop, 37
voltage gain, 58, 94
voltage regulation, 69
voltmeter, 135, 141-143, 242
VOM, 135
VTVM, 135, 143

W
wattmeter, 141, 142, 230
waveform, phase modulation in, 185
waveguides, 216, 217, 238, 255-257, 261, 266, 267
wavemeter, 202
weather observation messages, 4
wet cells, 207
Wheatstone bridge, 130
wideband, 186
wideband amplifiers, 97
windings, 41
wires, resistance in, 37, 38
words and phrases, 9, 15-20, 23
working frequency, 17, 21
World Administrative Radio Treaty of 1979, 16
WWV broadcast, 203